Body Image, Nutrition and Mental Health

Body Image, Nutrition and Mental Health

Editors

Hubertus Himmerich
Khadijeh Mirzaei

MDPI

Basel • Beijing • Wuhan • Barcelona • Belgrade • Novi Sad • Cluj • Manchester

Editors
Hubertus Himmerich
King's College London
London
UK

Khadijeh Mirzaei
Tehran University of Medical Sciences
(TUMS)
Tehran
Iran

Editorial Office
MDPI
St. Alban-Anlage 66
4052 Basel, Switzerland

This is a reprint of articles from the Special Issue published online in the open access journal *Nutrients* (ISSN 2072-6643) (available at: https://www.mdpi.com/journal/nutrients/special_issues/FOG560482C).

For citation purposes, cite each article independently as indicated on the article page online and as indicated below:

Lastname, A.A.; Lastname, B.B. Article Title. *Journal Name* **Year**, *Volume Number*, Page Range.

ISBN 978-3-7258-1303-2 (Hbk)
ISBN 978-3-7258-1304-9 (PDF)
doi.org/10.3390/books978-3-7258-1304-9

Cover image courtesy of Hubertus Himmerich and Holly Walker

Contents

About the Editors

Hubertus Himmerich

Hubertus Himmerich is a Reader in Eating Disorders at King's College London and a consultant psychiatrist in an inpatient ward for patients with eating disorders at the Bethlem Royal Hospital in London. His scientific and clinical interests include eating and weight disorders, psychopharmacology, psychoimmunology, and the therapeutic use of music. He has led and performed national and international scientific projects with researchers from Africa, Asia, Australia, America, and Europe.

Khadijeh Mirzaei

Khadijeh Mirzaei is an academic member of the Department of Community Nutrition, School of Nutritional Sciences and Dietetics, Tehran University of Medical Sciences (TUMS). She is currently the Vice Dean for international affairs at the School of Nutritional Sciences and Dietetics. Her work focusses on understanding the metabolic factors linking obesity to an increased disease risk during growth. She is also investigating the effects of nutrition on the body's ability to manage stressors and on mental well-being. She has been a visiting researcher at Harvard University and is currently involved in national and international population health and community research programs in mental health, lifestyle, obesity, and lipid disorders.

Preface

This Special Issue explores the complex relationships among body image, nutrition, and mental health.

The articles cover timely research topics regarding eating and weight disorders and their treatment, such as the regulation of emotions, appetite, and metabolism; social media, social networking sites, and sociocultural pressure; and the therapeutic use of smart toys and bariatric surgery. Therefore, this Special Issue provides a comprehensive overview of the current research ideas and activities in this area.

We are grateful to the team of diverse authors from Austria, Belgium, Chile, China, Colombia, France, Georgia, Germany, Iran, Italy, Japan, Lebanon, Lithuania, Poland, Saudi Arabia, Spain, Türkiye, the United Kingdom, and the United States of America.

We would also like to thank Holly Walker for the cover image that is a handmade embroidery depicting a skeleton drinking coffee.

We hope that this Special Issue will inform and entertain our readers.

Hubertus Himmerich and Khadijeh Mirzaei

Editors

nutrients

MDPI

Editorial

Body Image, Nutrition, and Mental Health

Hubertus Himmerich [1,*] and Khadijeh Mirzaei [2]

1 Department of Psychological Medicine, Institute of Psychiatry, Psychology & Neuroscience, King's College London, London SE5 8AF, UK
2 Department of Community Nutrition, School of Nutritional Sciences and Dietetics, Tehran University of Medical Sciences (TUMS), Tehran, Iran; mirzaei_kh@sina.tums.ac.ir
* Correspondence: hubertus.himmerich@kcl.ac.uk

Classical examples of disorders associated with body image disturbances are eating disorders (EDs) such as anorexia nervosa (AN) and bulimia nervosa (BN), as well as body dysmorphic disorder (BDD). Body image is a complex construct comprising thoughts, feelings, evaluations, and behaviors related to one's body. Body image disturbances are not exclusively found in EDs and BDD; they are also highly prevalent in people with other mental or physical health problems, e.g., depression or obesity, and in the general population [1].

Similarly, altered nutritional intake is not restricted to EDs and obesity. A change in food intake can be a symptom of a mental health disorder such as depression [2], or a consequence of psychopharmacological treatment [3]. Increased appetite and food intake lead to obesity, which is often associated with depression [4]. Vice versa, weight loss can help with depression in people with obesity if they lose weight under a calorie-restricted diet [5]. These examples indicate the close association between nutrition and mental health disorders.

This Special Issue examines and illuminates the complex relationships between body image, nutrition, and mental health. More specifically, it covers the psychological and social risk factors of body image disturbances and associated disorders, biological aspects of appetite regulation and the metabolic syndrome, and therapeutic approaches for EDs and weight disorders and their health consequences.

Psychosocial risk factors of body image disturbances and associated disorders are the first group of themes covered in this Special Issue.

Yumen et al. [6] and Karam et al. [7] examined the influence of social websites and social media use on body image and food intake. In a sample of young Japanese women, Yumen et al. found that longer social networking site use was associated with lower body weight and with a thinner body shape ideal [6]. Karam et al. researched the relationship between social media use and body image in Lebanese university students. Individuals with more social media use had higher odds of exhibiting body image concerns and were at risk for emotional overeating [7].

Witaszek et al. investigated anxiety and depression as risk factors of the use of food intake to regulate emotions and found that women with anxiety and depression showed higher scores for uncontrolled and emotional eating but lower scores of cognitive restraint [8]. As an additional result, they reported that the use of the glucagon-like peptide 1 (GLP-1) receptor agonists liraglutide and semaglutide was associated with increased cognitive restraint in the Three-Factor Eating Questionnaire.

In a study using a virtual reality (VR) kitchen as experimental paradigm, Bektas et al. observed that ED symptoms correlated positively with food-specific trait and state disgust [9].

Baceviciene et al. compared disordered eating, body image, and sociocultural and coach-related pressures between adolescent and adult and between male and female athletes. An interesting finding was that vomiting, laxative misuse, and excessive exercise

Citation: Himmerich, H.; Mirzaei, K. Body Image, Nutrition, and Mental Health. *Nutrients* **2024**, *16*, 1106. https://doi.org/10.3390/nu16081106

Received: 15 March 2024
Accepted: 3 April 2024
Published: 10 April 2024

were more prevalent in adolescent female athletes than adult females, while dietary restraint was more common in adult male athletes compared to adolescent males. Adolescent female athletes experienced a high sociocultural pressure from family and peers, and sport-related pressure from the coach [10].

Fabrig et al. examined a large sample of candidates for bariatric surgery and found a strong association between weight-related experienced stigmatization and depressive symptoms in candidates with high weight bias internalization [11].

Biological appetite regulation and markers of the metabolic syndrome constitute the second group of themes covered in this Special Issue.

In a sophisticated immunohistochemical study, Fichtner et al. provided evidence for a widespread distribution of the glial-cell-derived neurotrophic factor (GDNF) receptor alpha-like (GFRAL), which is the receptor for growth differentiation factor-15 (GDF-15). They found GFRAL in the prefrontal cortex, the hippocampus, the arcuate nucleus, and in peripheral tissues. GFRAL had already been implicated in food intake and body weight regulation, but these novel findings indicate a broader role of GFRAL in metabolism [12].

Bilska et al. investigated markers of metabolic syndrome and adipokines in a sample of patients with a depressive episode of bipolar disorder. They found that in women with bipolar depression, visfatin, S100B, and leptin concentrations correlated with metabolic syndrome, whereas adiponectin and leptin-receptor levels were negatively associated with it [13].

Therapeutic approaches for EDs and weight disorders and their health consequences are the third group of themes covered in this Special Issue.

Clemente-Suárez et al. summarized the non-pharmacological interventions in people with AN in a narrative review and thus covered nutritional interventions, psychological and family therapy, social media use management, and physical therapy interventions including relaxation, massages, and exercises [14].

The therapeutic use of the novel technology-enabled smart toy Purrble, which is designed for emotional regulation, was explored in a mixed-method analysis by Chubinidze et al. [15]. They found that it might be helpful particularly for patients with EDs and complex presentations.

A systematic review and meta-analysis of randomized controlled trials to treat obesity in military populations was performed by Gravina et al. [16]. They found that the current weight loss interventions are effective in military populations with a high level of evidence for physical activity, dietary and nutritional interventions, cognitive behavioral therapy, and structured outcome monitoring (clinical or self-monitoring).

Paszynska et al. summarized the biological, behavioral (binge eating episodes, vomiting, acidic diet, poor oral hygiene), and pharmacotherapeutic factors that may threaten oral health in people with EDs [17]. In their article, they advocate for early diagnosis, reductions in behaviors that are destructive for oral health, nutritional counselling, and medical interventions to treat and protect oral soft and hard tissues.

In summary, the use of social websites and social media, anxiety and depression, food-specific disgust, social pressure, physical fitness pressure, weight-related experienced stigmatization, and weight bias internalization were identified as risk factors for body image disturbances; disordered, restrictive, or over-eating; vomiting; laxative misuse; excessive exercise; and associated mental health problems [7–11]. The GDF-15 receptor GFRAL, which is known to be involved in appetite regulation, was found to be expressed across various central-nervous as well as peripheral tissues [12], and visfatin, S100B, and leptin are associated with the development of metabolic syndrome in people with bipolar depression [13]. Four articles of this Special Issue [14–17] summarized already available (e.g., nutritional interventions, psychological and family therapy) [14] and experimental (smart toy Purrble) [15] therapies for AN, effective therapies for the treatment of obesity (e.g., physical activity, dietary and nutritional intervention, cognitive behavioral therapy, clinical and self-monitoring) [16], as well as the preventive and therapeutic options for oral health consequences of EDs [17]. Witaszek et al.'s finding that liraglutide and semaglutide were associated with increased cognitive restraint [8] is also therapeutically relevant as these GLP-1 receptor agonists are approved for the treatment of obesity.

The mentioned findings reported in this Special Issue are clinically relevant and pave the way for future research in EDs and weight-related disorders.

For example, as weight gain is a clinically significant problem during the treatment with antidepressants and antipsychotics [3], prescription of the GLP-1 receptor agonists liraglutide and semaglutide alongside the treatment with weight-gain-inducing psychopharmacological medications might help to prevent or attenuate weight gain [18]. Witaszek et al.'s finding that liraglutide and semaglutide were associated with increased cognitive restraint [8] might help to explain weight loss during treatment with these GLP-1 receptor agonists.

The findings of Bilska et al. [13] that visfatin, S100B, and leptin concentrations correlated with the metabolic syndrome point to the involvement of the immune system in EDs and weight disorders, because visfatin, S100B, and leptin modulate the release of pro-inflammatory cytokines like tumor necrosis factor (TNF)-α and interleukin (IL)-6 [19–21]. TNF-α and IL-6 have been found elevated in people with obesity [22] but also in patients with AN [23]. Moreover, TNF-α and IL-6 inhibitors have been reported to lead to an increase in body weight and were therefore suggested as potential future treatments of AN [24,25]. Taking their influence on mood, cognition and behavior into account, these cytokines might not only be involved in the pathophysiology of EDs but also other mental health disorders [26]. Vice versa, psychopharmacological agents that treat mental health disorders like antipsychotics, antidepressants, and mood stabilizers have been shown to alter cytokine production in vivo and in vitro [27,28].

By using novel technical solutions such as VR [9] or a smart toy [15] as add-ons, therapists might enhance the success of their therapies in the future.

To conclude, this Special Issue covers psychological, social, and biological aspects of mental health problems associated with body image disturbances, and over- and undernutrition. It also provides novel ideas for psychometric and biological markers and therapeutic options for people with EDs and weight disorders such as the use of GLP-1 receptor agonists, GDF-15 and GFRAL signaling modification, and the enhancement of psychotherapies using VR and smart toys.

Author Contributions: H.H. and K.M. have written the manuscript of this editorial together. All authors have read and agreed to the published version of the manuscript.

Conflicts of Interest: The authors declare no conflicts of interest.

References

1. Hosseini, S.A.; Padhy, R.K. Body Image Distortion. In *StatPearls*; StatPearls Publishing: Treasure Island, FL, USA, 2024.
2. Kroemer, N.B.; Opel, N.; Teckentrup, V.; Li, M.; Grotegerd, D.; Meinert, S.; Lemke, H.; Kircher, T.; Nenadic, I.; Krug, A.; et al. Functional Connectivity of the Nucleus Accumbens and Changes in Appetite in Patients with Depression. *JAMA Psychiatry* **2022**, *79*, 993–1003. [CrossRef]
3. McIntyre, R.S.; Kwan, A.T.H.; Rosenblat, J.D.; Teopiz, K.M.; Mansur, R.B. Psychotropic Drug-Related Weight Gain and Its Treatment. *Am. J. Psychiatry* **2024**, *181*, 26–38. [CrossRef] [PubMed]
4. Blasco, B.V.; García-Jiménez, J.; Bodoano, I.; Gutiérrez-Rojas, L. Obesity and Depression: Its Prevalence and Influence as a Prognostic Factor: A Systematic Review. *Psychiatry Investig.* **2020**, *17*, 715–724. [CrossRef] [PubMed]
5. Patsalos, O.; Keeler, J.; Schmidt, U.; Penninx, B.W.J.H.; Young, A.H.; Himmerich, H. Diet, Obesity, and Depression: A Systematic Review. *J. Pers. Med.* **2021**, *11*, 176. [CrossRef]
6. Yumen, Y.; Takayama, Y.; Hanzawa, F.; Sakane, N.; Nagai, N. Association of Social Networking Sites Use with Actual and Ideal Body Shapes, and Eating Behaviors in Healthy Young Japanese Women. *Nutrients* **2023**, *15*, 1589. [CrossRef] [PubMed]
7. Karam, J.M.; Bouteen, C.; Mahmoud, Y.; Tur, J.A.; Bouzas, C. The Relationship between Social Media Use and Body Image in Lebanese University Students. *Nutrients* **2023**, *15*, 3961. [CrossRef] [PubMed]
8. Witaszek, T.; Babicki, M.; Brytek-Matera, A.; Mastalerz-Migas, A.; Kujawa, K.; Kłoda, K. Maladaptive Eating Behaviours, Generalised Anxiety Disorder and Depression Severity: A Comparative Study between Adult Women with Overweight, Obesity, and Normal Body Mass Index Range. *Nutrients* **2023**, *16*, 80. [CrossRef] [PubMed]

9. Bektas, S.; Natali, L.; Rowlands, K.; Valmaggia, L.; Di Pietro, J.; Mutwalli, H.; Himmerich, H.; Treasure, J.; Cardi, V. Exploring Correlations of Food-Specific Disgust with Eating Disorder Psychopathology and Food Interaction: A Preliminary Study Using Virtual Reality. *Nutrients* **2023**, *15*, 4443. [CrossRef] [PubMed]
10. Baceviciene, M.; Jankauskiene, R.; Rutkauskaite, R. The Comparison of Disordered Eating, Body Image, Sociocultural and Coach-Related Pressures in Athletes across Age Groups and Groups of Different Weight Sensitivity in Sports. *Nutrients* **2023**, *15*, 2724. [CrossRef] [PubMed]
11. Fabrig, A.; Schmidt, R.; Mansfeld, T.; Sander, J.; Seyfried, F.; Kaiser, S.; Stroh, C.; Dietrich, A.; Hilbert, A. Depressive Symptoms among Bariatric Surgery Candidates: Associations with Stigmatization and Weight and Shape Concern. *Nutrients* **2024**, *16*, 510. [CrossRef] [PubMed]
12. Fichtner, K.; Kalwa, H.; Lin, M.M.; Gong, Y.; Müglitz, A.; Kluge, M.; Krügel, U. GFRAL Is Widely Distributed in the Brain and Peripheral Tissues of Mice. *Nutrients* **2024**, *16*, 734. [CrossRef] [PubMed]
13. Bilska, K.; Dmitrzak-Węglarz, M.; Osip, P.; Pawlak, J.; Paszyńska, E.; Permoda-Pachuta, A. Metabolic Syndrome and Adipokines Profile in Bipolar Depression. *Nutrients* **2023**, *15*, 4532. [CrossRef] [PubMed]
14. Clemente-Suárez, V.J.; Ramírez-Goerke, M.I.; Redondo-Flórez, L.; Beltrán-Velasco, A.I.; Martín-Rodríguez, A.; Ramos-Campo, D.J.; Navarro-Jiménez, E.; Yáñez-Sepúlveda, R.; Tornero-Aguilera, J.F. The Impact of Anorexia Nervosa and the Basis for Non-Pharmacological Interventions. *Nutrients* **2023**, *15*, 2594. [CrossRef] [PubMed]
15. Chubinidze, D.; Li, Z.; Slovak, P.; Baudinet, J.; Dufour, E.; Tchanturia, K. Introducing a Smart Toy in Eating Disorder Treatment: A Pilot Study. *Nutrients* **2024**, *16*, 467. [CrossRef] [PubMed]
16. Gravina, D.; Keeler, J.L.; Akkese, M.N.; Bektas, S.; Fina, P.; Tweed, C.; Willmund, G.D.; Treasure, J.; Himmerich, H. Randomized Controlled Trials to Treat Obesity in Military Populations: A Systematic Review and Meta-Analysis. *Nutrients* **2023**, *15*, 4778. [CrossRef] [PubMed]
17. Paszynska, E.; Hernik, A.; Rangé, H.; Amaechi, B.T.; Gross, G.S.; Pawinska, M. Diet Traps during Eating Disorders among Dentate Patients at an Oral Health Glance. *Nutrients* **2023**, *15*, 4414. [CrossRef] [PubMed]
18. Vasiliu, O. Therapeutic management of atypical antipsychotic-related metabolic dysfunctions using GLP-1 receptor agonists: A systematic review. *Exp. Ther. Med.* **2023**, *26*, 355. [CrossRef] [PubMed]
19. Heo, Y.J.; Choi, S.E.; Jeon, J.Y.; Han, S.J.; Kim, D.J.; Kang, Y.; Lee, K.W.; Kim, H.J. Visfatin Induces Inflammation and Insulin Resistance via the NF-kappaB and STAT3 Signaling Pathways in Hepatocytes. *J. Diabetes Res.* **2019**, *2019*, 4021623. [CrossRef] [PubMed]
20. Piazza, O.; Leggiero, E.; De Benedictis, G.; Pastore, L.; Salvatore, F.; Tufano, R.; De Robertis, E. S100B induces the release of pro-inflammatory cytokines in alveolar type I-like cells. *Int. J. Immunopathol. Pharmacol.* **2013**, *26*, 383–391. [CrossRef] [PubMed]
21. Iikuni, N.; Lam, Q.L.; Lu, L.; Matarese, G.; La Cava, A. Leptin and Inflammation. *Curr. Immunol. Rev.* **2008**, *4*, 70–79. [CrossRef] [PubMed]
22. Kern, L.; Mittenbühler, M.J.; Vesting, A.J.; Ostermann, A.L.; Wunderlich, C.M.; Wunderlich, F.T. Obesity-Induced TNFα and IL-6 Signaling: The Missing Link between Obesity and Inflammation-Driven Liver and Colorectal Cancers. *Cancers* **2018**, *11*, 24. [CrossRef]
23. Dalton, B.; Bartholdy, S.; Robinson, L.; Solmi, M.; Ibrahim, M.A.A.; Breen, G.; Schmidt, U.; Himmerich, H. A meta-analysis of cytokine concentrations in eating disorders. *J. Psychiatr. Res.* **2018**, *103*, 252–264. [CrossRef]
24. Patsalos, O.; Dalton, B.; Leppanen, J.; Ibrahim, M.A.A.; Himmerich, H. Impact of TNF-α Inhibitors on Body Weight and BMI: A Systematic Review and Meta-Analysis. *Front. Pharmacol.* **2020**, *11*, 481. [CrossRef] [PubMed]
25. Patsalos, O.; Dalton, B.; Himmerich, H. Effects of IL-6 Signaling Pathway Inhibition on Weight and BMI: A Systematic Review and Meta-Analysis. *Int. J. Mol. Sci.* **2020**, *21*, 6290. [CrossRef] [PubMed]
26. Himmerich, H.; Berthold-Losleben, M.; Pollmächer, T. The relevance of the TNF-alpha system in psychiatric disorders. *Fortschr. Neurol. Psychiatr.* **2009**, *77*, 334–345. [CrossRef] [PubMed]
27. Müller, N. Immunological aspects of the treatment of depression and schizophrenia. *Dialogues Clin. Neurosci.* **2017**, *19*, 55–63. [CrossRef] [PubMed]
28. Himmerich, H.; Bartsch, S.; Hamer, H.; Mergl, R.; Schönherr, J.; Petersein, C.; Munzer, A.; Kirkby, K.C.; Bauer, K.; Sack, U. Modulation of cytokine production by drugs with antiepileptic or mood stabilizer properties in anti-CD3- and anti-Cd40-stimulated blood in vitro. *Oxid. Med. Cell Longev.* **2014**, *2014*, 806162. [CrossRef] [PubMed]

nutrients

MDPI

Article

Association of Social Networking Sites Use with Actual and Ideal Body Shapes, and Eating Behaviors in Healthy Young Japanese Women

Yukina Yumen [1,2,*], Yumi Takayama [1], Fumiaki Hanzawa [3], Naoki Sakane [4] and Narumi Nagai [1,3]

1 Laboratory of Nutritional Physiology, Graduate School of Human Science and Environment, University of Hyogo, Himeji City 670-0092, Hyogo, Japan; nagai.lab.uh@gmail.com (N.N.)
2 Kyoto College of Nutritional and Medical Sciences, Kyoto 616-8376, Kyoto, Japan
3 Department of Food Science and Nutrition, School of Human Science and Environment, University of Hyogo, Himeji City 670-0092, Hyogo, Japan; hanzawa@shse.u-hyogo.ac.jp
4 Division of Preventive Medicine, Clinical Research Institute for Endocrine and Metabolic Disease, National Hospital Organization, Kyoto Medical Center, Kyoto 612-8555, Kyoto, Japan; nsakane@gf6.so-net.ne.jp
* Correspondence: yukinayumen@gmail.com

Abstract: Recent reports have associated the use of social networking sites (SNS) with the drive for thinness in young women; however, its influence on their actual body shape and eating behaviors (EB) remains unclear. We aimed to examine the effect of SNS use on body mass index (BMI), body image (BI), and EB in young women. Participants included 196 healthy women (20–29 years) who answered questions about their SNS use, height, weight, BI and EB via a web-based survey. First, the correlation between time spent on SNS and each variable was determined. Participants were then divided into quartiles according to the duration of daily SNS use as long ($\geq$3 h, n = 52) and short (<1 h, n = 54), and the data were then compared between the groups. Correlation analysis showed that the longer the duration of daily SNS use, the significantly lower the BMI, the use of nutrition labels, and the frequency of consumption of milk and dairy products. The long group had significantly lower BMI and ideal BI than the short group. The results suggest that spending more time on SNS in young women may be associated with thinner actual and ideal body shapes and poorer access to health information and healthy foods.

Keywords: social media; body mass index; body image; nutrition labeling; healthy eating; thinness

Citation: Yumen, Y.; Takayama, Y.; Hanzawa, F.; Sakane, N.; Nagai, N. Association of Social Networking Sites Use with Actual and Ideal Body Shapes, and Eating Behaviors in Healthy Young Japanese Women. *Nutrients* **2023**, *15*, 1589. https://doi.org/10.3390/nu15071589

Academic Editors: Hubertus Himmerich and Leslie D. Frazier

Received: 1 March 2023
Revised: 19 March 2023
Accepted: 22 March 2023
Published: 24 March 2023

1. Introduction

Underweight (body mass index; BMI < 18.5 kg/m^2) is found in approximately 10% of adult women worldwide, and female thinness is prevalent not only in countries and regions with inadequate food supplies, but also in those without [1]. One of these countries, Japan, has the highest prevalence of underweight among adult women in the Organization for Economic Cooperation and Development (OECD) and among developed countries [1]. In particular, nearly 20% of young women are underweight [2]. Women in younger generations are more likely to engage in intentional dietary restriction for weight loss, or dieting, due to their greater drive for thinness and desire to achieve a slimmer body shape [3]. The increasing prevalence of thinness among young women has raised concerns about its possible adverse effects not only on women's own life-course [4], but also on the health of offspring across generations from the Developmental Origins of Health and Disease (DOHaD) perspective [5], highlighting the prevention of thinness among young women as an important public health nutrition issue.

The drive to be thin, which contributes to female thinness, is known to be enhanced by comparison with others [6]. The recent development of social networking sites (SNS), along with a constant stream of attractive platforms for posting images and videos, has dramatically increased the opportunities for comparison with others. The majority of SNS users belong to the younger generations [7]. Several studies have reported associations between SNS use and the drive for thinness in young women [8]. Tiggemann et al. [9,10] showed that the more time female high school students spent on SNS, the higher their drive for thinness. Cohen et al. [11,12] focused on behavioral outcomes associated with SNS use and found that appearance-focused activities on SNS among young adult women promoted thin-ideal internalization and body surveillance, thereby enhancing their drive for thinness. As described above, SNS use has been documented to increase the drive for thinness in adolescent and young adult women, though its influence on their actual and ideal body shapes remains unknown.

Furthermore, given the link between thinness and eating behaviors (EB), the relationship between SNS use and EB also needs to be addressed [13]. Yao et al. [14] reported that among female college students, appearance comparisons on SNS were associated with an increased tendency toward restrained eating. However, the relationships between SNS use and dietary habits and food choices have not been fully elucidated.

This study aimed to clarify the impact of SNS use on actual and ideal body shapes and EB in non-obese young Japanese women.

2. Materials and Methods

2.1. Participants and Procedure

This cross-sectional study was conducted using a web-based survey between December 2020 and February 2021. Participants were healthy Japanese women over the age of 20 who were recruited through flyers, social media, and emails promoting the study. Exclusion criteria were participants who (1) were pregnant and breastfeeding, (2) had any of the following weight-related medical conditions within the past 5 years: cancer, eating disorders, diabetes mellitus, or endocrine disorders. Participation was voluntary. The study was approved by the Research Committee of the School of Human Science and Environment, University of Hyogo (No.192, 7 December 2018).

Seven hundred women completed questions on demographics, body size, nutrition, and lifestyle (all required questions). However, only the objectified body consciousness questionnaires were non-required questions, resulting in missing values for these questions.

Age and employment status were considered to influence the duration of SNS use, so participants older than 30 years (n = 342) and with employment status (n = 155) were excluded from the analyses. In addition, seven women reported not having an SNS account (e.g., Facebook, Instagram, Twitter, or TikTok) and were excluded from the analyses, resulting in a final dataset of 196 women between the ages of 20 and 29 years (M = 21.1, SD = 1.8) with a mean reported BMI of 20.3 (SD = 2.3). The prevalence of being underweight (BMI < 18.5 kg/m^2) and overweight (BMI $\geq$ 25 kg/m^2) among all participants was 17.9% and 1.5%, respectively.

2.2. Measures

2.2.1. SNS Use

A questionnaire was developed based on a previous report [11]. Participants indicated the number of times they accessed/checked their accounts daily on a 7-point scale: hardly ever, 1 or 2 times, 3–5 times, 5–10 times, 11–15 times, 15–20 times, more times than I can count, and the average amount of time they spent on SNS a day on a 12-point scale: 0–15 min, 15–30 min, 1–2 h, 2–3 h, 3–4 h, 4–5 h, 5–6 h, 6–7 h, 7–8 h, 8–9 h, 9–10 h, 10 or more hours, each scored at the median. Participants also indicated how often they looked at other people's photos of themselves on a 5-point scale: almost never, rarely, sometimes, often, almost every time I log on. Participants were asked how many selfies they took and posted per day.

2.2.2. Anthropometric Assessments

Participants reported their age, sex, height, and weight. BMI (kg/m^2) was calculated from self-reported the height and weight data.

2.2.3. Body Image

Body image was assessed using the Japanese version of the Body Image Scale, which has 10 silhouettes ranging from thin (score 1) to obese (score 10). This scale was developed and its reliability and validity were assessed in the Japanese adult population [15]. Regarding reliability, good test–retest reliability was found for both men ($n = 35$, $\rho = 0.90$, $p < 0.01$) and women ($n = 257$, $\rho = 0.90$, $p < 0.01$). In terms of validity, there was a significant positive correlation between the body image score and BMI in both men ($n = 335$, $\rho = 0.82$, $p < 0.001$) and women ($n = 444$, $\rho = 0.81$, $p < 0.001$). The area under the curve (AUC) calculated by sensitivity and specificity was > 0.9 for thinness (BMI < 18.5 kg/m^2) and obesity (BMI $\geq$ 25 g/m^2) in both genders. Body image scores are reliable indicators of thinness or obesity.

Participants were asked to identify the silhouette that best represented their current body image (CBI) and ideal body image (IBI). The difference between CBI and IBI was calculated as body dissatisfaction discrepancy (CBI−IBI). To assess body perception, participants reported their own body shape on a 5-point scale: thin, slightly thin, normal, slightly overweight, or overweight. Body image distortion was scored based on participants' body perception and actual body shape: (1) underestimation: normal body shape perceived as thin or slightly thin, and overweight perceived as thin, slightly thin, or normal; (2) overestimation: underweight perceived as normal, slightly overweight, or overweight, and normal body shape perceived as slightly overweight or overweight; (3) no distortion: perceived body shape equal to actual body shape.

2.2.4. Objectified Body Consciousness

The Objectified Body Consciousness Scale [16] was used to assess objectified body consciousness tendencies. The Objectified Body Consciousness Scale consists of three subscales of eight items pertaining to body surveillance, body shame, and appearance control. Participants rate their level of agreement with 24 items (e.g., "I often worry about whether the clothes I am wearing make me look good") on a 5-point scale (1 = strongly disagree, 5 = strongly agree). McKinley and Hyde (1996) [16] reported good construct and discriminant validity in a sample of female undergraduates. In the present study, alpha was 0.75, 0.74, 0.60, and 0.66 for body surveillance, body shame, appearance control, and total score, respectively. The total score for each subscale (8 items) and the total score for all 24 items were calculated and used for scoring. To consider participants who did not want to answer sensitive questionnaires, only the Objectified body consciousness items were open-ended.

2.2.5. Subjective Feelings

Subjective health perceptions were assessed using a self-report questionnaire [17] that asked, "Do you feel that you are in good health?" Participants reported on a 4-point scale (1 = poor, 4 = excellent), with lower scores indicating lower subjective health perceptions.

The four-item Subjective Happiness Scale [18] was used to measure subjective happiness. Participants reported on a 7-point scale (1 = not a very happy person, 7 = a very happy person) with lower scores indicating lower subjective happiness. The scale has shown good internal consistency in a Japanese undergraduate sample [19]; in the present study, the alpha was 0.88. The average score for the four items was calculated and used for scoring.

2.2.6. Eating Behaviors

Dietary habits, including skipping breakfast, eating dinner within two hours of bedtime, and eating snacks after dinner were assessed using a self-report questionnaire [20]. Participants answered "yes" if they did so on three or more days per week, and "no" if they did so on two or fewer days per week. In addition, use of nutrition labels was assessed using a self-report questionnaire [21] that asked, "Do you usually refer to the nutrition label when you buy food?" Participants responded on a 4-point scale: hardly ever (0), rarely (1), often (2), and every time (3).

The frequency of food consumption of fruits, fish, milk and dairy products, and alcohol was assessed using a self-report questionnaire; the food items were selected from the healthy lifestyle behaviors that have been shown to be associated with cardiovascular mortality among Japanese people [22]. Participants reported on a 5-point scale: rarely (0), 1–2 days a month (1), 1–2 days a week (2), 3–4 days a week (3), almost every day (4). Each frequency weight was set to 0, 0.05 (1.5/30), 0.214 (1.5/7), 0.5 (3.5/7), and 1.0.

2.3. Statistical Analysis

Using all data (n = 196), we evaluated the association between the duration of SNS use and body image and eating behaviors using Spearman correlation coefficients. Based on previous reports [23,24], we compared participants in the highest (long: ≥75th percentile; ≥3 h per day) and lowest (short: <25th percentile; <1 h per day) quartiles of duration of SNS use. Comparisons between the two groups (long vs. short) were performed using the Mann–Whitney U or chi-squared test, as appropriate. Cases with missing data were removed from the relevant analysis and Little's missing completely at random (MCAR) test was used to demonstrate that the data were missing at random [25]. All statistical analyses were performed using the Statistical Package for the Social Sciences (SPSS for Windows™ ver. 28, IBM Inc., Tokyo, Japan). The significance level was set to alpha = 0.05.

3. Results

There were no missing values except for four items of the Objectified Body Consciousness. For the missing data, the number of missing data in the long and short groups was 17 (31.5%) and 15 (28.8%), respectively. Little's MCAR test showed that the missing data were missing completely at random (Chi-square = 0.09; p = 0.768).

3.1. SNS Use

Most participants (88.8%, n = 174) checked their SNS account at least 3–5 times per day. Approximately half of the participants (51.0%, n = 100) reported using SNS for 2 or more hours per day. Two-thirds of the participants (69.9%, n = 137) looked at other people's photos of themselves often or almost every time they logged on. Approximately one in five participants (17.9%, n = 35) habitually took selfies, and only 10 participants (5.1%) posted the photos on SNS (both > 0 times per day).

3.2. Relationship between SNS Use and Actual and Ideal Body Shapes, and Eating Behaviors

Figure 1 presents the correlation between the duration of SNS use and (a) BMI, (b) CBI, (c) IBI, and (d) body dissatisfaction of all data. BMI was negatively correlated with the duration of SNS use (r = −0.149, p = 0.037).

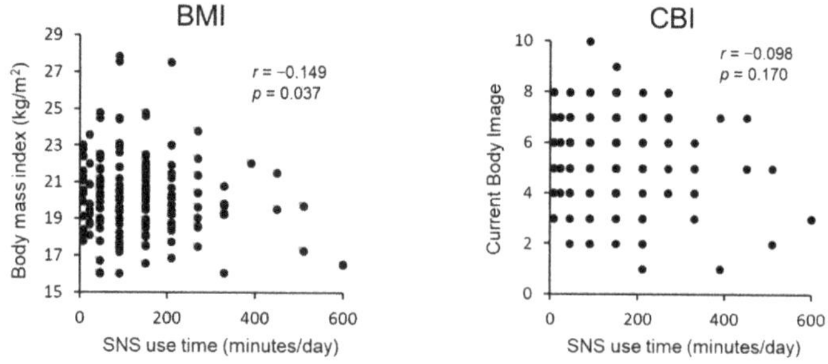

(a) (b)

Figure 1. *Cont.*

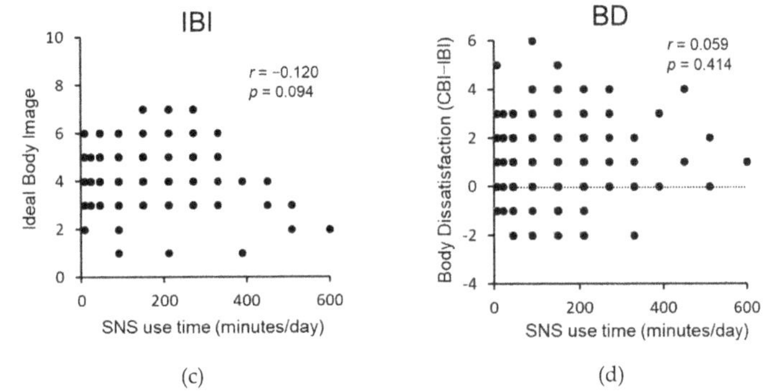

Figure 1. Correlations between the duration of SNS use and (**a**) body mass index, (**b**) current body image, (**c**) ideal body image, and (**d**) body dissatisfaction (all $n = 196$). Spearman's correlation. Body dissatisfaction was calculated as the difference between current and ideal body image. SNS = social networking sites.

Use of nutrition labels ($r = -0.159$, $p = 0.026$) and the frequency of consumption of milk and dairy products ($r = -0.195$, $p = 0.006$) were negatively correlated with the duration of SNS use. Frequency of consumption of fruit, fish, and alcohol was not significantly correlated with the duration of SNS use.

3.3. *Characteristics in the Long and Short Time Groups of SNS Use*

3.3.1. Daily SNS Use

Table 1 shows the daily SNS use, body shape, body image, Objectified Body Consciousness, and subjective feelings of the long and short groups. Compared to the short group, the long group accessed SNS more frequently ($p < 0.001$), took more selfies ($p = 0.001$), and posted more of them ($p = 0.001$) per day. There were no differences in the frequency of viewing others' photos of themselves (Table 1).

3.3.2. Body Shape

BMI was lower in the long group than in the short group ($p = 0.027$). Body weight status based on the BMI value did not differ significantly between the two groups.

3.3.3. Body Image

There was no difference in CBI between the two groups ($p = 0.182$). On the other hand, IBI was lower in the long group than in the short group ($p = 0.026$). There were no differences in body dissatisfaction, body perception, and body image distortion between the two groups (Table 1).

3.3.4. Objectified Body Consciousness

The total scores of Objectified Body Consciousness ($p = 0.032$) and body surveillance ($p = 0.002$) were higher in the long group than in the short group. There were no differences in body shame and appearance control between the two groups (Table 1).

3.3.5. Subjective Feelings

The subjective health score was higher in the long group than in the short group ($p = 0.038$). There was no difference in the subjective happiness score between the two groups (Table 1).

3.3.6. Eating Behaviors

Table 2 shows the eating behaviors of the long and short groups. There were significant differences in the frequency of consumption of fruits ($p = 0.017$), milk and dairy products ($p = 0.045$), and alcohol ($p = 0.043$) in the two groups. There was no difference in skipping breakfast, eating dinner within two hours of bedtime, and eating snacks after dinner between the two groups (Table 2).

Table 1. Daily SNS use, body shape, body image, objectified body consciousness, and subjective feelings by the duration of SNS use.

Variables	Long (n = 52)	Short (n = 54)	p-Value
Daily SNS use [1]			
Frequency of access [3]	4.9 ± 1.3	2.1 ± 1.3	**<0.001**
Frequency of viewing others' photos of themselves [4]	2.9 ± 1.0	2.8 ± 1.0	0.383
Number of selfies taken	0.4 ± 0.7	0.1 ± 0.4	**0.001**
Number of selfies posted	0.2 ± 0.4	0.0 ± 0.0	**0.001**
Body shape			
Body mass index (kg/m^2) [1]	19.6 ± 2.3	20.5 ± 2.0	**0.027**
Body weight status [2]			0.395
underweight (BMI < 18.5)	11 (21.2)	8 (14.8)	
normal (18.5 ≤ BMI < 25)	40 (76.9)	46 (85.2)	
overweight (25 ≤ BMI)	1 (1.9)	0 (0.0)	
Body Image			
Current body image (CBI) [1]	5.1 ± 1.8	5.6 ± 1.4	0.182
Ideal body image (IBI) [1]	4.0 ± 1.4	4.4 ± 1.1	**0.026**
Body dissatisfaction [1,5]	1.1 ± 1.6	1.2 ± 1.4	0.796
Body perception [2]			0.194
thin, slightly thin	9 (17.3)	4 (7.4)	
normal	10 (19.2)	21 (38.9)	
slightly overweight, overweight	33 (63.5)	29 (53.7)	
Body image distortion [2]			0.967
underestimation	1 (1.9)	1 (1.9)	
no distortion	18 (34.6)	20 (37.0)	
overestimation	33 (63.5)	33 (61.1)	
Objectified Body Consciousness [1,6]			
Body surveillance [7]	30.4 ± 5.5	26.7 ± 4.9	**0.002**
Body shame [7]	23.1 ± 6.0	21.8 ± 4.9	0.332
Appearance control [7]	26.3 ± 4.9	26.9 ± 3.9	0.551
Total	79.8 ± 10.1	75.3 ± 8.4	**0.032**
Subjective feelings [1]			
Health [8]	3.1 ± 0.5	2.8 ± 0.7	**0.038**
Happiness [9]	4.8 ± 1.2	4.7 ± 1.2	0.694

Note. SNS = social networking sites. [1] Values represent mean ± SD. p-values were analyzed via Mann–Whitney U. [2] Values represent n (%). p-values were analyzed by Chi-square test. [3] score range 0–6 (0 = hardly ever, 6 = more times than I can count). [4] score range 0–4 (0 = almost never, 4 = nearly every time I log on). [5] Body dissatisfaction represents the difference between CBI and IBI (IBI−CBI). [6] Long; n = 37, Short; n = 37 [7] score range 8–40 (8 items total score, 1 item score range 1 = strongly disagree, 5 = strongly agree). [8] score range 1–4 (1 = poor, 4 = excellent). [9] score range 1–7 (1 = not a very happy person, 7 = a very happy person). Bold: significant differences in p-Value.

Table 2. Eating behaviors by the duration of SNS use.

Variables	Long (n = 52)	Short (n = 54)	p-Value
Skipping breakfast (≥3 days/week)			0.157
Yes	34 (65.4)	42 (77.8)	
No	18 (34.6)	12 (22.2)	
Eating dinner within two hours of bedtime (≥3 days/week)			0.777
Yes	36 (69.2)	36 (66.7)	
No	16 (30.8)	18 (33.3)	
Eating snacks after dinner (≥3 days/week)			0.582
Yes	32 (61.5)	36 (66.7)	
No	20 (38.5)	18 (33.3)	
Nutrition label use			
hardly ever	9 (17.3)	6 (11.1)	0.059
Rarely	10 (19.2)	8 (14.8)	
Often	29 (55.8)	25 (46.3)	
every time	4 (7.7)	15 (27.8)	
Frequency of food consumption Fruits			**0.017**
Rarely	11 (21.2)	15 (27.8)	
1–2 days a month	17 (32.7)	9 (16.7)	
1–2 days a week	9 (17.3)	13 (24.1)	
3–4 days a week	12 (23.1)	5 (9.3)	
almost every day	3 (5.8)	12 (22.2)	
Fish			0.357
Rarely	10 (19.2)	6 (11.1)	
1–2 days a month	9 (17.3)	8 (14.8)	
1–2 days a week	27 (51.9)	29 (53.7)	
3–4 days a week	6 (11.5)	8 (14.8)	
almost every day	0 (0.0)	3 (5.6)	
Milk and Dairy products			**0.045**
Rarely	4 (7.7)	3 (5.6)	
1–2 days a month	8 (15.4)	3 (5.6)	
1–2 days a week	20 (38.5)	15 (27.8)	
3–4 days a week	10 (19.2)	8 (14.8)	
almost every day	10 (19.2)	25 (46.3)	
Alcohol			**0.043**
rarely	30 (57.7)	34 (63.0)	
1–2 days a month	11 (21.2)	10 (18.5)	
1–2 days a week	4 (7.7)	10 (18.5)	
3–4 days a week	3 (5.8)	0 (0.0)	
almost every day	4 (7.7)	0 (0.0)	

Note. SNS = social networking sites. Values represent n (%). p-values were analyzed using Chi-square test. Bold: significant differences in p-Value.

4. Discussion

This study had two main findings: (1) in non-obese young Japanese women, the longer the duration of daily SNS use, the significantly lower the BMI, use of nutrition labels, and frequency of consumption of milk and dairy products; and (2) when comparing subgroups of time spent on SNS, the long group had a lower BMI and IBI than the short group.

4.1. SNS Use and Actual and Ideal Body Shapes

A limited number of studies have investigated the association between time spent on SNS and physiques in young women. In contrast with the present finding that the more time young women spent on SNS, the lower their BMI, Wagner et al. [26] reported that there was no correlation between the frequency of posting selfies on SNS and BMI in female college students. This discrepancy is probably due to the methodological difference between the studies, in that Wagner et al. assessed the frequency of posting selfies, one form of SNS activity, whereas we examined the time spent engaging in SNS activity as a whole. This activity mainly consisted of posting and browsing, considering the importance of the total time spent on both components. In 2000, before SNS became an information source, it was suggested that the portrayal of thin models and celebrities on television and in fashion magazines would promote the drive for thinness in young women [27]. As their main source of information is currently shifting from television to SNS [7], our finding that a longer duration of daily SNS use can lead to a thinner physique in young women suggests that further attention needs to be paid to the modes of SNS use in order to reduce thinness among young women.

An interesting finding of the present study was that the long group had a thinner actual and ideal physiques than the short group. Tiggemann et al. [9,10] reported a positive correlation between time spent on SNS and the drive for thinness in adolescent girls. Fardouly et al. [28] found that the more female college students engaged in SNS use, as measured by the mean scores of time spent and frequency, the higher their drive for thinness and body dissatisfaction. The finding from these previous studies that SNS use is associated with increased drive for thinness in young women is in good agreement with our findings that women in the long group who spent more time on SNS had a thinner body size ideal. In the present study, however, no significant difference in body dissatisfaction was noted between the long and short groups. This could be explained by the fact that the long group actually had a thinner physique, with a mean BMI of 19.5 kg/m^2; therefore, they were less likely to develop greater body dissatisfaction.

4.2. SNS Use and Eating Behaviors

The analysis of time spent on SNS and EB revealed that the longer the duration of daily SNS use, the lower the frequency of referring to nutrition labels when purchasing food. The percentage of adult women in Japan [2] who cited SNS as a source of information influencing their diet was highest among those in their 20s (approximately 40%), which, together with our results, suggests that SNS information, rather than reliable information on nutrition labels, is used as a reference for food purchases in young women who spend more time on SNS.

Another interesting new finding of this study was that the longer the duration of daily SNS use, the lower frequency of consumption of milk and dairy products. Previous research on the associations between SNS use and EB found that scores on the Eating Attitudes Test-26 [29] and dieting behavior scores (which measure items such as the frequency of attempts to lose weight) [10] increased with the amount of time spent on SNS. However, to our knowledge, no studies have shown that SNS use affects specific food consumption. In light of this, why were the milk and dairy product intakes lower? In this regard, the study on food selection patterns in Japanese adolescent girls reported that body dissatisfaction was associated with lower frequency of milk consumption [30]. Additionally, a study of Japanese mothers of preschool children found that they were more likely to perceive that "milk and dairy products are high in energy due to their fat content and that consumption of these foods causes weight gain" [31]. In Japan, national surveys have shown a decline in dairy consumption over the past 20 years among non-obese women aged 20–39 years [32]. In reality, however, milk and dairy products are desirable for young women as they are rich sources of nutrients such as calcium and protein. Taken together with the finding on the use of nutrition labels, this finding indicates that the use of SNS as a source of information may have led to limited access to correct information, such as nutrition labels, leading to misconceived avoidance of healthy foods (milk and dairy products). It is plausible that overuse of SNS may be a barrier to access both health information and healthy foods.

4.3. Limitations

The limitations of this study were as follows: (1) A detailed dietary survey was not conducted, and data on participants' actual food or nutrient intake were not available. (2) SNS use was self-reported, so the possibility of underestimation cannot be excluded [33,34]. In the future, research using objective data logs and screen time applications to assess SNS use is needed [35]. (3) The survey focused primarily on time spent on SNS and not on content viewed and the use of individual application, all of which should be investigated in future studies. (4) The study design was cross-sectional, so cause and effect cannot be inferred. In addition, the possibility of reverse causation cannot be excluded [36]. Future longitudinal or interventional studies are required to confirm our findings. Despite the above limitations, to the best of our knowledge, the present study is the first report to demonstrate the negative impact of prolonged SNS use on both actual and ideal body shapes and healthy EB in non-obese young women.

5. Conclusions

The present results suggest that spending more time on SNS may be associated with thinner actual and ideal body shapes and poorer access to health information and healthy foods among young women. In order to prevent thinness and promote healthy EB among young women, it may be necessary to pay attention to the time spent on SNS.

Author Contributions: Conceptualization, Y.Y., N.S. and N.N.; methodology, Y.Y., N.S. and N.N.; validation, N.S.; formal analysis, Y.Y., N.S. and N.N.; investigation, Y.Y., Y.T. and N.N.; data curation, Y.Y., F.H. and N.N.; writing—original draft preparation, Y.Y.; writing—review and editing, Y.T., F.H., N.S. and N.N.; project administration, N.N.; funding acquisition, Y.Y. and N.N. All authors have read and agreed to the published version of the manuscript.

Funding: This research was funded by a grant from the Nakatani Suzuyo Memorial Fund for Nutrition and Dietetics and Health Labour Sciences Research Grant (22FA1023).

Institutional Review Board Statement: The study was conducted in accordance with the Declaration of Helsinki and approved by the School of Human Science and Environment Research Ethics Committee of the University of Hyogo (No. 192, 7 December 2018).

Informed Consent Statement: Informed consent was obtained from all subjects involved in the study.

Data Availability Statement: Raw data cannot be shared publicly as it is in a re-identifiable database. These restrictions were placed by the Research Committee of the School of Human Science and Environment, University of Hyogo.

Acknowledgments: The authors are grateful to Yuji Wada, Akiko Suganuma, and Kanako Ogura for assistance with the web survey.

Conflicts of Interest: The authors declare no conflict of interest.

References

1. NCD Risk Factor Collaboration: Adult Body-Mass Index. 2018. Available online: http://www.ncdrisc.org/data-downloads-adiposity.html (accessed on 26 February 2023).
2. Ministry of Health, Labour and Welfare. The National Health and Nutrition Survey in Japan. 2019. Available online: https://www.mhlw.go.jp/stf/seisakunitsuite/bunya/kenkou_iryou/kenkou/eiyou/r1-houkoku_00002.html (accessed on 27 February 2023).

3. Hamamoto, Y.; Suzuki, S.; Sugiura, M. Two components of body-image disturbance are differentially associated with distinct eating disorder characteristics in healthy young women. *PLoS ONE* **2022**, *17*, e0262513. [CrossRef]
4. Szegda, K.L.; Whitcomb, B.W.; Purdue-Smithe, A.C.; Boutot, M.E.; Manson, J.E.; Hankinson, S.E.; Rosner, B.A.; Bertone-Johnson, E.R. Adult adiposity and risk of early menopause. *Hum. Reprod.* **2017**, *32*, 2522–2531. [CrossRef]
5. Gluckman, P.D.; Hanson, M.A.; Pinal, C. The developmental origins of adult disease. *Matern. Child Nutr.* **2005**, *1*, 130–141. [CrossRef] [PubMed]
6. Jiotsa, B.; Naccache, B.; Duval, M.; Rocher, B.; Grall-Bronnec, M. Social Media Use and Body Image Disorders: Association between Frequency of Comparing One's Own Physical Appearance to That of People Being Followed on Social Media and Body Dissatisfaction and Drive for Thinness. *Int. J. Environ. Res. Public Health* **2021**, *18*, 2880. [CrossRef] [PubMed]
7. Ministry of Internal Affairs and Communications. Information and Communications in Japan. 2021. Available online: https://www.soumu.go.jp/johotsusintokei/whitepaper/eng/WP2021/2021-index.html (accessed on 27 February 2023).
8. Holland, G.; Tiggemann, M. A systematic review of the impact of the use of social networking sites on body image and disordered eating outcomes. *Body Image* **2016**, *17*, 100–110. [CrossRef]
9. Tiggemann, M.; Miller, J. The Internet and Adolescent Girls' Weight Satisfaction and Drive for Thinness. *Sex Roles* **2010**, *63*, 79–90. [CrossRef]
10. Tiggemann, M.; Slater, A.E. NetTweens: The Internet and body image concerns in preteenage girls. *J. Early Adolesc.* **2014**, *34*, 606–620. [CrossRef]
11. Cohen, R.; Newton-John, T.; Slater, A. The relationship between Facebook and Instagram appearance-focused activities and body image concerns in young women. *Body Image* **2017**, *23*, 183–187. [CrossRef]
12. Cohen, R.; Newton-John, T.; Slater, A. 'Selfie'-objectification: The role of selfies in self-objectification and disordered eating in young women. *Comput. Hum. Behav.* **2018**, *79*, 68–74. [CrossRef]
13. Rounsefell, K.; Gibson, S.; McLean, S.; Blair, M.; Molenaar, A.; Brennan, L.; Truby, H.; McCaffrey, T.A. Social media, body image and food choices in healthy young adults: A mixed methods systematic review. *Nutr. Diet.* **2020**, *77*, 19–40. [CrossRef]
14. Yao, L.; Niu, G.; Sun, X. Body Image Comparisons on Social Networking Sites and Chinese Female College Students' Restrained Eating: The Roles of Body Shame, Body Appreciation, and Body Mass Index. *Sex Roles* **2021**, *84*, 465–476. [CrossRef]
15. Yumen, Y.; Takayama, Y.; Suzuki, A.; Wada, Y.; Sakane, N.; Nagai, N. Development of the Japanese version of Body Image Scale, and assessment of its reliability and validity. *J. Jpn. Soc. Study Obes.* **2021**, *27*, 140–148. Available online: http://www.jasso.or.jp/contents/magazine/index.html (accessed on 26 February 2023). (In Japanese).
16. McKinley, N.M.; Hyde, J.S. The Objectified Body Consciousness Scale. *Psychol. Women Q.* **1996**, *20*, 181–215. [CrossRef]
17. Idler, E.L.; Angel, R.J. Self-rated health and mortality in the NHANES-I Epidemiologic Follow-up Study. *Am. J. Public Health* **1990**, *80*, 446–452. [CrossRef] [PubMed]
18. Lyubomirsky, S.; Lepper, H.S. A Measure of Subjective Happiness: Preliminary Reliability and Construct Validation. *Soc. Indic. Res.* **1999**, *46*, 137–155. [CrossRef]
19. Shimai, S.; Otake, K.; Utsuki, N.; Ikemi, A.; Lyubomirsky, S. Development of a Japanese version of the Subjective Happiness Scale (SHS), and examination of its validity and reliability. *[Nihon Koshu Eisei Zasshi] Jpn. J. Public Health* **2004**, *51*, 845–853. (In Japanese)
20. Ministry of Health, Labour and Welfare. Questionnaire Sheet for General Health Examinations. 2018. Available online: https://www.mhlw.go.jp/content/10900000/000496780.pdf (accessed on 27 February 2023).
21. Ministry of Health, Labour and Welfare. The National Health and Nutrition Survey in Japan. 2015. Available online: https://www.mhlw.go.jp/bunya/kenkou/eiyou/h27-houkoku.html (accessed on 27 February 2023).
22. Eguchi, E.; Iso, H.; Tanabe, N.; Wada, Y.; Yatsuya, H.; Kikuchi, S.; Inaba, Y.; Tamakoshi, A. Healthy lifestyle behaviours and cardiovascular mortality among Japanese men and women: The Japan collaborative cohort study. *Eur. Heart J.* **2012**, *33*, 467–477. [CrossRef]
23. Carbonell, X.; Panova, T. A critical consideration of social networking sites' addiction potential. *Addict. Res. Theory* **2016**, *25*, 48–57. [CrossRef]
24. Sidani, J.E.; Shensa, A.; Hoffman, B.; Hanmer, J.; Primack, B.A. The Association between Social Media Use and Eating Concerns among US Young Adults. *J. Acad. Nutr. Diet.* **2016**, *116*, 1465–1472. [CrossRef] [PubMed]
25. Baraldi, A.N.; Enders, C.K. An introduction to modern missing data analyses. *J. Sch. Psychol.* **2010**, *48*, 5–37. [CrossRef]
26. Wagner, C.N.; Aguirre Alfaro, E.; Bryant, E.M. The relationship between Instagram selfies and body image in young adult women. *First Monday* **2016**, *21*, 1–10. [CrossRef]
27. British Medical Association Board of Science and Education. *Eating Disorder, Body Image & the Media*; Morgan, D.R., Ed.; BMJ Publish Group: London, UK, 2000; pp. 1–66.
28. Fardouly, J.; Pinkus, R.T.; Vartanian, L.R. The impact of appearance comparisons made through social media, traditional media, and in person in women's everyday lives. *Body Image* **2017**, *20*, 31–39. [CrossRef]
29. Mabe, A.G.; Forney, K.; Keel, P.K. Do you "like" my photo? Facebook use maintains eating disorder risk. *Int. J. Eat. Disord.* **2014**, *47*, 516–523. [CrossRef]
30. Nakaoka, N.; Noda, S.; Yamada, A.; Togashi, Y.; Namiki, N.; Goseki-Sone, M. The Examination of Body Shape Misperception and Body Dissatisfaction of 5th and 6th Grade Students—Focusing on Physical Condition, Dietary/Exercise Habits, and Physical Strength. *J. Jpn. Soc. Shokuiku* **2020**, *14*, 27–39. (In Japanese) [CrossRef]

31. Sato, M. The Parents or Guardians' image of Milk, Cheese and Yogurt Is Influence Intake of their Calcium for Nursery School Children. *Jpn. Soc. Home Econ.* **2016**, *67*, 513–525. (In Japanese) [CrossRef]
32. Matsumoto, M.; Tajima, R.; Fujiwara, A.; Yuan, X.; Okada, E.; Takimoto, H. Trends in Food Group Intake According to Body Size among Young Japanese Women: The 2001–2019 National Health and Nutrition Survey. *Nutrients* **2022**, *14*, 4078. [CrossRef] [PubMed]
33. Ellis, D.A.; Davidson, B.I.; Shaw, H.; Geyer, K. Do smartphone usage scales predict behavior? *Int. J. Hum.-Comput. Stud.* **2019**, *130*, 86–92. [CrossRef]
34. Mahalingham, T.; McEvoy, P.M.; Clarke, P.J. Assessing the validity of self-report social media use: Evidence of No relationship with objective smartphone use. *Comput. Hum. Behav.* **2023**, *140*, 107567. [CrossRef]
35. Kaye, L.K.; Orben, A.; Ellis, D.A.; Hunter, S.C.; Houghton, S. The Conceptual and Methodological Mayhem of "Screen Time". *Int. J. Environ. Res. Public Health* **2020**, *17*, 3661. [CrossRef]
36. Hartanto, A.; Quek, F.Y.X.; Tng, G.Y.Q.; Yong, J.C. Does Social Media Use Increase Depressive Symptoms? A Reverse Causation Perspective. *Front. Psychiatry* **2021**, *12*, 641934. [CrossRef] [PubMed]

nutrients MDPI

Article

The Relationship between Social Media Use and Body Image in Lebanese University Students

Joanne M. Karam [1,2,3], Carol Bouteen [3], Yara Mahmoud [3], Josep A. Tur [4,5,6,*] and Cristina Bouzas [4,5,6]

1 Institut National de Santé Publique, d'Épidémiologie Clinique et de Toxicologie-Liban (INSPECT-LB), Beirut 1103 2180, Lebanon
2 Department of Natural Sciences, Lebanese American University, Beirut 1102 2801, Lebanon
3 School of Health Sciences, Modern University for Business and Science, Damour 113-7501, Lebanon
4 Research Group on Community Nutrition and Oxidative Stress, University of the Balearic Islands-IUNICS, E-07122 Palma de Mallorca, Spain; cristina.bouzas@uib.es
5 Health Research Institute of Balearic Islands (IdISBa), E-07120 Palma, Spain
6 CIBEROBN (Physiopathology of Obesity and Nutrition CB12/03/30038), Instituto de Salud Carlos III, E-28029 Madrid, Spain
* Correspondence: pep.tur@uib.es; Tel.: +34-971-1731; Fax: +34-971-173184

Abstract: Well-being is not only defined as being physically healthy; multiple factors can affect a person's well-being. Social media is strongly correlated with the body dissatisfaction of an individual. High exposure to lean and toned body shapes has created new standards and "idealized" body types. The aim of this article was to assess the relationship between social media and body image among university students in Lebanon. Data were obtained from 292 university students (median age: 22 years), selected from different Lebanese regions by using convenience sampling. Demographic data, social media addiction, body satisfaction, levels of physical activity, eating behaviors, and ultimate well-being were expressed as median and interquartile range. People who relied more on social media were younger than those who did not. Individuals addicted to social media had higher odds of having moderate and marked body image concerns. A significant association was found between social media addiction and emotional overeating, food responsiveness, and feeling hunger. These findings stress the need for rising regional and national awareness among social media users, especially the younger ones, and the implementation of intervention and prevention techniques to help prevent body image dissatisfaction, disordered eating patterns, and the alteration of overall well-being.

Keywords: body image; self-perception; social media addiction; social media use; university students; Lebanon

Citation: Karam, J.M.; Bouteen, C.; Mahmoud, Y.; Tur, J.A.; Bouzas, C. The Relationship between Social Media Use and Body Image in Lebanese University Students. *Nutrients* **2023**, *15*, 3961. https://doi.org/10.3390/nu15183961

Academic Editors: Hubertus Himmerich and Khadijeh Mirzaei

Received: 24 July 2023
Revised: 11 September 2023
Accepted: 12 September 2023
Published: 13 September 2023

1. Introduction

The World Health Organization (WHO) defines health as a "state of complete physical, mental, and social well-being" and not merely the absence of disease or infirmity [1]. Human beings seek to feel well in their lives, which makes well-being a vital human aim [2]. The concept of well-being describes how people feel about their lives, which may or may not be strongly correlated with the unchanging realities of people's lives [3]. According to the center for disease control, CDC, well-being is associated with job, family, health, and economically related benefits [2]. Psychological well-being specifically is a main building block of mental health and may be defined as enjoyment and pleasure on one hand and meaning and fulfillment on the other hand, as well as resilience (coping, emotional regulation, and healthy problem solving) [4].

Among university students, well-being is specifically important as it is a critical contributor to lifelong fulfilment [5], but it is at risk since the students encounter many academic difficulties, as well as social, emotional, and psychological ones [6].

Lebanon is undergoing several crises arising due to a massive economic collapse and the tragic August 4 Beirut port blast [7]. Young adults representing 28% of the total population were found to be most affected by these crises because they are the population most vulnerable to stressors since their future is at stake, and it is their time to shape their lives in a country that is crippling them [8].

Social media has become the primary source of communication for young people, and its usage has increased substantially among university students, who mainly use it for non-academic purposes [9]. It presents significant challenges for the younger generation since it allows them to discover the world as a filtered image rather than portraying the real world [10]. Social media is responsible for the new perception of what is called an "ideal man" or "ideal woman". Those terms refer to men or women socially perceived as attractive, according to sociocultural norms [11]. Moreover, this is considered the primary cause of the exertion of extra pressure on the public, which leads to body dissatisfaction due to the unattainable standards that are set in society [12]. The social comparison theory might explain this association as it asserts that individuals tend to compare themselves to others who are similar [13]. This theory posits that individuals evaluate themselves by comparing their abilities and opinions to those of others, which influences their self-perception and emotional well-being [14].

People frequently display an idealized picture of themselves on social media, adding only the most appealing photographs of themselves to their profile [15]; those pictures might be even changed using applications and filters. Even though social media provides images of a variety of individuals (e.g., friends, family, strangers, and celebrities), it is mainly used to communicate with one's peers [16].

Evidence shows that peer comparisons may have a particularly strong impact on body image [17]. Physique dissatisfaction is frequently tested by asking people to rate their actual body with their ideal form; as with social comparison theory, the gap between the two is the level of body dissatisfaction [12]. Body image and psychological well-being are inextricably linked; body dissatisfaction is mainly based on a person's negative thoughts about their appearance, which are frequently influenced by social experiences like media representations [18]. In recent years, body dissatisfaction has become a serious worry for both men and women, particularly women [19,20]. Body dissatisfaction has previously been related to unhealthy lifestyles, whereas body satisfaction has been related to healthier lifestyles [20]. Therefore, authors hypothesize that the usage of social media could be linked to body dissatisfaction and to lifestyle.

Understanding and investigating students' body dissatisfaction as influenced by the effects of social media is significantly important in assessing the need to develop appropriate education on health awareness and the appropriate use of social media. Such studies are lacking, especially among university students in Lebanon during the crisis. This study aims to assess the relationship between social media and body image among university students in Lebanon.

2. Methods

2.1. Design and Sample Size

This study is a cross-sectional survey. Participants were asked to complete an online questionnaire including six sections. This survey was designed to assess the effect of social media on body image. For data collection purposes, the online survey was distributed throughout different social media platforms. The inclusion criterion for subjects to participate in this study consisted of being Lebanese university students registered in different universities across all Lebanese regions. Exclusion criteria included being a Lebanese student outside Lebanon and being a non-Lebanese student in a university in Lebanon. The subject population included both genders regardless of the educational level or occupation of the participants. A convenience sampling technique was used in this research. The sample size was 292 participants. Data collection took place in November and December 2021.

This study received approval from the ethical committee of the Modern University of Business and Science (approval reference MU-20211105-27). The survey included a consent form. Subjects had the full right to drop out of the study at any time they wanted without giving any reason. Participants were asked to complete a 10 min questionnaire. All questionnaire responses were confidential and anonymous, and the analyzed data were non-identifiable.

2.2. Assessment Tools

The questionnaire, distributed in English, included 6 sections: (1) sociodemographic. (2) The Bergen social media usage scale (BSMAS), a six-item scale used to assess addiction to social media [21], from which scores were calculated according to a certain ranking (any level above 18 was considered an addiction to social media). In the present study, the internal consistency of BSMAS was quite acceptable (Cronbach $\alpha = 0.762$). (3) The Body Shape Questionnaire 16-Item Version (BSQ-16), a self-reported measurement of body shape concerns [22]; scores were also recorded and classified in a system split into four categories: 'no concern' (scoring less than 38), 'mild concern' (scoring between 38 and 51), 'moderate concern' (scoring between 52 and 66), and 'marked concern with shape' (scoring over 66) (Cronbach $\alpha = 0.945$). (4) The Godin leisure scale, a four-item scale used to assess physical activity [23], in which scoring was as follows: active (scoring 24 units or more), moderately active (scoring between 14 and 23 units), and insufficiently active/sedentary (less than 14 units), and "how often they work up a sweat", to which the possible answers were often, sometimes, and never (Cronbach $\alpha = 0.649$). (5) The Adult Eating Behavior Questionnaire (AEBQ), used to assess positive and negative beliefs about food and eating [24] and which consists of 35 items, was rated on a 5-point Likert-type scale (1–strongly disagree; 5–strongly agree); these items were split into sub-categories which are as follows: hunger, food responsiveness, emotional eating, enjoyment of food, satiety responsiveness, emotional undereating, food fussiness, and slow eating (Cronbach α for each sub category = 0.700–0.895). (6) The last part of the questionnaire was the World Health Organization-5 Well-being Index (WHO-5) validated in Lebanon and used to assess psychological well-being [25,26], which consisted of five statements, rated according to the following scale: all of the time 5, most of the time 4, more than half of the time 3, less than half of the time 2, some of the time 1, at no time 0 (Cronbach $\alpha = 0.843$). The total raw score, ranging from 0 to 25, was multiplied by 4 to give the final score, with 0 representing the worst imaginable well-being and 100 representing the best imaginable well-being. All variables were calculated as the authors of the questionnaire and the validation study have stated.

Addiction status was classified according to the total score of BSMAS. The median value was 18 points; therefore, a total score > 18 was an addiction indicator, while a total score ≤ 18 was a no-addiction indicator. All variables from BSQ-16, weekly leisure-time activity, and WHO-5 that were not dichotomous were transformed considering each specific study issue as a cutoff point. All variables from AEBQ that were not dichotomous were transformed taking the median as a cutoff point.

2.3. Statistics

The SPSS statistical software package version 27.0 (SPSSS Inc., Chicago, IL, USA) was used to perform analyses. The distribution of variables was assessed by performing the Kolmogorov–Smirnov test. The data are shown as median and interquartile range (IQR), and differences among groups were analyzed by performing the Mann–Whitney U test because the variables did not follow a normal distribution. The prevalence is shown as sample size and percentage. The difference in prevalence among groups was analyzed using χ^2 (all p values are two-tailed). The association between body shape concern, physical activity level, well-being status, and adult eating behaviors (dependent variables) and addiction status (independent variables) was analyzed by calculating the odds ratio (OR). For each item, two ORs were calculated: crude and adjusted by sociodemographic characteristics (age, gender, and university student). Results were considered statistically significant when the p-value < 0.05.

3. Results

Sociodemographic characteristics according to social media addiction (no addiction: total BSMAS score $\leq$ 18; addiction: total BSMAS score > 18) are shown in Table 1. Participants with a major social media addiction were slightly younger (21 years) than those who did not have an addiction.

Table 1. Sociodemographic characteristics of participants according to social media addiction.

	No Addiction (n = 166)	Addiction (n = 154)	p-Value
Age (years) *	22.0 (8.0)	21.0 (4.0)	0.017
Gender (female; n; %)	138 (83.1)	124 (80.5)	0.544
Province (n; %)			0.916
Beirut	17 (10.3)	16 (10.5)	
Mount Lebanon	119 (72.1)	108 (70.6)	
North of Lebanon	2 (1.2)	1 (0.7)	
South of Lebanon	10 (6.1)	10 (6.5)	
Begaa	16 (9.7)	15 (9.8)	
Nabatiyeh	1 (0.6)	3 (2.0)	

* Values are median (IQR: interquartile range). Differences between groups were tested by performing the Mann–Whitney U test. Differences in prevalence across groups were examined using χ^2.

Body shape concern, weekly leisure-time activity, and well-being index according to social media addiction are available in Table 2. Most people without social media addiction are not concerned with their body shape (63.9%). In the social media addiction group, the percentage of participants who had moderate and marked concern with their body shape was notably high (31.2%).

Table 2. Body shape concern, weekly leisure-time activity, and well-being index according to social media addiction.

	No Addiction (n = 166)	Addiction (n = 154)	p-Value
BSQ-16 (n; %)			0.001
No concern with shape	106 (63.9)	74 (48.1)	
Mild concern with shape	37 (22.3)	32 (20.8)	
Moderate and marked concern with shape	23 (13.9)	48 (31.2)	
Weekly Leisure-Time Activity (n; %)			0.563
Insufficiently Active/Sedentary	67 (40.4)	55 (35.7)	
Moderately Active	24 (14.5)	28 (18.2)	
Active	75 (45.2)	71 (46.1)	
WHO-5 (n; %)			0.137
Worst possible well-being	2 (1.2)	6 (3.9)	
Poor well-being	44 (26.5)	51 (33.1)	
Mild well-being	66 (39.8)	60 (39.0)	
Good and best possible well-being	54 (32.5)	37 (24.0)	

Abbreviations: BSQ-16: Body Shape Questionnaire 16-Item Version; WHO-5: World Health Organization-5 Well-being Index. Differences in prevalence were examined using χ^2.

Eating behaviors according to social media addiction are shown in Table 3. Regarding the enjoyment of food, participants obtained the same score in both groups (4.0). Nevertheless, the scores referring to emotional overeating, food responsiveness, and hunger were higher in the social media addiction group participants, at (3.2)/(3.3)/(3.4), respectively.

Table 3. Eating behaviors according to social media addiction.

	No Addiction (*n* = 166)	Addiction (*n* = 154)	*p*-Value
AEBQ *			
Enjoyment of food (EF)	4.0 (1.0)	4.0 (1.0)	0.028
Emotional overeating (EOE)	2.6 (1.4)	3.2 (1.6)	<0.001
Emotional undereating (EUE)	3.2 (1.4)	3.0 (1.6)	0.409
Food fussiness (FF)	2.2 (1.0)	2.4 (1.1)	0.245
Food responsiveness (FR)	3.0 (1.0)	3.3 (0.8)	<0.001
Hunger (H)	3.0 (0.8)	3.4 (1.0)	<0.001
Slowness in eating (SE)	2.8 (1.3)	2.8 (1.5)	0.705
Satiety responsiveness (SR)	2.8 (1.0)	3.0 (1.2)	0.067

* Values are median (IQR: interquartile range). Abbreviation: AEBQ: Adult Eating Behavior Questionnaire. Differences between groups were tested by performing the Mann–Whitney U test.

In Table 4, crude and adjusted ORs were calculated to find an association between body shape concern, weekly leisure-time activity, well-being index, and social media addiction. No addiction was established as the reference. Crude and adjusted analysis showed that students addicted to social media had higher odds of reporting moderate and marked concern with shape (adjusted OR: 0.48; 95% CI: 0.30–0.77). Accordingly, students with social media addiction were less likely to report no concerns with body shape than students without addiction (adjusted OR: 3.01; 95% CI: 1.70–5.33).

Table 4. Association between body shape concern, weekly leisure-time activity, and well-being index and social media addiction.

		No Addiction (*n* = 166) OR (95% CI)	Addiction (*n* = 154) OR (95% CI)	*p*-Value
BSQ-16				
No concern with shape	Crude OR	1.00 (ref.)	0.52 (0.34–0.82)	0.005
	Adjusted OR	1.00 (ref.)	0.48 (0.30–0.77)	0.002
Mild concern with shape	Crude OR	1.00 (ref.)	0.91 (0.54–1.56)	0.743
	Adjusted OR	1.00 (ref.)	0.94 (0.54–1.64)	0.830
Moderate and marked concern with shape	Crude OR	1.00 (ref.)	2.82 (1.61–4.91)	<0.001
	Adjusted OR	1.00 (ref.)	3.01 (1.70–5.33)	<0.001
Weekly Leisure-Time Activity				
Insufficiently Active/Sedentary	Crude OR	1.00 (ref.)	0.82 (0.52–1.29)	0.393
	Adjusted OR	1.00 (ref.)	0.88 (0.55–1.39)	0.576
Moderately Active	Crude OR	1.00 (ref.)	1.32 (0.73–2.39)	0.368
	Adjusted OR	1.00 (ref.)	1.40 (0.76–2.58)	0.288
Active	Crude OR	1.00 (ref.)	1.04 (0.67–1.61)	0.868
	Adjusted OR	1.00 (ref.)	0.95 (0.60–1.49)	0.818
WHO-5				
Worst well-being	Crude OR	1.00 (ref.)	3.32 (0.66–16.73)	0.145
	Adjusted OR	1.00 (ref.)	2.68 (0.53–13.64)	0.235
Poor well-being	Crude OR	1.00 (ref.)	1.37 (0.85–2.22)	0.197
	Adjusted OR	1.00 (ref.)	1.45 (0.88–2.37)	0.143
Mild well-being	Crude OR	1.00 (ref.)	0.97 (0.62–1.52)	0.884
	Adjusted OR	1.00 (ref.)	0.95 (0.60–1.49)	0.807
Good and best well-being	Crude OR	1.00 (ref.)	0.66 (0.40–1.07)	0.093
	Adjusted OR	1.00 (ref.)	0.66 (0.40–1.08)	0.098

Values are OR (95% CI). Abbreviations: BSQ-16: Body Shape Questionnaire 16-Item Version; CI: confidence interval; OR: odds ratio; adjusted OR: odds ratio adjusted by sociodemographic characteristics (age, gender, university student); WHO-5: World Health Organization-5 Well-being Index.

Table 5 shows crude and adjusted ORs for association between eating behaviors and social media addiction. No addiction was established as the reference (1.00). In all cases, crude and adjusted analyses showed that the OR for the addiction group was higher than for the no-addiction group for "Emotional overeating" (adjusted OR: 2.2; 95% CI: 1.42–3.54), "Food responsiveness" (adjusted OR: 2.13; 95% CI: 1.35–3.36), and "Hunger" (adjusted OR: 2.16; 95% CI: 1.37–3.41) items. Hence, students with social media addiction were more likely to report emotional overeating, food responsiveness, and hunger than students without social media addiction.

Table 5. Association between eating behaviors and social media addiction.

		No Addiction (n = 166) OR (95% CI)	Addiction (n = 154) OR (95% CI)	p-Value
Adult Eating Behavior Questionnaire (AEBQ)				
Enjoyment of food (EF)	Crude OR	1.00 (ref.)	1.33 (0.85–2.07)	0.209
	Adjusted OR	1.00 (ref.)	1.25 (0.79–1.97)	0.343
Emotional overeating (EOE)	Crude OR	1.00 (ref.)	2.24 (1.44–3.51)	<0.001
	Adjusted OR	1.00 (ref.)	2.25 (1.42–3.54)	0.001
Emotional undereating (EUE)	Crude OR	1.00 (ref.)	0.76 (0.49–1.19)	0.230
	Adjusted OR	1.00 (ref.)	0.73 (0.46–1.15)	0.176
Food fussiness (FF)	Crude OR	1.00 (ref.)	1.10 (0.70–1.73)	0.672
	Adjusted OR	1.00 (ref.)	1.09 (0.69–1.72)	0.710
Food responsiveness (FR)	Crude OR	1.00 (ref.)	2.31 (1.47–3.61)	<0.001
	Adjusted OR	1.00 (ref.)	2.13 (1.35–3.36)	0.001
Hunger (H)	Crude OR	1.00 (ref.)	2.18 (1.39–3.41)	0.001
	Adjusted OR	1.00 (ref.)	2.16 (1.37–3.41)	0.001
Slowness in eating (SE)	Crude OR	1.00 (ref.)	1.02 (0.66–1.58)	0.934
	Adjusted OR	1.00 (ref.)	0.99 (0.63–1.56)	0.954
Satiety responsiveness (SR)	Crude OR	1.00 (ref.)	1.48 (0.93–2.35)	0.097
	Adjusted OR	1.00 (ref.)	1.47 (0.91–2.37)	0.118

Values are OR (95% CI). Abbreviations: CI: confidence interval; OR: odds ratio; adjusted OR: odds ratio adjusted by sociodemographic characteristics (age, gender, university student).

4. Discussion

The current study explored the effect of social media on body image among university students in Lebanon. Younger participants displayed major social media addiction compared to older participants. Social media addiction was also more dominant among young women (80.5%) than among men. Similar findings were obtained in Turkey [27] and Romania [28]. These results are also consistent with prior reports of social media addiction, where findings showed that young people [29–35] and females [36] were more prone to social media addiction. The tendency of the young population to be addicted to social media can be attributed to their need to express their personality, achieve dominance, escape family pressure, overcome loneliness, and earn social approval, in addition to coping with psychological disorders, economic problems, and physical inabilities [27]. Negative mood states are also considered causes of social media addiction among young people; these include timidity, depression, social phobia, and worry about the future. Hence, a strong body of evidence relates youth and social media addiction. However, other studies have shown significant relationships between males and older adults with problematic social media use or who have not displayed any relationship [37–40], which reflects that there is no agreement among studies on the association of a particular sociodemographic with social media addiction.

A positive association between no addiction to social media and no concerns about body shape was determined in this current study. Moreover, the percentage of participants having moderate and marked concerns with body shape was significantly higher among the social media addiction group compared to the group not addicted to social media. Previous studies examining the relationship between social media addiction and body image concerns yielded comparable results [41,42], especially among females [12,43–45].

As has been pointed out [46], the digital filters that hide flaws set an unrealistic standard of beauty. Moreover, studies have shown that internalizing the portrayed images exposed by the media leads to a desire to achieve ideal beauty [47,48]. This association can be explained by the social comparison theory, suggesting that individuals tend to compare themselves to others who are similar to them [13]. In fact, the significant Western influence in the media in Lebanon has spread perceptions of a 'perfect body' and thus led to increased body concerns [45]. However, these results contradict other studies that did not find a direct association between social media addiction and body concerns or dissatisfaction [49–52].

The social media addiction group obtained higher scores for emotional overeating, food responsiveness, and hunger. These current results confirm previous studies among Lebanese university students, where eating disorders (mainly emotional and restrained eating) were significantly linked to social media influence and pressure [53], and separation anxiety from technological devices was associated with the risk of eating disorders [54]. Similar findings were obtained in a recent study evaluating the relationship between social media addiction and eating behavior and disorder risk in university students during the pandemic period in Turkey, where emotional eating behavior was higher among social media addiction participants [55]. This was also supported by the literature on the impact of social media addiction on eating behaviors [56–59]. Research in America has also shown that social media use can affect eating choices and increase individuals' unaware food intake despite their not feeling hungry [60]. In fact, the use of social media can influence eating choices and food intake due to product marketing and excessive energy-dense food exposure [61], in addition to peer pressure, as young adults tend to pay increased attention to products recommended by friends on social media [62]. Emotional eating can also result in overeating [63] and higher energy-dense food intake in attempts to rapidly reduce negative moods [64].

Strengths and Limitations

The current study addresses a currently hot topic that affects health and well-being. Moreover, it addresses students, who are at most risk of developing social media addiction, specifically in the context of Lebanon, where it might become an escape mechanism due to the escalating humanitarian crisis [7], with no previous studies having tackled this subject since the crisis boomed in 2019. Tools used for the current research were previously validated, providing the article with scientific soundness. Nevertheless, this study also has limitations. It was conducted using online self-reported data, which introduces a risk of memory bias. Moreover, the studied population is limited to Lebanese university students, which might be a hindrance in extending the findings to other populations. In addition, not all the instruments used were previously validated in Lebanon. Last but not least, due to the cross-sectional nature of the study, there is a potential reverse causation, as its cross-sectional design does not allow it to explore whether the behaviors are an outcome or an antecedent of social media use [65]. Therefore, the use of social media and body concerns or behaviors could have an inherent relationship that cannot be ascertained due to the study's design.

5. Conclusions

Social media addiction among Lebanese students is related to body image concerns and several behaviors/feelings such as emotional overeating, food responsiveness, and feeling hunger. Such associations increase the risk of unhealthy behaviors, especially those related to food intake, which will directly affect health.

These findings should raise awareness among social media users, especially younger users, of the need to avoid the detrimental effects on health of social media overuse. These findings should be taken into consideration by authorities regulating social media access and when developing programs to address eating behavior disorders.

Longitudinal research would be advisable to be able to better establish causality and to check possible interventions to improve such associations.

Author Contributions: J.M.K., J.A.T. and C.B. (Cristina Bouzas) provided literature searches and reviews and prepared the main outline of the manuscript and approved the final manuscript. All authors (J.M.K., C.B. (Carol Bouteen), Y.M., J.A.T. and C.B. (Cristina Bouzas)) contributed substantially to the acquisition of data or analysis and interpretation of data. All authors revised the article critically for important intellectual content. All authors have read and agreed to the published version of the manuscript.

Funding: Instituto de Salud Carlos III through CIBEROBN CB12/03/30038, which is co-funded by the European Regional Development Fund. Other funding received: IdISBa grants (FOLIUM, PRIMUS, SYNERGIA, and LIBERI). The funding sponsors had no role in the design of the study; in the collection, analyses, or interpretation of the data; in the writing of the manuscript; or in the decision to publish the results.

Institutional Review Board Statement: Research Ethics Committees from all recruitment centers approved the study protocol, according to the ethical standards of the Declaration of Helsinki. This study received approval from the ethical committee of the Modern University of Business and Science (approval reference MU-20211105-27). All participants were informed of the purpose and the implications of the study, and all provided the written informed consent to participate.

Informed Consent Statement: The results and writing of this manuscript followed the Committee on Publication Ethics (COPE) guidelines on how to deal with potential acts of misconduct, maintaining integrity of the research and its presentation and following the rules of good scientific practice, with trust in the journal, in the professionalism of scientific authorship, and in the entire scientific endeavor. Written informed consent has been obtained from the participant(s) to publish this paper if applicable.

Data Availability Statement: There are restrictions on the availability of the data in the research, due to the signed consent agreements around data sharing, which only allow access to external researchers for studies following the project purposes. Requestors wishing to access the data used in this study can make a request to the last author: cristina.bouzas@uib.es. The request will then be passed to the authors for deliberation.

Acknowledgments: The authors especially thank Eman Swaidan, student at the Modern University for Business and Science, Damour, Lebanon, and Marina Yammine, student at the Department of Nutritional Sciences, Faculty of Health Sciences, University of Balamand, Koura, Lebanon for their exceptional effort.

Conflicts of Interest: The authors declare no conflict of interest.

Abbreviations

AEBQ: Adult Eating Behavior Questionnaire; BSMAS: Bergen social media usage scale; BSQ-16: Body Shape Questionnaire 16-Item Version; IQR: interquartile range; OR: odds ratio; WHO-5: World Health Organization-5 Well-being Index.

References

1. World Health Organization. Who Remains Firmly Committed to the Principles Set out in the Preamble to the Constitution. 2023. Available online: https://www.who.int/about/governance/constitution (accessed on 7 July 2023).
2. Centers for Disease Control and Prevention. Well-Being Concepts. 2018. Available online: https://www.cdc.gov/hrqol/wellbeing.htm (accessed on 7 July 2023).
3. Abdullahi, A.M.; Orji, R.; Rabiu, A.M.; Kawu, A.A. Personality and Subjective Well-Being: Towards Personalized Persuasive In-terventions for Health and Well-Being. *Online J. Public Health Inform.* **2020**, *12*, e1. [CrossRef] [PubMed]
4. Tang, Y.-Y.; Tang, R.; Gross, J.J. Promoting Psychological Well-Being Through an Evidence-Based Mindfulness Training Program. *Front. Hum. Neurosci.* **2019**, *13*, 237. [CrossRef] [PubMed]
5. Ross, D.A.; Hinton, R.; Melles-Brewer, M.; Engel, D.; Zeck, W.; Fagan, L.; Herat, J.; Phaladi, G.; Imbago-Jácome, D.; Anyona, P.; et al. Adolescent Well-Being: A Definition and Conceptual Framework. *J. Adolesc. Health* **2020**, *67*, 472–476. [CrossRef] [PubMed]
6. Visser, M.; Law-Van, W.E. University Students' Mental Health and Emotional Wellbeing during the COVID-19 Pandemic and Ensuing Lockdown. *S. Afr. J. Psychol.* **2021**, *51*, 229–243. [CrossRef]
7. Abouzeid, M.; Halwani, D.A.; Mokdad, A.H.; Habib, R.R. A generation at risk: The impacts of lebanon's escalating humanitarian crisis on children. *Front. Public Health* **2021**, *9*, 704678. [CrossRef] [PubMed]
8. Moussa, S.; Malaeb, D.; Obeid, S.; Hallit, S. Correlates of Positivity Among a Sample of Lebanese University Students. *Front. Psychol.* **2022**, *13*, 880437. [CrossRef]

9. Kolhar, M.; Kazi, R.N.A.; Alameen, A. Effect of social media use on learning, social interactions, and sleep duration among university students. *Saudi J. Biol. Sci.* **2021**, *28*, 2216–2222. [CrossRef]
10. Bozzola, E.; Spina, G.; Agostiniani, R.; Barni, S.; Russo, R.; Scarpato, E.; Di Mauro, A.; Di Stefano, A.V.; Caruso, C.; Corsello, G.; et al. The Use of Social Media in Children and Adolescents: Scoping Review on the Potential Risks. *Int. J. Environ. Res. Public Health* **2022**, *19*, 9960. [CrossRef]
11. Buote, V.M.; Wilson, A.E.; Strahan, E.J.; Gazzola, S.B.; Papps, F. Setting the bar: Divergent sociocultural norms for women's and men's ideal appearance in real-world contexts. *Body Image* **2011**, *8*, 322–334. [CrossRef]
12. Aparicio-Martinez, P.; Perea-Moreno, A.-J.; Martinez-Jimenez, M.P.; Redel-Macías, M.D.; Pagliari, C.; Vaquero-Abellan, M. Social Media, Thin-Ideal, Body Dissatisfaction and Disordered Eating Attitudes: An Exploratory Analysis. *Int. J. Environ. Res. Public Health* **2019**, *16*, 4177. [CrossRef]
13. Festinger, L. A Theory of social comparison processes. *Hum. Relat.* **1954**, *7*, 117–140. [CrossRef]
14. Powdthavee, N. Social Comparison Theory. In *Encyclopedia of Well-Being and Well-Being Research*; Michalos, A.C., Ed.; Springer: Dordrecht, The Netherlands, 2014; pp. 6028–6029. [CrossRef]
15. Pedalino, F.; Camerini, A.-L. Instagram Use and Body Dissatisfaction: The Mediating Role of Upward Social Comparison with Peers and Influencers among Young Females. *Int. J. Environ. Res. Public Health* **2022**, *19*, 1543. [CrossRef] [PubMed]
16. Rounsefell, K.; Gibson, S.; McLean, S.; Blair, M.; Molenaar, A.; Brennan, L.; Truby, H.; McCaffrey, T.A. Social media, body image and food choices in healthy young adults: A mixed methods systematic review. *Nutr. Diet.* **2020**, *77*, 19–40. [CrossRef]
17. Papageorgiou, A.; Fisher, C.; Cross, D. "Why don't I look like her?" How adolescent girls view social media and its connection to body image. *BMC Womens Health* **2022**, *22*, 261. [CrossRef]
18. Mahon, C.; Hevey, D. Processing Body Image on Social Media: Gender Differences in Adolescent Boys' and Girls' Agency and Active Coping. *Front. Psychol.* **2021**, *12*, 626763. [CrossRef]
19. Quittkat, H.L.; Hartmann, A.S.; Düsing, R.; Buhlmann, U.; Vocks, S. Body Dissatisfaction, Importance of Appearance, and Body Appreciation in Men and Women Over the Lifespan. *Front. Psychiatry* **2019**, *10*, 864. [CrossRef] [PubMed]
20. Bouzas, C.; Bibiloni, M.d.M.; Tur, J.A. Relationship between Body Image and Body Weight Control in Overweight $\geq$ 55-Year-Old Adults: A Systematic Review. *Int. J. Environ. Res. Public Health* **2019**, *16*, 1622. [CrossRef] [PubMed]
21. Andreassen, C.S.; Billieux, J.; Griffiths, M.D.; Kuss, D.J.; Demetrovics, Z.; Mazzoni, E.; Pallesen, S. The relationship between addictive use of social media and video games and symptoms of psychiatric disorders: A large-scale cross-sectional study. *Psychol. Addict. Behav.* **2016**, *30*, 252–262. [CrossRef]
22. Evans, C.; Dolan, B. Body Shape Questionnaire: Derivation of shortened "alternate forms". *Int. J. Eat Disord.* **1993**, *13*, 315–321. [CrossRef]
23. Godin, G.; Shephard, R.J. A simple method to assess exercise behavior in the community. *Can. J. Appl. Sport Sci.* **1985**, *10*, 141–146.
24. Hunot, C.; Fildes, A.; Croker, H.; Llewellyn, C.H.; Wardle, J.; Beeken, R.J. Appetitive traits and relationships with BMI in adults: Development of the Adult Eating Behaviour Questionnaire. *Appetite* **2016**, *105*, 356–363. [CrossRef] [PubMed]
25. World Health Organization. Regional Office for Europe. Wellbeing Measures in Primary Health Care/the DepCare Project: Report on a WHO Meeting: Stockholm, Sweden, 12–13 February 1998. World Health Organization. Regional Office for Europe. 1998. Available online: https://apps.who.int/iris/handle/10665/349766 (accessed on 7 July 2023).
26. Sibai, A.M.; Chaaya, M.; Tohme, R.A.; Mahfoud, Z.; Al-Amin, H. Validation of the Arabic version of the 5-item WHO well being index in elderly population. *Int. J. Geriatr. Psychiatry* **2009**, *24*, 106–107. [CrossRef] [PubMed]
27. Kırık, A.M.; Arslan, A.; Çetinkaya, A.; Gül, M. A Quantitative Research on the Level of Social Media Addiction Among Young People in Turkey. *Int. J. Sci. Cult. Sport* **2015**, *3*, 108–122.
28. Stănculescu, E.; Griffiths, M.D. Social media addiction profiles and their antecedents using latent profile analysis: The contri-bution of social anxiety, gender, and age. *Telemat. Inf.* **2022**, *74*, 101879. [CrossRef]
29. Andreassen, C.S.; Torsheim, T.; Brunborg, G.S.; Pallesen, S. Development of a Facebook Addiction Scale. *Psychol. Rep.* **2012**, *110*, 501–517. [CrossRef] [PubMed]
30. Andreassen, C.S.; Griffiths, M.D.; Gjertsen, S.R.; Krossbakken, E.; Kvam, S.; Pallesen, S. The relationships between behavioral ad-dictions and the five-factor model of personality. *J. Behav. Addict.* **2013**, *2*, 90–99. [CrossRef]
31. Dailey, S.L.; Howard, K.; Roming, S.M.P.; Ceballos, N.; Grimes, T. A biopsychosocial approach to understanding social media addiction. *Hum. Behav. Emerg. Technol.* **2020**, *2*, 158–167. [CrossRef]
32. Griffiths, M.D.; Kuss, D.J.; Demetrovics, Z. Social networking addiction: An overview of preliminary findings. In *Behavioral Addictions: Criteria, Evidence and Treatment*; Ros-enberg, K., Feder, L., Eds.; Elsevier: New York, NY, USA, 2014; pp. 119–141.
33. Monacis, L.; De Palo, V.; Griffiths, M.D.; Sinatra, M. Social networking addiction, attachment style, and validation of the Italian version of the Bergen Social Media Addiction Scale. *J. Behav. Addict.* **2017**, *6*, 178–186. [CrossRef]
34. Shensa, A.; Escobar-Viera, C.G.; Sidani, J.E.; Bowman, N.D.; Marshal, M.P.; Primack, B.A. Problematic social media use and depressive symptoms among U.S. young adults: A nationally-representative study. *Soc. Sci. Med.* **2017**, *182*, 150–157. [CrossRef]
35. Turel, O.; Serenko, A. The benefits and dangers of enjoyment with social networking websites. *Eur. J. Inf. Syst.* **2012**, *21*, 512–528. [CrossRef]
36. Kircaburun, K.; Griffiths, M.D. Instagram addiction and the Big Five of personality: The mediating role of self-liking. *J. Behav. Addict.* **2018**, *7*, 158–170. [CrossRef] [PubMed]
37. Çam, E.; Isbulan, O. A new addiction for teacher candidates: Social networks. *Turk. Online J. Educ. Technol.-TOJET* **2012**, *11*, 14–19.

38. Floros, G.; Siomos, K. The relationship between optimal parenting, Internet addiction and motives for social networking in adolescence. *Psychiatry Res.* **2013**, *209*, 529–534. [CrossRef] [PubMed]
39. Koc, M.; Gulyagci, S. Facebook addiction among Turkish college students: The role of psychological health, demographic, and usage characteristics. *Cyberpsychology Behav. Soc. Netw.* **2013**, *16*, 279–284. [CrossRef]
40. Tang, J.; Chang, Y.; Aggarwal, C.; Liu, H. A survey of signed network mining in social media. *ACM Comput. Surv.* **2016**, *49*, 1–37. [CrossRef]
41. Ryding, C.F.; Kuss, D.J. The use of social networking sites, body image dissatisfaction and Body Dysmorphic Disorder: A review of psychological research. *Psychol Pop Media Cult.* **2020**, *9*, 412–435. [CrossRef]
42. Yurdagül, C.; Kircaburun, K.; Emirtekin, E.; Wang, P.; Griffiths, M.D. Psychopathological consequences related to problematic Instagram use among adolescents: The mediating role of body image dissatisfaction and moderating role of gender. *Int. J. Ment. Health Addict.* **2021**, *19*, 1385–1397. [CrossRef]
43. Hong, M.Y.; Shepanski, T.L.; Gaylis, J.B. Majoring in nutrition influences BMI of female college students. *J. Nutr. Sci.* **2016**, *5*, e8. [CrossRef]
44. Mingoia, J.; Hutchinson, A.D.; Wilson, C.; Gleaves, D.H. The Relationship between Social Networking Site Use and the Internalization of a Thin Ideal in Females: A Meta-Analytic Review. *Front Psychol.* **2017**, *8*, 1351. [CrossRef]
45. Zeeni, N.; Kharma, J.A.; Mattar, L. Social media use impacts body image and eating behavior in pregnant women. *Curr. Psychol.* **2021**, *42*, 4948–4955. [CrossRef]
46. Adam, A. The ugly side of Snapchat and Instagram Filters. 2019. Available online: https://globalnews.ca/news/5676392/ugly-side-filters-snapchat-instagram-selfie-dysmorphia/ (accessed on 7 July 2023).
47. Carrotte, E.R.; Prichard, I.; Lim, M.S. "Fitspiration" on Social Media: A Content Analysis of Gendered Images. *J. Med. Internet Res.* **2017**, *19*, e95. [CrossRef]
48. de Vries, D.A.; Vossen, H.G.M.; van der Kolk-van der Boom, P. Social Media and Body Dissatisfaction: Investigating the Attenuating Role of Positive Parent-Adolescent Relationships. *J. Youth Adolesc.* **2019**, *48*, 527–536. [CrossRef] [PubMed]
49. Aparicio-Martínez, P.; Ruiz-Rubio, M.; Perea-Moreno, A.-J.; Martínez-Jiménez, M.P.; Pagliari, C.; Redel-Macías, M.D.; Vaquero-Abellán, M. Gender differences in the addiction to social networks in the Southern Spanish university students. *Telemat. Inform.* **2020**, *46*, 101304. [CrossRef]
50. Al Saud, D.F.; Alhaddab, S.A.; Alhajri, S.M.; Alharbi, N.S.; Aljohar, S.A.; Mortada, E.M. The association between body image, body mass index and social media addiction among female students at a Saudi Arabia Public University. *Malays. J. Med. Health Sci.* **2019**, *15*, 16–22.
51. Delgado-Rodríguez, R.; Linares, R.; Moreno-Padilla, M. Social network addiction symptoms and body dissatisfaction in young women: Exploring the mediating role of awareness of appearance pressure and internalization of the thin ideal. *J. Eat. Disord.* **2022**, *10*, 117. [CrossRef]
52. Guizzo, F.; Canale, N.; Fasoli, F. Instagram Sexualization: When posts make you feel dissatisfied and wanting to change your body. *Body Image* **2021**, *39*, 62–67. [CrossRef]
53. Sanchez-Ruiz, M.J.; El-Jor, C.; Kharma, J.A.; Bassil, M.; Zeeni, N. Personality, emotion-related variables, and media pressure predict eating disorders via disordered eating in Lebanese university students. *Eat. Weight. Disord. Stud. Anorex. Bulim. Obes.* **2019**, *24*, 313–322. [CrossRef]
54. Zeeni, N.; Doumit, R.; Kharma, J.A.; Sanchez-Ruiz, M.-J. Media, Technology Use, and Attitudes: Associations with Physical and Mental Well-Being in Youth with Implications for Evidence-Based Practice. *Worldviews Evid. Based Nurs.* **2018**, *15*, 304–312. [CrossRef]
55. Ayyıldız, F.; Şahin, G. Effect of social media addiction on eating behavior, body weight and life satisfaction during pandemic period. *Br. Food J.* **2022**, *124*, 2980–2992. [CrossRef]
56. Boswell, R.G.; Kober, H. Food cue reactivity and craving predict eating and weight gain: A meta-analytic review. *Obes. Rev.* **2016**, *17*, 159–177. [CrossRef]
57. Collier, D.A.; Treasure, J.L. The aetiology of eating disorders. *Br. J. Psychiatry* **2004**, *185*, 363–365. [CrossRef]
58. Karmila, S.; Aritonang, E.Y.; Sudaryati, E. The relationship of the duration of social media Instagram usage and student's eating behavior in university of sumatera utara. *Br. Int. Humanit. Soc. Sci.* **2020**, *2*, 289–295. [CrossRef]
59. Rodgers, R.F.; Lombardo, C.; Cerolini, S.; Franko, D.L.; Omori, M.; Fuller-Tyszkiewicz, M.; Linardon, J.; Courtet, P.; Guillaume, S. The impact of the COVID -19 pandemic on eating disorder risk and symptoms. *Int. J. Eat. Disord.* **2020**, *53*, 1166–1170. [CrossRef]
60. Vaterlaus, J.M.; Patten, E.V.; Roche, C.; Young, J.A. Getting healthy: The perceived influence of social media on young adult health behaviors. *Comput. Hum. Behav.* **2015**, *45*, 151–157. [CrossRef]
61. Muthia, A.S.; Margawati, A.; Fitranti, D.Y.; Dieny, F.F.; Hananingtyas, A. Correlation between Eating Behavior and Use of Social Media with Energy-Dense Food Intake Based on Gender among Students in Semarang, Indonesia. *Open Access Maced. J. Med. Sci.* **2022**, *10*, 602–610. [CrossRef]
62. Chen, H. College-aged young consumers' perceptions of social media marketing: The story of Instagram. *J. Curr. Issues Res. Advert.* **2021**, *39*, 22–36. [CrossRef]
63. van Strien, T.; Ouwens, M.A. Effects of distress, alexithymia and impulsivity on eating. *Eat Behav.* **2007**, *8*, 251–257. [CrossRef]

64. Camilleri, G.M.; Méjean, C.; Kesse-Guyot, E.; Andreeva, V.A.; Bellisle, F.; Hercberg, S.; Péneau, S. The associations between emotional eating and consumption of energy-dense snack foods are modified by sex and depressive symptomatology. *J. Nutr.* **2014**, *144*, 1264–1273. [CrossRef]
65. Hartanto, A.; Quek, F.Y.X.; Tng, G.Y.Q.; Yong, J.C. Does Social Media Use Increase Depressive Symptoms? A Reverse Causation Perspective. *Front. Psychiatry* **2021**, *12*, 641934. [CrossRef]

nutrients MDPI

Article

Maladaptive Eating Behaviours, Generalised Anxiety Disorder and Depression Severity: A Comparative Study between Adult Women with Overweight, Obesity, and Normal Body Mass Index Range

Tomasz Witaszek [1], Mateusz Babicki [2,*], Anna Brytek-Matera [3], Agnieszka Mastalerz-Migas [2], Krzysztof Kujawa [4] and Karolina Kłoda [5]

1 Hipokrates Przychodnia Lekarzy Rodzinnych i Specjalistów, ul. Powstańców WIelkopolskich 1a, 64-330 Opalenica, Poland; witaszek.tomasz@gmail.com
2 Department of Family Medicine, Wroclaw Medical University, 50-367 Wrocław, Poland; agnieszka.mastalerz-migas@umw.edu.pl
3 Eating Behavior Laboratory (EAT Lab), Institute of Psychology, University of Wroclaw, ul. J. Dawida 1, 50-527 Wrocław, Poland; anna.brytek-matera@uwr.edu.pl
4 Statistical Analysis Centre, Wroclaw Medical University, 50-368 Wrocław, Poland
5 MEDFIT Karolina Kłoda, ul. Narutowicza 13E/11, 70-240 Szczecin, Poland; wikarla@gazeta.pl
* Correspondence: ma.babicki@gmail.com

Abstract: (1) Background: Causes of obesity are multifactorial and include genetic predisposition as well as behavioural, psychological, social, and hormonal influences. We aimed to compare adult women with normal weight, overweight, and obesity, with a focus on maladaptive eating behaviours, the presence of generalised anxiety disorder, and the severity of depression. Additionally, we explored the context of anti-obesity pharmacotherapy and the status of bariatric surgery. (2) Methods: The sample was composed of 1105 adult women. The following measures, through the Computer-Assisted Web Interview (CAWI), were used in the present study: the Three Factor Eating Questionnaire (TFEQ-R18), the 7-item Generalised Anxiety Disorders Scale (GAD-7), and the 9-item Patient Health Questionnaire (PHQ-9). (3) Results: All domains of the TFEQ-R18 had correlations with Body Mass Index (BMI). There was a weak negative association between BMI and Cognitive Restraint ($r = -0.172$, $p < 0.001$) and a weak positive relationship between BMI and Uncontrolled as well as Emotional Eating ($r = 0.165$, $p < 0.001$; $r = 0.191$, $p < 0.001$, respectively). Women who screened positive for anxiety scored lower in the Cognitive Restraint domain (10.11 ± 3.58, $p = 0.042$) and higher in the Uncontrolled Eating (12.69 ± 6.04, $p < 0.001$) and Emotional Eating (5.29 ± 2.75, $p < 0.001$) domains. Similarly, women screening positive for depression had lower scores in Cognitive Restraint (9.88 ± 3.61, $p < 0.001$) and higher scores in Uncontrolled Eating (12.64 ± 6.09, $p < 0.001$) and Emotional Eating (5.31 ± 2.71, $p < 0.001$). A significant association between liraglutide and semaglutide administration and Cognitive Restraint was observed. (4) Conclusions: Individualised treatment for obesity should consider the existing and confirmed association between maladaptive eating behaviours and generalised anxiety disorder, as well as the severity of depression influencing the BMI altogether. The use of anti-obesity pharmacotherapy needs further exploration because the evidence for the use of liraglutide and semaglutide in terms of positive associations with eating behaviours is encouraging.

Keywords: obesity; maladaptive eating behaviours; anxiety; depression; women health; BMI

Citation: Witaszek, T.; Babicki, M.; Brytek-Matera, A.; Mastalerz-Migas, A.; Kujawa, K.; Kłoda, K. Maladaptive Eating Behaviours, Generalised Anxiety Disorder and Depression Severity: A Comparative Study between Adult Women with Overweight, Obesity, and Normal Body Mass Index Range. *Nutrients* **2024**, *16*, 80. https://doi.org/10.3390/nu16010080

Academic Editors: Hubertus Himmerich, Khadijeh Mirzaei and Aline Meirhaeghe

Received: 23 November 2023
Revised: 21 December 2023
Accepted: 22 December 2023
Published: 26 December 2023

1. Introduction

Since 1975, there has been a threefold increase in obesity and overweight prevalence. In 2019, an estimated 38.2 million children under the age of 5 were overweight or obese, which is an unfavourable prognosis for the future. It is estimated that around 55% of children with obesity will become adolescents with obesity, and out of them, around 80% will become adults living with this disease [1,2]. Obesity is the result of complex relationships between epigenetic, genetic, psychological, socioeconomic, and cultural factors [3,4]. Due to its common occurrence, obesity is now an emerging topic of scientific study, not only in terms of pathogenesis but also because of its economic and societal impact [5]. Positive energy balance in obesity is related to numerous factors, including eating patterns affected by maladaptive eating behaviours or potentially occurring eating disorders. In the literature, three specific dimensions of eating behaviour are presented: cognitive restraint (placing restrictions on food without taking into account the sensations of hunger and satiety), uncontrolled eating (characterised by a disordered perception of hunger that results in excessive food intake), and emotional eating (where food is consumed in response to various negative emotions) [4,6]. The Three Factor Eating Questionnaire (TFEQ) is a widely used method that quantifies these three dimensions [7,8]. In the case of the Polish population, it was found that the three-factor structure of the TFEQ-R18 (18-items) is invariant at each level–configural, metric, scalar, and strict [6]. Several studies have examined the associations between body weight and TFEQ-R18 scores. They demonstrated that gaining weight is significantly associated with uncontrolled eating and emotional eating and that higher scores in cognitive restraint and emotional eating are found to be associated with a higher Body Mass Index (BMI) in women [9–11].

Apart from the fact that psychological factors influence the development of obesity, this disease is also the cause of mental disorders. Anxiety and depression have a high prevalence among people with obesity, especially females [12]. According to a recent meta-analysis by Jung et al. (2017), obesity increases the risk of depression [13]. A bidirectional link between these two diseases has been observed [14]. Individuals with more severe obesity (BMI $\geq$ 40 kg/m^2) exhibited a stronger association with depression [13,14]. In some studies, the link between being overweight and the onset of depression was found in women but not in men [13]. For women with a baseline BMI of 30 kg/m^2 or higher, the odds of a major depressive disorder (MDD) are significantly higher and independent of other risk factors such as age, education, prior depressive symptoms, marital status, chronic disease, low social support, and financial strain [15]. A connection was also found between obesity and general anxiety disorder (GAD), but the risk between obesity and the onset of anxiety disorders (i.e., panic disorder, agoraphobia without panic disorder, social phobia, specific phobia, and generalised anxiety disorder) has been less investigated so far [16]. Women who were either overweight or obese were significantly more likely to have the three mental health issues simultaneously—current depression; lifetime diagnosed depression; and anxiety disorder—than women with a normal BMI range [17].

Achieving long-term weight loss maintenance in patients with obesity through current treatment methods continues to pose a clinical challenge. This challenge is particularly evident in the varied responses to obesity interventions, including changes in eating patterns, medication use, and surgical procedures [4]. Lifestyle interventions for weight loss are currently not personalised to address the underlying pathophysiology and behavioural traits associated with obesity [18]. There is a need to develop a valid classification for this heterogeneous patient population and assess its effectiveness in obesity management to provide a more individualised approach. One of the hypothesised obesity phenotypes is emotional eating, which could be identified by screening patients for depression, anxiety, and emotional eating behaviour using validated tools [4,18]. This approach is especially promising because a phenotype-targeted pharmacological treatment of obesity resulted in a mean weight loss of 4.1% higher after 6 months compared to the non-phenotype-targeted approach. Additionally, the proportion of patients who lost >10% at 1 year was 79%, compared with 34% with non-phenotype-guided treatment, with no increase in the adverse effect ratio [4].

The aim of the present study was to compare adult women with normal weight, overweight, and obesity in relation to maladaptive eating behaviours, generalised anxiety disorder, and depression severity using the TFEQ-R18, the 7-item Generalised Anxiety

Disorders Scale (GAD-7) and the 9-item Patient Health Questionnaire (PHQ-9). As this topic is severely understudied, we have also explored the associations between anti-obesity pharmacotherapy and undergoing bariatric surgery in the past with TFEQ-R18 domain scores. We have put forward the following hypotheses: 1. There is an association between BMI and the results of the above-mentioned questionnaires, and the significance of this relationship will increase with weight. 2. There is a correlation between the TFEQ-R18 scales and the PHQ-9 and GAD-7. 3. The use of particular medications influences the scores obtained on the subscales of the TFEQ-R18 scale.

2. Materials and Methods

2.1. Participants and Recruitment

This cross-sectional survey was conducted between 4 September 2023 and 19 October 2023 using a computer-assisted web interviewing (CAWI) technique. A proprietary questionnaire was used and uploaded to Google Forms, making it available to any device connected to the internet. The survey was then distributed through social media platforms like Instagram and Facebook within support groups for individuals with obesity, undergoing pharmacological obesity treatment, or undergoing bariatric surgery. Before proceeding with the questionnaire, participants were informed about the research objectives and methodology, after which they gave their informed consent to participate in this study. Respondents had the option to discontinue their participation without explanation at any point. In order to maintain anonymity, no personal data were collected. This study's inclusion criteria were being female, being over 18 years old, residing in Poland, and having internet access.

This study was conducted in accordance with the Declaration of Helsinki and approved by the Bioethics Committee of the Wroclaw Medical University, Poland (approval number: 349/2023N).

This study questionnaire included a series of questions aimed at evaluating the sociodemographic profile of the participants. These questions covered topics such as age, gender, height, current weight, and the highest weight ever recorded. Following this section, participants were asked about the presence of any chronic diseases, whether they were undergoing pharmacological treatment for obesity, or if they had undergone bariatric treatment. In the event of a positive response, participants were presented with an additional multiple-choice question to specify the particular co-morbid condition, specific anti-obesity medication taken, and type of bariatric procedure. The second part of the survey consisted of standardised psychometric tools: the Three Factor Eating Questionnaire (TFEQ-R18), the Patient Health Questionnaire (PHQ-9), and the General Anxiety Disorder 7-Item (GAD-7). The questionnaire required all questions to be answered. Failure to answer at least one question prevented progression to the next stage.

2.2. TFEQ-R18 Scale

The Three Factor Eating Questionnaire (TFEQ-R18) was developed by Stunkard and Messick [7] as a 51-item questionnaire. It was further modified by Karlsson et al. [19], who abbreviated the measure to 18 items (TFEQ-R18). The Polish version of the TFEQ-R18 was validated [6]. The questionnaire refers to current dietary practise and measures three domains of eating behaviour: cognitive restraint (conscious restriction of food intake in order to control body weight), uncontrolled eating (excessive food intake due to a loss of control accompanied by subjective feelings of hunger), and emotional eating (inability to resist emotional cues). The TFEQ-R18 consists of 17 items on a four-point Likert response scale (ranging from "definitely true" to "definitely false") and an additional item to rate, on an eight-point Likert scale, how often respondents restrain their eating. For the purpose of analysis, this particular item was re-coded into four categories, with responses 1 and 2 combined into category 1, 3 and 4 into category 2, 5 and 6 into category 3, and 7 and 8 into category 4. The range of points that can be obtained in each subscale is: 0–18 for cognitive restraint, 0–27 for uncontrolled eating, and 0–9 for emotional eating. The reliability of the tool, as measured by Cronbach's alpha coefficient, was 0.763. Additionally, the Cronbach's alpha coefficient for the individual subscales was, respectively, 0.799 for cognitive restraint, 0.889 for uncontrolled eating, and 0.883 for uncontrolled eating.

2.3. PHQ-9 Scale

The Patient Health Questionnaire (PHQ-9) is an instrument used to diagnose criteria-based depressive disorders commonly encountered in primary care settings. The PHQ-9 asks patients to rate, on a four-point scale ranging from "not at all" to "most days", the frequency with which they have experienced certain depressive symptoms in the preceding 2 weeks. Items are rated from 0 to 3 accordingly. The possible point range to be scored is between 0 and 27 points. Researchers have confirmed the validity and reliability of the PHQ-9 [20]. The Polish version of the questionnaire has been validated [21]. In the validation of the tool in the Polish population, a cutoff value of 12 points was calculated [21]. The reliability of the tool, as measured by Cronbach's alpha coefficient, was 0.872.

2.4. GAD-7 Scale

The General Anxiety Disorder 7-Item (GAD-7) Scale is commonly used to assess general anxiety symptoms in various settings and populations. The GAD-7 has shown good reliability and construct validity [22]. It comprises seven items that measure worry and anxiety symptoms. Each item is rated on a four-point Likert scale (0–3), with total scores ranging from 0 to 21, where higher scores indicate greater anxiety severity. Scores exceeding 10 are considered to be within the clinical range. In this study, we used the Polish translation of the GAD-7 provided by the MAPI Research Institute [23]. A high internal consistency for Cronbach's alpha was found (0.924).

For a better understanding of the research methodology, the English version of the survey is attached in Supplementary Table S1. The English Version of this Study Questionnaire.

2.5. Statistical Analysis

The variables in this study had both qualitative and quantitative characteristics. To assess the normality of the distribution, the Shapiro–Wilk test was used. Qualitative variable comparisons were conducted using the chi-squared test. For quantitative variables, non-parametric tests such as the Kruskal–Wallis H Test or the Mann–Whitney U test were used. The level of correlation between variables was assessed with the Spearman correlation test coefficient. The BMI was assumed to mediate the effect of eating behaviour (TFEQ-R18 subscales) on anxiety (GAD-7) and depression (PHQ-9), because it can be affected by the eating behavior, and on the other hand, it can affect anxiety or health. Based on this assumption, the associations between eating behavior, BMI, anxiety, and health were analysed using the mediation model. The model enables assessing the average direct effect (ADE) of a given factor (subscale of TFEQ-R18) and the average causal mediation effect (ACME) of a mediator (BMI). To test hypothesis 1, the Kruskal–Wallis H Test and Mann–Whitney U test were used. Additionally, the mediation model was used. To test hypothesis 2, the Spearman correlation test was used, and to test hypothesis 3, the Mann–Whitney U test was used.

The hypotheses verified based on a common group of patients were considered a family of hypotheses, and the Bonferroni correction was used for controlling I-type error. Therefore, a corrected alpha level amounted to 0.008 (0.05/6) (with respect to scales GAD-7 and PHQ-9), or to 0.017 (0.05/3) (for Age and BMI). The analysis was conducted using Statistica 13.0 (TIBCO Software, Palo Alto, CA, USA). The mediation analysis was performed using the command 'mediate' from the R-package version 4.0.4 'mediation'.

3. Results

3.1. Characteristics of the Study Group

This study involved 1105 female participants. The average age was 38.89 years (SD = 9.0). The mean BMI was 33.32 kg/m^2 (SD = 6.67). Overall, 44.2% of women (n = 488) were undergoing pharmacological treatment for obesity, with liraglutide being the most common (21.3%), and 11.9% of women (n = 131) had undergone surgical treatment for obesity. Among all participants, 29.6% had one or more chronic diseases. Out of 68.7% of women with obesity (n = 759), only 44.8% (n = 496) recognised it as a chronic disease. Approximately 46.4% of the women (n = 513) screened positively for anxiety, and 50.3% (n = 556) screened positively for depression. A detailed overview of the characteristics of the study group is presented in Table 1, and a summary of the results and interpretation of the Three-Factor Eating Questionnaire, Generalised Anxiety Disorder, and Patient Health Questionnaire-9 is presented in Table 2.

Table 1. Characteristics of the study group.

Variable		N/M ± SD
Age [years]		38.89 ± 9.00
Body Mass Index [kg/m^2]		33.32 ± 6.67
Body Mass Index	Underweight	0 (0.0)
	Normal weight	92 (8.3)
	Overweight	254 (23.0)
	Obesity I	366 (33.1)
	Obesity II	247 (22.4)
	Obesity III	146 (13.2)
Pharmacological treatment of obesity		488 (44.2)
Pharmacological treatment of obesity	Semaglutide	232 (21.0)
	Bupropion + Naltrexone	41 (3.7)
	Liraglutide	235 (21.3)
Surgical treatment of obesity		131 (11.9)
Chronic diseases		327 (29.6)
Chronic diseases	Hypertension	220 (19.9)
	Cardiovascular diseases other than hypertension	57 (5.2)
	Obesity	496 (44.8)
	Hypothyroidism	387 (35.0)
	Diabetes mellitus type 2	116 (10.5)
	Osteoarthritis	76 (6.9)
	Depression	211 (19.1)
	Anxiety	153 (13.8)
	Dyslipidemia	101 (9.1)
	Fatty liver disease	112 (10.1)
	Other	367 (33.2)

M—mean; SD—Standard deviation; N—number,, kg/m^2—kilogram(s) per square metre.

3.2. Correlations between the Three Domains of TFEQ-R18 and GAD-7, as Well as PHQ-9 Scores in Women with Normal Weight, Overweight, and Obesity

All domains of TFEQ-R18 were associated with BMI. There were weak negative correlations between BMI and Cognitive Restraint, (rho = −0.172, $p < 0.001$) and weak positive relationships between BMI and Uncontrolled as well as Emotional Eating (rho = 0.165, $p < 0.001$; rho = 0.191, $p < 0.001$, respectively). Emotional Eating was the only domain that was weakly negatively linked to age (rho = −0.086, $p = 0.004$).

Table 2. Summary of results and interpretation of the Three-Factor Eating Questionnaire, Generalised Anxiety Disorder, and Patient Health Questionnaire-9.

Variable		N/M ± SD
TFEQ-R18	Cognitive Restraint	10.33 ± 3.51
	Uncontrolled Eating	11.16 ± 5.85
	Emotional Eating	4.60 ± 2.76
GAD-7		9.57 ± 6.10
GAD-7 Interpretation	Anxiety	513 (46.4)
PHQ-9		10.6 ± 6.36
PHQ-9 Interpretation	Depression	556 (50.3)

M—mean; SD—Standard deviation; N—number; TFEQ-R18 Three-Factor Eating Questionnaire, GAD-7—Generalised anxiety disorder; PHQ-9—Patient Health Questionnaire-9.

Analysis of correlations between three domains of TFEQ-R18 and GAD-7 as well as PHQ-9 scores (as continuous values) revealed no significance for GAD-7 and PHQ-9 between Uncontrolled and Emotional Eating in women with normal weight. In overweight women, relationships were moderately significantly positive for both the GAD-7 and PHQ-9 scores in relation to Uncontrolled Eating (rho = 0.357, $p < 0.001$, and rho = 0.360, $p < 0.001$, respectively) and Emotional Eating (rho = 0.313, $p < 0.001$, and rho = 0.349, $p < 0.001$, respectively). Finally, we found weak negative significant correlations among women with class I obesity between GAD-7 and PHQ-9 scores and Cognitive Restraint (rho = −0.136, $p = 0.008$ and rho = −0.189, $p < 0.001$, respectively) and weak positive relationship between GAD-7 and PHQ-(score and Uncontrolled Eating (rho = 0.283, $p < 0.001$ and rho = 0.291 and $p < 0.001$, respectively) and Emotional Eating (rho = 0.286, $p < 0.001$ and rho = 0.314, $p < 0.001$, respectively). All analysed correlations are presented in Table 3.

Table 3. Correlations between the three domains of TFEQ-R18, GAD-7, and PHQ-9 scores (as a continuous value) in different BMI subgroups.

Variable	Cognitive Restraint		Uncontrolled Eating		Emotional Eating	
	rho	*p* *	rho	*p* *	rho	*p* *
Age [years]	0.055	0.067	−0.428	0.154	−0.086	**0.004**
BMI [kg/m^2]	−0.172	**<0.001**	0.165	**<0.001**	0.191	**<0.001**
Whole group (N = 1105)						
GAD-7	−0.091	**0.003**	0.299	**<0.001**	0.293	**<0.001**
PHQ-9	−0.154	**<0.001**	0.338	**<0.001**	0.329	**<0.001**
Normal weight (N = 92)						
GAD-7	0.048	0.648	0.179	0.086	0.149	0.155
PHQ-9	0.008	0.937	0.268	**0.008**	0.208	0.046
Overweight (N = 254)						
GAD-7	−0.040	0.524	0.357	**<0.001**	0.313	**<0.001**
PHQ-9	−0.112	0.074	0.360	**<0.001**	0.349	**<0.001**
Obesity I (N = 366)						
GAD-7	−0.136	**0.008**	0.283	**<0.001**	0.286	**<0.001**
PHQ-9	−0.189	**<0.001**	0.291	**<0.001**	0.314	**<0.001**
Obesity II (N = 247)						
GAD-7	−0.002	0.981	0.221	**<0.001**	0.215	**<0.001**
PHQ-9	0.005	0.934	0.207	**0.002**	0.191	**0.003**

Table 3. *Cont.*

Variable	Cognitive Restraint		Uncontrolled Eating		Emotional Eating	
	rho	*p* *	rho	*p* *	rho	*p* *
			Obesity III (N = 146)			
GAD-7	−0.118	0.156	0.291	**<0.001**	0.318	**<0.001**
PHQ-9	−0.316	**<0.001**	0.473	**<0.001**	0.435	**<0.001**

N—number; GAD-7—Generalised anxiety disorder; PHQ-9—Patient Health Questionnaire-9; * Spearman's rank correlation. Significant effects (<0.017 for BMI and age) and (<0.008) for the rest are marked in bold.

Associations between the three domains of TFEQ-R18 and GAD-7, as well as PHQ-9 scores (grouped by the screening cut-off threshold) revealed that women who screened positively for anxiety were scoring higher in the Uncontrolled Eating (12.69 ± 6.04, $p < 0.001$) and Emotional Eating (5.29 ± 2.75, $p < 0.001$) domains when compared to those that did not meet the screening threshold.

A similar tendency was noted for women who screened positively for depression. They were scoring lower in the Cognitive Restraint domain (9.88 ± 3.61, $p < 0.001$) while scoring higher in the Uncontrolled Eating (12.64 ± 6.09, $p < 0.001$) and Emotional Eating (5.31 ± 2.71, $p < 0.001$) domains.

After dividing the study group into BMI subgroups, it was found that women with normal weight presented no association between GAD-7 and PHQ-9 scores and TFEQ-R18 Cognitive Restraint scores, whereas in women with class I obesity, an association between the PHQ-9 score and Cognitive Restraint was observed. Moreover, in women with class I and class III obesity, the PHQ-9 score was significantly associated with all three domains of TFEQ-R18. A detailed overview of the results of GAD-7 and PHQ-9 is presented in Table 4, distinguishing between different BMI subgroups.

Table 4. Association between the three domains of TFEQ-R18, GAD-7, and PHQ-9 scores (grouped by the screening cut-off threshold).

Variable		Cognitive Restraint		Uncontrolled Eating		Emotional Eating	
		M ± SD	*p* *	M ± SD	*p* *	M ± SD	*p* *
				Whole group (N = 1105)			
GAD-7 Interpretation	Anxiety	10.11 ± 3.58	0.042	12.69 ± 6.04	**<0.001**	5.29 ± 2.75	**<0.001**
	No anxiety	10.53 ± 3.43		9.82 ± 5.32		3.99 ± 2.61	
PHQ-9 Interpretation	Depression	9.88 ± 3.61	**<0.001**	12.64 ± 6.09	**<0.001**	5.31 ± 2.71	**<0.001**
	No depression	10.80 ± 3.33		9.64 ± 5.17		3.88 ± 2.61	
				Normal weight (N = 92)			
GAD-7 Interpretation	Anxiety	11.34 ± 3.44	0.844	11.09 ± 6.02	0.016	4.28 ± 3.05	0.214
	No anxiety	11.22 ± 3.56		7.92 ± 4.41		3.33 ± 2.31	
PHQ-9 Interpretation	Depression	11.10 ± 3.59	0.861	11.63 ± 6.17	**0.003**	4.66 ± 3.04	0.018
	No depression	11.34 ± 3.48		7.76 ± 4.19		3.18 ± 2.25	
				Overweight (N = 254)			
GAD-7 Interpretation	Anxiety	10.76 ± 3.21	0.689	12.03 ± 6.01	**<0.001**	4.82 ± 2.79	**<0.001**
	No anxiety	10.90 ± 3.42		8.69 ± 5.16		3.57 ± 2.60	
PHQ-9 Interpretation	Depression	10.52 ± 3.56	0.233	11.70 ± 6.03	**<0.001**	4.77 ± 2.83	**0.001**
	No depression	11.09 ± 3.07		8.92 ± 5.27		3.59 ± 2.57	

Table 4. *Cont.*

Variable		Cognitive Restraint		Uncontrolled Eating		Emotional Eating	
		M ± SD	*p* *	M ± SD	*p* *	M ± SD	*p* *
Obesity I (N = 366)							
GAD-7 Interpretation	Anxiety	10.30 ± 3.42	0.183	12.63 ± 5.95	**<0.001**	5.11 ± 2.77	**<0.001**
	No anxiety	10.75 ± 3.28		10.13 ± 5.28		3.99 ± 2.63	
PHQ-9 Interpretation	Depression	10.04 ± 3.44	**0.007**	12.39 ± 6.18	**<0.001**	5.13 ± 2.64	**<0.001**
	No depression	11.02 ± 3.19		10.12 ± 5.03		3.88 ± 2.71	
Obesity II (N = 247)							
GAD-7 Interpretation	Anxiety	9.92 ± 3.44	0.566	13.01 ± 5.92	0.031	5.51 ± 0.26	0.018
	No anxiety	9.61 ± 3.42		11.39 ± 5.48		4.72 ± 2.64	
PHQ-9 Interpretation	Depression	9.78 ± 3.49	0.821	13.00 ± 6.01	0.028	5.42 ± 2.70	0.033
	No depression	9.77 ± 3.37		11.11 ± 5.16		4.72 ± 2.56	
Obesity III (N = 146)							
GAD-7 Interpretation	Anxiety	8.71 ± 3.96	0.084	13.78 ± 6.37	**0.002**	6.33 ± 4.35	**<0.001**
	No anxiety	9.89 ± 3.95		10.46 ± 5.49		4.35 ± 2.44	
PHQ-9 Interpretation	Depression	8.61 ± 3.91	0.016	13.94 ± 5.94	**<0.001**	6.28 ± 2.45	**<0.001**
	No depression	10.33 ± 3.91		9.48 ± 5.67		4.02 ± 2.47	

M—mean; SD—Standard deviation; N—number; GAD-7—Generalised anxiety disorder; PHQ-9—Patient Health Questionnaire-9; * Mann–Whitney U test Significant effects (<0.008) are marked in bold.

3.3. *Maladaptive Eating Behaviours among Adult Women with Overweight, Obesity, and Normal Weight*

After dividing participants into subgroups based on their BMI, we found an association between BMI and all three TFEQ-R18 domains. We observed increasing scores of Uncontrolled and Emotional Eating and decreasing scores of Cognitive Restraint in women with an increasing degree of obesity ($p < 0.001$).

Patients who are undergoing pharmacological treatment for obesity scored significantly higher in the Cognitive Restraint domain ($p < 0.001$). When considering every class of medication separately, this relationship was stronger with semaglutide ($p < 0.001$) than with liraglutide ($p < 0.016$). There was no observed relationship between cognitive restraint and the use of Bupropion/Naltrexone. However, it is the only medication for which a relationship with Emotional Eating was noted ($p < 0.001$). The status of bariatric surgery had significant differences in all three domains ($p < 0.001$). An in-depth overview of the results of the TFEQ-R18 is presented in Table 5.

Table 5. Associations between studied variables and the TFEQ-R18 domain scores.

Variable		Cognitive Restraint		Uncontrolled Eating		Emotional Eating	
		M ± SD	*p*	M ± SD	*p*	M ± SD	*p*
BMI	Normal weight (N = 92)	11.26 ± 3.49	**<0.001** ^	9.02 ± 5.22	**<0.001** ^	3.66 ± 2.62	**<0.001** ^
	Overweight (N = 254)	10.84 ± 3.30		10.14 ± 5.78		4.11 ± 2.75	
	Obesity I (N = 366)	10.56 ± 3.34		11.16 ± 5.69		4.45 ± 2.74	
	Obesity II (N = 247)	9.78 ± 3.43		12.28 ± 5.78		5.16 ± 2.66	
	Obesity III (N = 146)	9.21 ± 3.98		12.35 ± 6.21		5.48 ± 2.68	

Table 5. *Cont.*

Variable		Cognitive Restraint		Uncontrolled Eating		Emotional Eating	
		M ± SD	*p*	M ± SD	*p*	M ± SD	*p*
Pharmacological treatment of obesity	Yes (N = 488)	10.96 ± 3.27	**<0.001 ***	11.30 ± 5.91	0.492 *	4.68 ± 2.77	0.387 *
	No (N = 617)	9.83 ± 3.60		11.04 ± 5.92		4.54 ± 2.76	
Semaglutide	Yes (N = 232)	11.19 ± 3.21	**<0.001 ***	10.78 ± 3.92	0.223 *	4.47 ± 2.70	0.441 *
	No (N = 873)	10.11 ± 3.54		11.25 ± 5.82		4.63 ± 2.77	
Liraglutide	Yes (N = 235)	10.84 ± 3.25	**0.016 ***	11.56 ± 5.41	0.125 *	4.69 ± 2.79	0.588 *
	No (N = 870)	10.20 ± 3.56		11.04 ± 5.96		4.58 ± 2.75	
Bupropion + Naltrexone	Yes (N = 41)	10.61 ± 3.71	0.401 *	12.98 ± 5.67	0.067 *	6.13 ± 2.39	**<0.001 ***
	No (N = 1064)	10.32 ± 3.51		11.08 ± 5.84		4.54 ± 2.75	
Surgical treatment of obesity	Yes (N = 131)	11.78 ± 3.57	**<0.001 ***	7.83 ± 5.35	**<0.001 ***	3.28 ± 2.54	**<0.001 ***
	No (N = 974)	10.14 ± 3.45		11.60 ± 5.77		4.77 ± 2.73	

M—mean; SD—Standard deviation; N—number; BMI—body mass index; kg/m^2—kilogram(s) per square meter; ^ Kruskal–Wallis H Test; * Mann–Whitney U test; Significant effects (<0.017) are marked in bold.

3.4. Mediation Effect of BMI in Associations between the TFEQ-R18 and Anxiety and Health

Although the mediation effect of BMI was statistically significant in all the analyses performed, the effect size of the mediation was low. The rate of the mediated effect on the total effect of the analysed factors amounted to 0.04–0.06 (4–6%) for uncontrolled eating and emotional eating and to 0.20 and 0.16 (20 and 16%) for cognitive restraint (Table 6).

Table 6. The analysis of the mediation effect of BMI in associations between the TFEQ-R18 subscales and anxiety (GAD-7) and health (PHQ-9).

Explained Variable	Factor	Effect Statistics	Coeff.	Lower Limit of 95% CI	Upper Limit of 95% CI	*p*
GAD-7 score	Cognitive Restraint	ACME	−0.03	−0.06	−0.01	**<0.001**
		ADE	−0.13	−0.22	−0.01	0.028
		Total effect	−0.16	−0.26	−0.04	0.016
		Prop. med. effect	0.20	0.28	163.79	0.016
	Uncontrolled Eating	ACME	0.01	0.00	0.02	**0.006**
		ADE	0.31	0.25	0.37	**<0.001**
		Total effect	0.33	0.27	0.38	**<0.001**
		Prop. med. effect	0.04	0.01	0.08	**0.006**
	Emotional Eating	ACME	0.03	0.01	0.06	0.026
		ADE	0.63	0.49	0.76	**<0.001**
		Total effect	0.66	0.52	0.79	**<0.001**
		Prop. med. effect	0.04	0.01	0.10	0.026

Table 6. *Cont.*

Explained Variable	Factor	Effect Statistics	Coeff.	Lower Limit of 95% CI	Upper Limit of 95% CI	p
PHQ-9 score	Cognitive Restraint	ACME	−0.04	−0.07	−0.02	**<0.001**
		ADE	−0.25	−0.35	−0.13	**<0.001**
		Total effect	−0.29	−0.39	−0.18	**<0.001**
		Prop. med. effect	0.15	0.08	0.31	**<0.001**
	Uncontrolled Eating	ACME	0.02	0.01	0.03	**<0.001**
		ADE	0.38	0.32	0.44	**<0.001**
		Total effect	0.40	0.34	0.46	**<0.001**
		Prop. med. effect	0.05	0.02	0.09	**<0.001**
	Emotional Eating	ACME	0.05	0.02	0.08	**<0.001**
		ADE	0.72	0.59	0.86	**<0.001**
		Total effect	0.77	0.63	0.90	**<0.001**
		Prop. med. effect	0.06	0.03	0.12	**<0.001**

ACME—average causal mediation effect; ADE—average direct effect; Prop. Med. Effect—proportion of mediated effect (ACME/Total effect); GAD-7—Generalized anxiety disorder; PHQ-9—Patient Health Questionnaire-9; Significant effects (<0.008) are marked in bold.

4. Discussion

The aim of this study was to compare women with normal body weight, overweight, and obesity in relation to maladaptive eating behaviours, generalised anxiety disorder, and depression severity using the TFEQ-R18, the GAD-7, and the PHQ-9 scales. We have also hypothesised that there will be an association between body weight and scores on the aforementioned questionnaires and that the significance of this association will increase with increasing BMI.

Indeed, we found that BMI correlated significantly positively with Uncontrolled and Emotional Eating and significantly negatively with Cognitive Restraint. This was confirmed in a separate analysis of associations between the studied variables and TFEQ-R18 domain scores. An analysis of existing literature showed that this relationship may differ in relation to the studied population. For example, in the Finnish study on 2 997 females, higher BMI was associated with higher levels of Cognitive Restraint ($p < 0.001$) and Emotional Eating ($p < 0.001$), but not with Uncontrolled Eating [10]. Whereas, in the Arabic population, BMI was positively correlated with Uncontrolled Eating and Emotional Eating ($p < 0.001$ respectively), but not with Cognitive Restraint [8]. Another piece of evidence that this study population is of great importance gives a sample of 4377 Swedish, middle-aged men and women living with obesity in whom the original factor structure of TFEQ was not replicated at all [19]. Not without significance is also the mean age of the studied population, because it can be correlated with TFEQ-R18 domain scores. In the case of our study, a significant negative correlation was found between age and Emotional Eating. The aforementioned Finnish population consisted of very young women, with a mean age of 18.6 years, whereas our population consisted of mostly middle-aged women, with a mean age of 38.9 years. Our results also vary in the context of Cognitive Restraint, which is usually associated positively with a higher BMI [10,24] or is not associated with body weight [5,25]. Interestingly, a German sample of middle-aged adults exhibited a reversed U-shaped association between cognitive restraint and BMI, where BMI was high in subjects with medium cognitive restraint [24]. Population diversity in terms of the association between BMI and TFEQ-R18 is an interesting aspect that requires further research.

Our study revealed significant correlations between PHQ-9 scores and all three domains of TFEQ-R18. Also, when dividing the population into groups by BMI, relationships were noted for Uncontrolled Eating and Emotional Eating across all BMI subgroups. A similar observation was made in a study of 238 participants undergoing a behavioural weight loss intervention in Germany. PHQ-9 scores were positively correlated with Hunger (Emotional Eating) ($r = 0.245$, $p < 0.001$) and Disinhibition (Uncontrolled Eating) ($r = 0.353$,

$p < 0.05$). It is worth noting that the original version of the 51-item TFEQ was used. No significant correlation was observed between PHQ-9 scores and Cognitive Restraint [26]. Individuals with obesity and eating disorders are characterised by impulsivity, anxiety, and depression. However, even without the presence of eating disorders, people living with obesity are affected by psychological factors [27,28]. The linkage between weight gain and depression is considered to be formed on the basis of pathological eating behaviour, including emotional eating [29]. Thus, the results observed in the 1105 population of women in this study are in concordance with former reports and confirm the previous scientific findings.

The aforementioned study of the German population undergoing behavioural weight loss intervention also reveals correlations between GAD-7 scores and TFEQ-R18 Domains. GAD-7 scores are positively correlated with Hunger (Emotional Eating) ($r = 0.311$, $p < 0.001$) and Disinhibition (Uncontrolled Eating) ($r = 0.245$, $p < 0.001$) [26]. Another study on 129 Ghanaian students reveals a positive correlation between screening positively for anxiety with GAD-7 and Emotional Eating ($r = 0.471$, $p < 0.001$), but only for female participants [30]. Similar to depression, anxiety, and perceived stress affect individuals with obesity. Especially, perceived stress is considered a linkage between body image, stigma, and depressive symptoms, as well as food addiction [5,31–33]. There is an association between weight management and stress management. Optimal stress coping strategies result in weight reduction and changes in pathologic eating behaviours in women [34].

In one of the studies, TFEQ-R18 scores were measured before and after liraglutide treatment. After treatment with liraglutide, the Uncontrolled Eating score decreased from 36.8 ± 24.5 to 19.6 ± 18.4 ($p < 0.001$), and the Emotional Eating score decreased from 49.9 ± 33.3 to 28.5 ± 26.9 ($p < 0.001$). Scores for Cognitive Restraint were not changed [35]. What is interesting is that in our study, a significant association between liraglutide administration and Cognitive Restraint was observed. We can assume that participants already treated with liraglutide (before the study) had undergone lifestyle interventions and education, which are provided with obesity pharmacotherapy in Poland, thus their Cognitive Restraint obtained higher scores. A similar result was observed in the case of semaglutide.

We also found a relationship between the use of Bupropion/Naltrexone and the Emotional Eating domain of TFEQ-R18 ($p < 0.001$). However, it should be suspected that the presence of emotional eating is the reason why the participant received this medication in the first place. The Polish Society for the Treatment of Obesity recommends considering the use of this medication in patients exhibiting emotional eating patterns and significant cravings, especially with coexisting depression [36]. In one of the studies about phenotype-targeted pharmacological treatment, the dominance of the emotional eating domain in TFEQ was an indication for choosing this particular class of anti-obesity medication and led to a higher efficacy of treatment [4].

The status of bariatric surgery had significant differences in all three domains of TFEQ-R18 ($p < 0.001$). Those who underwent the surgery scored higher in Cognitive Restraint, while receiving lower scores in Emotional Eating and Uncontrolled Eating. This is reflected in the available literature. In one of the studies, 204 adults with severe obesity from three countries were followed 1 year after metabolic surgical procedures. The original version of the 51-item TFEQ was used. After 12 months, there were statistically significant increases in restraint and decreases in disinhibition and hunger [37].

The authors are aware of the limitations of this study, which are undoubtedly the data collection methodology. Firstly, the authors did not analyse socio-economic data, which can play a very large role in the parameters analysed. Furthermore, the use of self-reported height and weight to calculate BMI is controversial, as there are numerous studies that suggest self-reported measures are biassed as people tend to provide overestimates of their height and underestimates of their weight [38,39]. Additionally, the use of an online questionnaire and its distribution through social networks carry a risk of group selection. Only people using the Internet and social networks could participate in this study. However, on the other hand, this type of methodology allows a large number of people from different parts of the country to be reached in a quick way. In addition, research suggests that CAWI-type surveys are associated with a greater likelihood of providing truthful answers than socially acceptable ones. Participants also experience lower levels of stress when taking part in this type of survey. Another limitation of the survey is the lack of representativeness of the analysed group in Polish society. It should also be borne in mind that the authors do not know the number of people reached by the survey and cannot assess the response rate. More research on eating-related behaviours among patients is needed to understand the multifaceted nature of these disorders and to offer new solutions with a more individual approach.

5. Conclusions

Individualised treatment for obesity should consider the existing and confirmed association between maladaptive eating behaviours, and generalised anxiety disorder, as well as the severity of depression influencing the BMI altogether. The use of anti-obesity pharmacotherapy needs further exploration because the evidence for the use of liraglutide and semaglutide in terms of positive associations with eating behaviours is encouraging.

Supplementary Materials: The following supporting information can be downloaded at: https://www.mdpi.com/article/10.3390/nu16010080/s1. Table S1: The English Version of the Study Questionnaire and the distribution of responses among participants in the TFEQ-R18, GAD-7, and PHQ-9 questionnaires; Table S2: The distribution of the responses among the participants in the Three Factor Eating Questionnaire-18 (TFEQ-R18); Table S3: The distribution of the responses among the participants in the Patient Health Questionnaire-9 (PHQ-9); and Table S4: The distribution of the responses among the participants in the General Anxiety Disorder-7 Questionnaire (GAD-7).

Author Contributions: T.W.: Conceptualization, Writing—original draft, Writing—review and editing, Methodology, Validation. M.B.: Conceptualization, Formal analysis, Investigation, Methodology, Project administration, Supervision, Writing—original draft, Writing—review, and editing. A.B.-M.: Conceptualization, Supervision, Writing—original draft, Writing—review and editing. A.M.-M.: Funding acquisition, Supervision, Writing—original draft, Writing—review, and editing. K.K. (Karolina Kłoda): Conceptualization, Investigation, Methodology, Project administration, Validation, Visualization, Writing—original draft, Writing—review and editing. K.K. (Krzysztof Kujawa): Formal analysis; Writing—review and editing. All authors have read and agreed to the published version of the manuscript.

Funding: This research was funded by the Wroclaw Medical University, SUBZ.C290.23.069. A.B.-M. acknowledges the financial support provided by the "Excellence Initiative—Research University" programme for the years 2020–2026 at the University of Wroclaw.

Institutional Review Board Statement: This study was conducted in accordance with the Declaration of Helsinki and approved by the Bioethics Committee of the Wroclaw Medical University, Poland (approval number: 349/2023N, 26 October 2023).

Informed Consent Statement: Informed consent was obtained from all subjects involved in this study.

Data Availability Statement: The data presented in this study are available on request from the corresponding author. The data are not publicly available due to lack of a suitable platform.

Conflicts of Interest: The authors declare no conflicts of interest.

References

1. Obesity and Overweight. Available online: https://www.who.int/news-room/fact-sheets/detail/obesity-and-overweight (accessed on 21 November 2023).
2. Simmonds, M.; Llewellyn, A.; Owen, C.G.; Woolacott, N. Predicting Adult Obesity from Childhood Obesity: A Systematic Review and Meta-Analysis: Adult Obesity from Childhood Obesity. *Obes. Rev.* **2016**, *17*, 95–107. [CrossRef] [PubMed]
3. Apovian, C.M. Obesity: Definition, Comorbidities, Causes, and Burden. *Am. J. Manag. Care* **2016**, *22*, s176–s185. [PubMed]
4. Acosta, A.; Camilleri, M.; Dayyeh, A.; Calderon, B.; Gonzalez, G.; Mcrae, D.; Rossini, A.; Singh, W.; Burton, S.; Clark, D. Selection of Antiobesity Medications Based on Phenotypes Enhances Weight Loss: A Pragmatic Trial in an Obesity Clinic. *Obesity* **2021**, *29*, 662–671. [CrossRef] [PubMed]

5. Świder, K.; Baska, A.; Babicki, M.; Mastalerz-Migas, A.; Kłoda, K. Weight Stigma and Fat Phobia in Poland—Attitudes towards People Living with Obesity and the Level of Knowledge about Obesity among the Social Media Internet Respondents and Medical Professionals. *Front. Nutr.* **2023**, *10*, 1287783. [CrossRef] [PubMed]
6. Brytek-Matera, A.; Rogoza, R.; Czepczor-Bernat, K. The Three-Factor Eating Questionnaire-R18 Polish Version: Factor Structure Analysis among Normal Weight and Obese Adult Women. *Arch. Psyc. Psychother.* **2017**, *19*, 81–90. [CrossRef]
7. Stunkard, A.J.; Messick, S. The Three-Factor Eating Questionnaire to Measure Dietary Restraint, Disinhibition and Hunger. *J. Psychosom. Res.* **1985**, *29*, 71–83. [CrossRef] [PubMed]
8. Alhebshi, S.; Hilary, S.; Safi, S.K.H.; Ali, H.I.; Cheikh Ismail, L.; Al Dhaheri, A.; Stojanovska, L. Validity and Reliability of the Arabic Version of the Three-Factor Eating Questionnaire-R18. *Heliyon* **2023**, *9*, e17623. [CrossRef] [PubMed]
9. Tan, S.T.; Kannan, T.M. Comparison of Eating Behaviour by Relative Weight Change Status of Young Adults throughout the COVID-19 Pandemic. *Dialogues Health* **2023**, *2*, 100094. [CrossRef]
10. Anglé, S.; Engblom, J.; Eriksson, T.; Kautiainen, S.; Saha, M.-T.; Lindfors, P.; Lehtinen, M.; Rimpelä, A. Three Factor Eating Questionnaire-R18 as a Measure of Cognitive Restraint, Uncontrolled Eating and Emotional Eating in a Sample of Young Finnish Females. *Int. J. Behav. Nutr. Phys. Act.* **2009**, *6*, 41. [CrossRef]
11. Elfhag, K.; Linné, Y. Gender Differences in Associations of Eating Pathology between Mothers and Their Adolescent Offspring. *Obes. Res.* **2005**, *13*, 1070–1076. [CrossRef]
12. Kornetov, N.; Yarosh, O.; Sitnikova, E. Depressive and Anxiety Disorders among Women with Obesity. *Eur. Psychiatry* **2021**, *64*, S339. [CrossRef]
13. Jung, S.J.; Woo, H.-T.; Cho, S.; Park, K.; Jeong, S.; Lee, Y.J.; Kang, D.; Shin, A. Association between Body Size, Weight Change and Depression: Systematic Review and Meta-Analysis. *Br. J. Psychiatry* **2017**, *211*, 14–21. [CrossRef] [PubMed]
14. Luppino, F.S.; de Wit, L.M.; Bouvy, P.F.; Stijnen, T.; Cuijpers, P.; Penninx, B.W.J.H.; Zitman, F.G. Overweight, Obesity, and Depression: A Systematic Review and Meta-Analysis of Longitudinal Studies: A Systematic Review and Meta-Analysis of Longitudinal Studies. *Arch. Gen. Psychiatry* **2010**, *67*, 220–229. [CrossRef] [PubMed]
15. Kasen, S.; Cohen, P.; Chen, H.; Must, A. Obesity and Psychopathology in Women: A Three Decade Prospective Study. *Int. J. Obes.* **2008**, *32*, 558–566. [CrossRef] [PubMed]
16. de Wit, L.; Have, M.T.; Cuijpers, P.; de Graaf, R. Body Mass Index and Risk for Onset of Mood and Anxiety Disorders in the General Population: Results from the Netherlands Mental Health Survey and Incidence Study-2 (NEMESIS-2). *BMC Psychiatry* **2022**, *22*, 522. [CrossRef] [PubMed]
17. Zhao, G.; Ford, E.S.; Dhingra, S.; Li, C.; Strine, T.W.; Mokdad, A.H. Depression and Anxiety among US Adults: Associations with Body Mass Index. *Int. J. Obes.* **2009**, *33*, 257–266. [CrossRef] [PubMed]
18. Cifuentes, L.; Ghusn, W.; Feris, F.; Campos, A.; Sacoto, D.; De la Rosa, A.; McRae, A.; Rieck, T.; Mansfield, S.; Ewoldt, J.; et al. Phenotype Tailored Lifestyle Intervention on Weight Loss and Cardiometabolic Risk Factors in Adults with Obesity: A Single-Centre, Non-Randomised, Proof-of-Concept Study. *EClinicalMedicine* **2023**, *58*, 101923. [CrossRef]
19. Karlsson, J.; Persson, L.-O.; Sjöström, L.; Sullivan, M. Psychometric Properties and Factor Structure of the Three-Factor Eating Questionnaire (TFEQ) in Obese Men and Women. Results from the Swedish Obese Subjects (SOS) Study. *Int. J. Obes.* **2000**, *24*, 1715–1725. [CrossRef]
20. Cameron, I.M.; Crawford, J.R.; Lawton, K.; Reid, I.C. Psychometric Comparison of PHQ-9 and HADS for Measuring Depression Severity in Primary Care. *Br. J. Gen. Pract.* **2008**, *58*, 32–36. [CrossRef]
21. Kokoszka, A.; Jastrzębski, A.; Obrębski, M. Ocena Psychometrycznych Właściwości Polskiej Wersji Kwestionariusza Zdrowia Pacjenta-9 Dla Osób Dorosłych. *Psychiatria* **2016**, *13*, 187–193.
22. Johnson, S.U.; Ulvenes, P.G.; Øktedalen, T.; Hoffart, A. Psychometric Properties of the General Anxiety Disorder 7-Item (GAD-7) Scale in a Heterogeneous Psychiatric Sample. *Front. Psychol.* **2019**, *10*, 1713. [CrossRef] [PubMed]
23. Phqscreeners. Available online: https://www.phqscreeners.com/ (accessed on 21 November 2023).
24. Löffler, A.; Luck, T.; Then, F.S.; Sikorski, C.; Kovacs, P.; Böttcher, Y.; Breitfeld, J.; Tönjes, A.; Horstmann, A.; Löffler, M.; et al. Eating Behaviour in the General Population: An Analysis of the Factor Structure of the German Version of the Three-Factor-Eating-Questionnaire (TFEQ) and Its Association with the Body Mass Index. *PLoS ONE* **2015**, *10*, e0133977. [CrossRef] [PubMed]
25. Malkki-Keinänen, K.; Lankinen, M.; Karhunen, L.; Schwab, U. Psychometric Evaluation of Three-Factor Eating Questionnaire -R18 in Aging Finnish Men with Increased Risk for Type 2 Diabetes. *Nutr. Health* **2022**, 2601060221112178. [CrossRef] [PubMed]
26. Pruccoli, J.; Mack, I.; Klos, B.; Schild, S.; Stengel, A.; Zipfel, S.; Giel, K.E.; Schag, K. Mental Health Variables Impact Weight Loss, Especially in Patients with Obesity and Binge Eating: A Mediation Model on the Role of Eating Disorder Pathology. *Nutrients* **2023**, *15*, 3915. [CrossRef]
27. Gerlach, G.; Herpertz, S.; Loeber, S. Personality Traits and Obesity: A Systematic Review. *Obes. Rev.* **2015**, *16*, 32–63. [CrossRef]
28. Schag, K.; Schönleber, J.; Teufel, M.; Zipfel, S.; Giel, K.E. Food-Related Impulsivity in Obesity and Binge Eating Disorder—A Systematic Review: Food-Related Impulsivity. *Obes. Rev.* **2013**, *14*, 477–495. [CrossRef]
29. van Strien, T.; Konttinen, H.; Homberg, J.R.; Engels, R.C.M.E.; Winkens, L.H.H. Emotional Eating as a Mediator between Depression and Weight Gain. *Appetite* **2016**, *100*, 216–224. [CrossRef]
30. Amoako, M.; Amoah-Agyei, F.; Du, C.; Fenton, J.I.; Tucker, R.M. Emotional Eating among Ghanaian University Students: Associations with Physical and Mental Health Measures. *Nutrients* **2023**, *15*, 1526. [CrossRef]

31. Ziser, K.; Finklenburg, C.; Behrens, S.C.; Giel, K.E.; Becker, S.; Skoda, E.-M.; Teufel, M.; Mack, I.; Zipfel, S.; Junne, F. Perceived Stress Mediates the Relationship of Body Image and Depressive Symptoms in Individuals with Obesity. *Front. Psychiatry* **2019**, *10*, 852. [CrossRef]
32. Vidmar, A.P.; Yamashita, N.; Fox, D.S.; Hegedus, E.; Wee, C.P.; Salvy, S.J. Can a Behavioral Weight-Loss Intervention Change Adolescents' Food Addiction Severity? *Child. Obes.* **2022**, *18*, 206–212. [CrossRef]
33. Babicki, M.; Kotowicz, K.; Piotrowski, P.; Stramecki, F.; Kobyłko, A.; Rymaszewska, J. Areas of Stigma and Discrimination of Mentally Ill People among Internet Respondents in Poland. *Psychiatr. Pol.* **2018**, *52*, 93–102. [CrossRef]
34. Christaki, E.; Kokkinos, A.; Costarelli, V.; Alexopoulos, E.C.; Chrousos, G.P.; Darviri, C. Stress Management Can Facilitate Weight Loss in Greek Overweight and Obese Women: A Pilot Study. *J. Hum. Nutr. Diet.* **2013**, *26*, 132–139. [CrossRef] [PubMed]
35. Jensterle, M.; Kocjan, T.; Kravos, N.A.; Pfeifer, M.; Janez, A. Short-Term Intervention with Liraglutide Improved Eating Behavior in Obese Women with Polycystic Ovary Syndrome. *Endocr. Res.* **2015**, *40*, 133–138. [CrossRef] [PubMed]
36. Polskiego, S.; Otyłości, T.L.; Szulińska, N.M.M.; Walczak, N.M.M.; Mariusz Wyleżoł, N.M.; Psychologii, Z.; Katedra, N.; Śląski, U.; Dietetyki, K.; Szkoła, W.; et al. Zalecenia Kliniczne Dotyczące Postępowania u Chorych Na Otyłość 2022. Available online: https://ptlo.org.pl/resources/data/sections/114/ws_otylosc.pdf (accessed on 21 November 2023).
37. Story, N.M.; Lumby, J.; Fethney, J.; Waters, D. A Comparative Study of Eating Behaviours within and between Conventional Metabolic (Bariatric) Surgery Procedures. *J. Clin. Nurs.* **2021**, *30*, 3342–3354. [CrossRef]
38. Jacobson, B.H.; DeBock, D.H. Comparison of Body Mass Index by Self-Reported versus Measured Height and Weight. *Percept. Mot. Skills* **2001**, *92*, 128–132. [CrossRef]
39. Niedhammer, I.; Bugel, I.; Bonenfant, S.; Goldberg, M.; Leclerc, A. Validity of Self-Reported Weight and Height in the French GAZEL Cohort. *Int. J. Obes.* **2000**, *24*, 1111–1118. [CrossRef]

nutrients

MDPI

Article

Exploring Correlations of Food-Specific Disgust with Eating Disorder Psychopathology and Food Interaction: A Preliminary Study Using Virtual Reality

Sevgi Bektas [1,2,*], Ludovica Natali [3], Katie Rowlands [1], Lucia Valmaggia [4,5,6], Jerome Di Pietro [4], Hiba Mutwalli [1,7], Hubertus Himmerich [1,5], Janet Treasure [1,5] and Valentina Cardi [1,3]

1 Centre for Research in Eating and Weight Disorders (CREW), Institute of Psychiatry, Psychology and Neuroscience, King's College London, London SE5 8AF, UK; katie.rowlands@kcl.ac.uk (K.R.); hiba.mutwalli@kcl.ac.uk (H.M.); hubertus.himmerich@kcl.ac.uk (H.H.); janet.treasure@kcl.ac.uk (J.T.); valentina.cardi@unipd.it (V.C.)
2 Department of Psychology, Hacettepe University, Ankara 06800, Türkiye
3 Department of General Psychology, University of Padova, 35122 Padova, Italy
4 Department of Psychology, Institute of Psychiatry, Psychology & Neuroscience, King's College London, London SE5 8AF, UK; lucia.valmaggia@kcl.ac.uk (L.V.); jerome.dipietro@kcl.ac.uk (J.D.P.)
5 South London and Maudsley NHS Foundation Trust, London SE5 8AB, UK
6 Department of Psychiatry, KU Leuven, 3000 Leuven, Belgium
7 Department of Clinical Nutrition, College of Applied Medical Sciences, Imam Abdulrahman Bin Faisal University, Dammam 34212, Saudi Arabia
* Correspondence: sevgi.bektas1@kcl.ac.uk

Abstract: The emotion of disgust is thought to play a critical role in maintaining restrictive eating among individuals with anorexia nervosa. This exploratory cross-sectional study examined correlations between food-specific trait and state disgust, eating disorder psychopathology, illness severity (body mass index: BMI), and interactions with virtual foods in people with anorexia nervosa. Food-specific trait disgust and eating disorder symptoms were measured before exposure to virtual foods in one of three virtual reality (VR) kitchens to which participants were randomly allocated. Food interactions (eye gaze and reaching towards virtual foods) were measured during the VR exposure. Food-specific state disgust ratings were collected after the VR exposure. In the entire sample, eating disorder symptoms correlated positively with food-specific trait disgust (rs (68) = 0.45, $p < 0.001$). We also found a significant association between food-specific state disgust and eating disorder symptoms in each virtual kitchen scenario: virtual kitchen only (rs (22) = 0.40, $p = 0.05$), virtual kitchen plus pet (rs (22) = 0.80, $p < 0.001$), and virtual kitchen plus avatar (rs (20) = 0.78, $p < 0.001$). No significant correlation was observed for the link between food-specific disgust measures and food-related touch. Correlations between food-specific trait disgust and food-related eye gaze differed across scenarios. The current experimental paradigm needs to be improved to draw firm conclusions. Both food-specific trait and state disgust are associated with eating disorder psychopathology, and therefore, effective strategies are warranted to attenuate food-specific disgust.

Keywords: anorexia nervosa; disgust; food; eye gaze; touch; virtual reality

Citation: Bektas, S.; Natali, L.; Rowlands, K.; Valmaggia, L.; Di Pietro, J.; Mutwalli, H.; Himmerich, H.; Treasure, J.; Cardi, V. Exploring Correlations of Food-Specific Disgust with Eating Disorder Psychopathology and Food Interaction: A Preliminary Study Using Virtual Reality. *Nutrients* **2023**, *15*, 4443. https://doi.org/10.3390/nu15204443

Academic Editor: Antonios Dakanalis

Received: 13 September 2023
Revised: 2 October 2023
Accepted: 12 October 2023
Published: 19 October 2023

1. Introduction

Anorexia nervosa (AN) is a psychiatric disorder defined by a persistent and pervasive tendency of food avoidance. Exposure to food- or eating-related cues is associated with expected negative consequences (e.g., losing control, weight gain) and aversive emotions [1]. Much of the literature in this field has focused on the emotions of fear and anxiety, whereas only a few studies have investigated the role of disgust in maintaining eating disorder behaviours. A recent systematic review and meta-analysis of disgust in eating disorders [2] indicated that disgust in response to high-calorie food images was significantly elevated in people with eating disorders, whereas generic disgust sensitivity was lower. Individual differences in food disgust sensitivity compared to generic disgust sensitivity may be a more salient concept in people with AN since they tend to report disgust towards specific foods or food characteristics. Previous studies have examined state disgust, such as disgust reactions in response to food pictures in AN [2]. However, a pervasive pattern of food-specific disgust as a potential trait of individuals with AN has not yet been the focus of research.

Both state and trait forms of food-specific disgust may play a key role in shaping cognitive and behavioural mechanisms underlying AN symptomatology. For example, attentional or behavioural bias for food in AN has been studied using behavioural tasks and eye-tracking paradigms [3,4]. A recent systematic and methodological review of this literature [5] highlighted the variations in paradigms or stimuli used across studies. These variations have limited researchers' ability to conclude on the direction of biases (hypervigilance or avoidance of the stimuli). Nevertheless, it has been reported that attentional and/or behavioural avoidance of food, especially foods with high fat content, may be a common phenomenon in AN. It might be also possible that people with AN who experience greater food disgust may show enhanced attention and preparedness or greater avoidance of certain information or signals associated with danger (e.g., foods of unknown caloric density or changes in appearance), as suggested in the vigilance and avoidance model of attentional bias [6,7].

It is critical to explore relevant attentional and behavioural patterns linked to food disgust and eating disorder symptoms in AN for the development of treatment plans targeting disgust as well as anxiety or fear. Recently, virtual reality has been utilised as a practical research tool in the assessment of food-related attentional and behavioural mechanisms—for example, in people with binge eating disorders [8]. However, to our knowledge, data on disgust in AN using this technology are non-existent. Hence, the aim of this research is to test correlations between food-specific disgust, eating disorder psychopathology, and food interaction across different virtual kitchen scenarios in people with AN. No specific hypotheses were formulated due to the exploratory nature of this study.

2. Materials and Methods

2.1. Participants

English-speaking participants under treatment for AN were recruited from South London and Maudsley NHS Foundation Trust, through social media, and via email circulars at King's College London (KCL) between September 2022 and March 2023. The exclusion criteria were (1) age under 14 years; (2) body mass index (BMI) > 25; (3) self-reported diagnosis of neurological disorders, psychosis, or substance abuse disorders; and (4) visual/hearing impairments not corrected by glasses/hearing aids. All participants provided informed consent. The study was approved by the Research Ethics Committee North West—Liverpool East (reference number: 18/NW/0853) in the UK.

2.2. Apparatus

The virtual reality (VR) environment was developed by the KCL VR Research Lab for an ongoing large study led by V.C. and supported by the Medical Research Council [9]. The VR environment was built using the Unity3D game engine, Oculus Integration SDK for Unity, and Oculus Quest 2 headsets. A head-mounted display (HMD) headset and two Oculus Touch controllers were used. The headset has six degrees of freedom (6DoF) technology, which provides continuous rotational and positional tracking. The screen had a resolution of 1832 × 1920 pixels per eye.

The five-minute 3D scenario displayed a virtual kitchen, which stored foods of various calorie contents. A full list of the virtual foods is provided in Table S1. Three different versions of the virtual kitchen scenario were used: (1) a virtual kitchen only, (2) a virtual kitchen plus a virtual pet to induce a positive mood, and (3) a virtual kitchen plus an avatar to induce social support (see Text S1). Pre-recorded vocal instructions were used to guide participants on freedom to move in the kitchen, to look around, to open the fridge/cupboards, and to reach out to touch and hold the food using the hand controllers (for pictures of the virtual kitchen environment, see Figure S1).

2.3. Procedure

After providing written consent, participants completed a Qualtrics survey consisting of demographic and baseline questionnaires (related to eating disorder psychopathology and disgust sensitivity). Subsequently, participants were invited to complete the VR exposure session at KCL. During their lab visit, they were individually guided to familiarise themselves with the Oculus Quest 2 headset and controllers in a quiet room and then entered one of three virtual kitchen scenarios selected at random for each participant.

The use of components that increase the presence of positive emotional experiences can be beneficial for psychological treatment adjuncts [10]. So far, it has been demonstrated that both positive mood induction and recovery-focused support from peers are associated with increased calorie intake during a standard test meal and decreased attention bias to food after the test meal compared to a control condition [11–13]. In addition, it has been reported that encouraging support from people with lived experiences is highly endorsed by patients over time and associated with fewer eating disorder symptoms [12]. Based on this evidence, a positive mood induction condition and a social support condition were combined with the food exposure in the virtual kitchen scenarios with the goal of exploring whether these forms of support might impact participants' emotions/behaviours in the virtual kitchen compared to the food-only condition.

In the virtual kitchen plus pet scenario, participants entered the kitchen with a virtual pet, a pink elephant making soft gurgling noises and following the participant around the environment. In the virtual kitchen plus avatar scenario, participants were accompanied in the kitchen by a virtual avatar (they could choose between four different avatars, see Figure S2) who spoke from a supportive and motivational script (e.g., "It is normal to feel anxious, you are doing the best you can right now. Nobody is here to judge what you will do in the kitchen, which foods you will look at or grab. Take a few deep breaths, slow your mindset down, and think of the bigger picture. Each minute more is a minute easier"), encouraging participants to challenge the eating disorder voice and approach the virtual foods.

After the completion of the exposure session, participants were asked to rate their disgust at the thought of eating the high-calorie or palatable foods that were shown in the virtual kitchen. Participants were debriefed and reimbursed with a GBP 20 Amazon voucher upon completing the survey and VR exposure.

2.4. Measures

The following data were collected:

Before VR exposure

The socio-demographic and clinical variables collected were age, gender, ethnicity, years of education, illness duration, self-reported weight, and height (used to calculate BMI as kg/m^2). At baseline, participants also completed the Food Disgust Scale [14], a 32-item scale measuring eight domains of food disgust (Cronbach's alpha in this study = 0.93) as a parameter reflecting food-specific trait disgust, and the Eating Disorders Examination Questionnaire [15], a 28-item scale for the assessment of eating disorder symptom severity (Cronbach's alpha in this study = 0.94), with four dimensions: dietary restraint, shape concern, weight concern, and eating concern.

During VR exposure

The frequency of eye gazes towards and touching of virtual foods were automatically recorded in the VR environment by the software. The application used a narrow BoxCast, a rectangular ray fired along the Quest HDM's z-axis (see Figure 1, produced by J.D.P.), to approximate eye gaze direction. The head movement can, in this way, be used to determine where the participant is looking at the task, despite its limitation to perfectly analogous and accurate eye tracking.

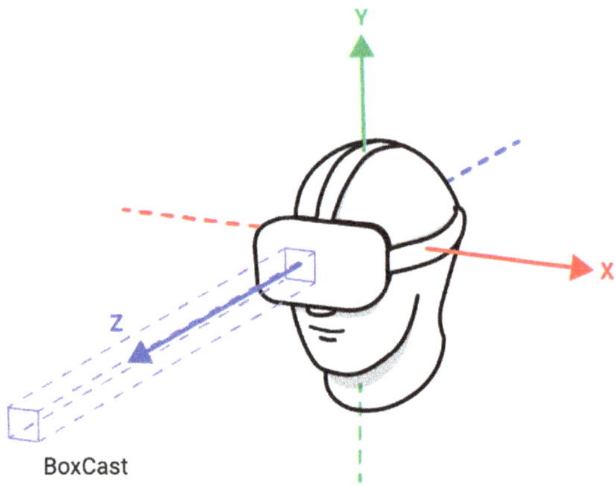

Figure 1. A diagram of the virtual reality device.

After VR exposure

As a parameter reflecting state disgust, momentary disgust reactions towards pictures of high-calorie foods that were shown in the virtual kitchen were collected using a Likert scale ranging from 0 (not at all) to 7 (extremely disgusted) with one question ("How disgusted do you feel now about eating high-calorie or palatable foods (e.g., pizza, chocolate, crisps)?").

For a summary of measures collected at different time points, see Table 1.

Table 1. Summary of measures collected in the present study.

Before the VR kitchen	During the VR kitchen (Participants were randomly allocated to different scenarios: (1) virtual kitchen only, (2) virtual kitchen + pet, (3) virtual kitchen + avatar	After the VR kitchen
Food-specific trait disgust measure: Food Disgust Scale, a 32-item scale from 1 (not disgusting at all) to 6 (extremely disgusting) for food disgust sensitivity.	**Food interaction measure:** The frequency of food-related eye gaze towards virtual food items.	**Food-specific state disgust measure:** A Likert scale with one question ("How disgusted do you feel now about eating high-calorie or palatable foods (e.g., pizza, chocolate, crisps)?") for post-exposure disgust towards high-calorie foods that were shown in the virtual kitchen.
Eating disorder psychopathology measure: Eating Disorder Examination Questionnaire, a 28-item scale for eating disorder symptom severity.	**Food interaction measure:** The frequency of food-related touching of virtual food items.	
Illness severity measure: body mass index (BMI) as kg/m^2		

2.5. Statistical Analyses

All statistical analyses were performed with SPSS v.28 (IBM Corp., Armonk, NY, USA). The descriptive analyses provided an overview of the demographic and clinical variables of the entire sample. Data were non-parametrically distributed. Thus, median and interquartile ranges were reported. The Pearson Chi-square test and the Kruskal–Wallis one-way analysis of variance with a post hoc Dunn test was used to compare the participants' characteristics among different virtual kitchen scenarios since the criteria for homogeneity of variance was not met. Spearman rank–order correlation coefficients were used to identify the strength and directions of relationships between the variables of interest. Significance was determined with p-values ≤ 0.05. We did not control multiple measurements, as this was an exploratory study.

3. Results

A total of 70 participants with AN with an average age of 25.7 (SD = 7) years completed the study. Most participants were female (94.3%) and White British (88.6%). The overall mean BMI was 16.8 (SD = 2.5) kg/m^2.

The Kruskal–Wallis test revealed significant differences between the three virtual scenarios only for illness duration. A follow-up Dunn test revealed that the length of illness duration of participants in the virtual kitchen plus avatar scenario was significantly lower than those in the kitchen-only scenario (H (2) = 7.91, p = 0.02). Descriptive statistics and group comparisons are reported in Table 2.

Table 2. Descriptive statistics of the demographic and clinical variables and group comparisons.

	N (%) or Median (IQR)				
Variable	**Whole Sample (n = 70)**	**Kitchen (n = 24)**	**Kitchen + Pet (n = 24)**	**Kitchen + Avatar (n = 22)**	**Test of Group Differences**
Age	23.5 (21.00–29.25)	26.0 (21.00–29.75)	24.5 (16.0–40.0)	22.5 (20.00–26.25)	H (2) = 3.15, p = 0.21
Sex					X2 (4) = 6.17, p = 0.19
Female	66 (94.3%)	21 (87.5%)	24 (100%)	21 (95.5%)	
Male	2 (2.9%)	1 (4.2%)	N/A	1 (4.5%)	
Other	2 (2.9%)	2 (8.3%)	N/A	N/A	
Ethnicity					X2 (6) = 10.42, p = 0.11
Asian British	2 (2.9%)	2 (8.3%)	N/A	N/A	
White British	62 (88.6%)	19 (79.2%)	22 (91.7%)	21 (95.5%)	
Mixed race	2 (2.9%)	N/A	1 (4.2%)	1 (4.5%)	
Other (non-British)	4 (5.7%)	3 (12.5%)	1 (4.2%)	N/A	
Years of education	16.0 (14.00–18.00)	16.0 (15.00–18.00)	15.5 (13.00–18.00)	16.0 (14.00–18.00)	H (2) = 0.24, p = 0.89
Illness duration	8.0 (4.75–12.00)	10.0 (7.13–13.00)	8.0 (2.88–13.75)	6.5 (3.50–9.00)	H (2) = 7.91, p = 0.02 * Kitchen > kitchen + avatar, p = 0.02
BMI (kg/m2)	16.6 (15.34–18.47)	16.5 (14.90–18.24)	16.5 (14.96–17.64)	17.3 (17.29–19.61)	H (2) = 3.40, p = 0.18
EDE-Q					H (2) = 0.25, p = 0.88
EDE-Q global	3.6 (2.72–4.73)	3.8 (2.92–4.84)	3.5 (2.50–4.72)	4.0 (2.21–4.88)	
EDE-Q restraint	3.9 (2.35–4.65)	3.7 (2.10–4.75)	3.9 (2.45–4.60)	4.2 (2.00–4.85)	
EDE-Q eating concern	2.6 (1.55–4.20)	2.8 (1.80–4.70)	2.6 (1.45–3.70)	2.3 (1.30–4.40)	
EDE-Q weight concern	4.0 (2.80–5.40)	4.1 (2.90–5.40)	4.0 (3.00–5.20)	4.0 (2.50–5.40)	
EDE-Q shape concern	4.5 (3.35–5.63)	4.6 (3.53–5.60)	4.4 (3.38–5.38)	4.8 (2.88–5.91)	
FDS					H (2) = 0.05, p = 0.98
FDS global	4.0 (3.27–4.69)	4.0 (3.30–4.85)	4.2 (3.10–4.71)	3.8 (3.30–4.53)	

Table 2. *Cont.*

Variable	N (%) or Median (IQR)				Test of Group Differences
	Whole Sample (n = 70)	Kitchen (n = 24)	Kitchen + Pet (n = 24)	Kitchen + Avatar (n = 22)	
FDS animal meat	4.8 (3.50–5.50)	4.9 (3.38–5.94)	4.9 (3.56–5.50)	4.5 (3.44–5.31)	
FDS poor hygiene	5.6 (4.95–6.00)	5.8 (4.85–6.00)	5.5 (4.45–5.95)	5.7 (5.30–5.85)	
FDS human contamination	3.0 (1.94–4.50)	3.4 (2.00–4.50)	3.5 (1.81–6.00)	2.6 (1.75–3.63)	
FDS mold	4.0 (4.00–5.50)	4.0 (2.06–5.25)	3.9 (1.81–6.00)	4.9 (2.38–5.50)	
FDS decaying fruit	2.5 (1.25–4.00)	2.5 (1.31–4.00)	2.1 (1.38–4.00)	2.9 (1.19–3.75)	
FDS fish	3.4 (2.25–5.00)	3.3 (2.50–5.50)	3.8 (1.81–5.00)	3.3 (2.25–4.31)	
FDS decaying vegetables	3.4 (2.25–4.56)	3.0 (2.25–4.19)	3.4 (2.25–4.44)	4.1 (2.81–4.75)	
FDS living contamination	6.0 (4.92–6.00)	6.0 (5.08–6.00)	6.0 (4.67–6.00)	5.7 (4.92–6.00)	
Food interaction					
VR eye gazes	290.5 (229.50–353.25)	287.0 (233.25–379.75)	292.0 (216.75–357.75)	292.5 (216.50–347.00)	H (2) = 0.03, p = 0.99
VR touches	19.0 (12.00–25.25)	21.5 (15.00–29.75)	18.5 (12.00–22.00)	15.0 (9.50–26.50)	H (2) = 5.26, p = 0.07
Post-VR disgust	5.0 (4.00–6.00)	5.0 (4.00–6.00)	6.0 (4.00–6.50)	5.0 (3.00–6.25)	H (2) = 0.36, p = 0.84

Abbreviations. IQR, interquartile range; N/A, not applicable; BMI, body mass index; EDE-Q, Eating Disorder Examination Questionnaire; FDS, Food Disgust Scale; VR eye gazes, the frequencies of eye gazes towards virtual foods; VR touches, the frequencies of touching virtual foods; post-VR disgust, momentary disgust reactions towards virtual foods with high-calorie content measured following the VR exposure; H, Kruskal–Wallis test; X^2, Pearson Chi-square test. * $p \leq 0.05$.

3.1. *Correlations between Eating Disorder Psychopathology and Food-Specific Disgust Measures*

The associations between food disgust sensitivity (food-specific trait disgust), post-exposure disgust towards high-calorie foods alone (food-specific state disgust), eating disorder severity (EDE-Q global and BMI), and food interactions (eye gazes towards and touching virtual foods) are presented in Table 3. The SPSS output for Spearman rank-order correlations and scatterplots are available in the Supplementary Materials (see Figures S3 and S4).

In the entire sample, the correlation between food disgust sensitivity (food-specific trait disgust) and the EDE-Q global score was 0.45 ($p < 0.001$, 95% CI [0.24, 0.63]), indicating a moderate relationship. There were no statistically significant correlations of food-specific disgust measures and BMI.

The correlation between post-exposure disgust ratings (food-specific state disgust) and the EDE-Q global scores was also significant in each virtual kitchen scenario: virtual kitchen only (rs = (22) = 0.40, $p = 0.05$, 95% CI [−0.02, 0.70]), virtual kitchen plus pet (rs = (22) = 0.80, $p < 0.001$, 95% CI [0.56, 0.92]), and virtual kitchen plus avatar (rs = (20) = 0.78, $p < 0.001$, 95% CI [0.52, 0.92]). However, there were no statistically significant correlations between post-exposure disgust ratings (food-specific state disgust) and BMI in the three virtual kitchen scenarios.

3.2. *Correlations between Food Interaction and Food-Specific Disgust Measures in the Virtual Kitchen Scenarios*

Food disgust sensitivity (food-specific trait disgust) was significantly and positively correlated with frequency of eye gazes towards virtual foods (rs = (22) = 0.66, $p < 0.001$, 95% CI [0.34, 0.84]) in the virtual kitchen plus pet scenario. There were no statistically significant correlations between food-specific disgust measures and food interaction measures in the virtual kitchen-only scenario or in the virtual kitchen plus pet condition.

Table 3. Spearman rank–order correlations for variables of interest in people with anorexia nervosa.

VR Scenarios	Variables					
	I FDS_Global	II Post-VR Disgust	III EDE-Q Global	IV BMI	V VR Eye Gazes	VI VR Touches
			All Sample (n = 70)			
I FDS global	1.00					
II Post-VR-Disgust	0.16	1.00				
III EDE-Q global	0.45 **	0.66 **	1.00			
IV BMI	0.15	−0.01	−0.00	1.00		
V VR eye gazes	0.25 *	−0.13	−0.16	0.27 *	1.00	
VI VR touches	−0.22	−0.17	−0.35 **	−0.07	0.33 **	1.00
			Kitchen (n = 24)			
I FDS global	1.00					
II post-VR-disgust	0.14	1.00				
III EDE-Q global	0.53 **	0.40 *	1.00			
IV BMI	0.05	−0.04	0.02	1.00		
V VR eye gazes	0.11	−0.27	−0.33	0.12	1.00	
VI VR touches	−0.36	−0.14	−0.31	−0.36	0.33	1.00
			Kitchen + Pet (n = 24)			
I FDS global	1.00					
II post-VR disgust	0.16	1.00				
III EDE-Q global	0.33	0.80 **	1.00			
IV BMI	0.10	−0.36	−0.42 *	1.00		
V VR eye gazes	0.66 **	0.17	0.09	0.35	1.00	
VI VR touches	0.15	−0.15	−0.47 *	0.34	0.37	1.00
			Kitchen + Avatar (n = 22)			
I FDS global	1.00					
II post-VR disgust	0.19	1.00				
III EDE-Q global	0.45 *	0.78 **	1.00			
IV BMI	0.29	0.24	0.28	1.00		
V VR eye gazes	0.02	−0.26	−0.23	0.25	1.00	
VI VR touches	−0.33	−0.20	−0.40	−0.21	0.28	1.00

Abbreviations. IQR, interquartile range; BMI, body mass index; EDE-Q, Eating Disorder Examination Questionnaire; FDS, Food Disgust Scale; VR eye gazes, the frequency of eye gazes towards virtual foods; VR touches, the frequency of touching virtual foods; post-VR disgust, momentary disgust reactions towards virtual foods with high-calorie content measured following the VR exposure. * $p \leq 0.05$; ** $p < 0.001$.

4. Discussion

This explorative study assessed correlations between food-specific trait and state disgust, eating disorder psychopathology (EDE-Q global and BMI), and food interaction measures (eye gaze and touch) in people with AN. We collected food disgust sensitivity (trait disgust) and eating disorder psychopathology data at baseline before participants were randomly assigned to the different virtual kitchen scenarios in which food interaction measures were obtained. Participants were then asked to rate their disgust reactions (state disgust) to high-calorie foods. The correlations involving variables collected during or after the VR paradigm were examined for each scenario rather than the whole sample. There are two main findings in the present study. The first finding is that food-specific trait and state disgust were significantly and positively associated with eating disorder symptom severity. The second finding is that inconsistent correlations between food disgust sensitivity and food-related eye gaze in different virtual kitchen scenarios were observed.

The first finding of an association between trait and state disgust with eating disorder psychopathology can be interpreted as an association of eating disorder severity with food-specific disgust. This is in line with a recent systematic review and meta-analysis that found that the levels of generic disgust sensitivity, self-disgust sensitivity, and momentary disgust to food images were significantly higher in people with eating disorders than in healthy controls [2]. Food disgust sensitivity has recently been investigated in non-clinical populations, and findings indicate that food disgust sensitivity may potentially influence food- or eating-related behaviours (e.g., texture-based food rejection, picky eating, selective eating, and variety-seeking in food) [16,17]. Surprisingly, there has been no research in patients with eating disorders so far in this regard.

In addition, food-specific state disgust correlated with eating disorder severity in the three different virtual kitchen scenarios. However, there was no difference among the scenarios regarding the question of "How disgusted do you feel now about eating high-calorie or palatable foods (e.g., pizza, chocolate, crisps)?" This indicates that there is no post-exposure effect on disgust ratings. However, this question is not suitable for investigating the impact of the kitchen scenarios comprehensively on the food-specific state of disgust. In addition, the experimental paradigm was primarily designed to test whether induced social support or positive mood would enhance the impact of virtual food exposure on food-related anxiety in people with AN [9]. It has been reported that disgust is more resistant to change compared to anxiety and fear [18], and therefore the development of treatment strategies directly targeting disgust in response to food or eating is necessary. Improvements in the current VR paradigm could be considered to ameliorate food-specific state disgust in the future. The availability of compassionate social support and social networks has a key role in the treatment of individuals with AN in terms of developing new positive associations with food stimuli [19]. Thus, an exposure to food-related environments providing social contact in VR settings may motivate individuals with AN to approach disgust-eliciting foods. For example, not only the company of the virtual avatar but also a positive interaction/dialogue in the virtual kitchen may have a potential therapeutic value of improving positive emotions (e.g., joy, connectedness, optimism) and diminishing disgust. Developing tailored and personally relevant scripts addressing disgust feelings in response to high-calorie foods and relevant negative expectations (e.g., body signals, taste pleasantness, or weight gain) and motivations (e.g., avoidance) might be useful.

Our second finding suggests that correlations between food disgust sensitivity and food-related eye gaze differed across the three virtual kitchen scenarios. A positive correlation between food disgust sensitivity and food-related eye gaze was observed for people with AN in the virtual kitchen plus pet scenario. Interestingly, this relationship was non-existent in the other two conditions. The variations between the conditions might partially explain the discrepancies in the results. Specifically, participants in the virtual kitchen plus avatar condition had a higher BMI and lower illness duration in comparison with those in the virtual kitchen plus pet condition. Lower BMI and longer duration of eating disorder diagnosis have previously been associated with non-response to treatment [20] or worse short-term treatment outcomes [21] in people with AN. The relationship between food disgust sensitivity and eye gaze towards virtual food items might be more evident in people with greater eating disorder severity. On the other hand, if the illness duration is critical to seeing a statistically significant correlation, the absence of this relationship in the kitchen-only scenario for participants with the longest illness duration is puzzling. This raises the question of whether the presence of a pet might motivate people with greater food disgust sensitivity to look at food items more often in the virtual kitchen environment. The inconsistent findings might also be explained by the low power due to the small sample size across the virtual kitchen scenarios. The correlation of food-specific trait and state disgust with food interaction measures should be tested in a larger sample of AN patients to better understand the relationship between food disgust sensitivity and food-related eye gaze. To prevent variations between conditions for illness duration, stratified randomisation may be considered.

5. Clinical Implications

In order to draw clinical conclusions, longitudinal studies on the relevance of disgust for the development of eating disorder pathology are needed. If this causal connection is confirmed, then strategies that enable people to modify their perception of food-specific disgust could be useful for the prevention and treatment of AN and other eating disorders.

So far, most interventions for AN have focused on food-related fears rather than disgust around food or eating. Our study highlights the importance of targeting food-specific disgust as well. Behavioural strategies—for example, exposure and counterconditioning of US revaluation—are recommended to reduce extinction-resistant disgust responses in the literature on obsessive-compulsive disorders [18]. Future research could test whether increasing the frequency and duration of exposure-based techniques leads to more effective and longer-term effects in reducing disgust-based avoidance. Cognitive techniques may also be of use to evaluate thoughts related to their perceptions of unpleasant or threatening experiences of food-specific disgust.

6. Strengths and Limitations

This study has two main strengths. First, to the best of our knowledge, this is the first study to examine the correlation between food-disgust sensitivity and eating disorder symptom severity in people with AN. Second, the VR paradigm enabled aspects of attention and approaches towards food to be measured.

There are some limitations in the present study to acknowledge. The first limitation is the lack of a healthy control group. Including a healthy control group would be valuable for supporting the hypothesis that food-specific disgust is critical to target in the treatment of anorexia and when investigating aberrant patterns in food-related eye gaze and touch. The second is the smaller number of participants from ethnic minority backgrounds, limiting the generalisability. Possible explanations could be that young White females, the majority of the participants, have more cultural pressure on body image, greater awareness of the illness, or access to better medical care [22,23]. Further studies may have a healthy control group and may implement new recruitment strategies to increase ethnic diversity in clinical anorexia research [24]. Future studies could also examine differences between individuals with diverse types of food-related avoidance (e.g., avoidant–restrictive food intake disorder, restrictive and binge–purge-type AN) to address important research questions of clinical significance. In the current sample, autistic traits were not evaluated. However, in patients with AN who have autistic traits, expectation about the impact of food on the body with aversive feelings or physical sensations appear to be more frequent and stronger than in non-autistic patients and more likely to generalise to other foods. Elevated food disgust sensitivity may cause susceptible individuals to experience disgust when encountering combined or modified edible foods (e.g., bread with seeds, wonky strawberries, chocolate with nuts) in a similar way to encountering inedible foods. Thus, having participants who are more similar (e.g., individuals with both AN and autism spectrum traits) might be useful in minimising the within-subject variability and being able to evaluate the effects of different virtual kitchen scenarios. Improving the existing food list with the more complex aforementioned foods could help with the assessment of food disgust sensitivity for people with more complex sensory difficulties related to food (e.g., some autistic individuals).

The current experimental paradigm also has limitations. For example, food-specific trait and state disgust were measured at different time points. Only one question was utilised to measure disgust reactions to a group of high-calorie foods that were shown in the virtual kitchen. Food-specific state disgust responses collected before and after the virtual kitchen exposure via VR would help to evaluate the effects of different virtual kitchen scenarios. Exposure to a virtual kitchen with a wider range of foods (e.g., decaying melon, wrinkled tomato, chocolate with chocolate bloom, snails, oysters, and grasshoppers) might be more likely to elicit disgust, especially in people with a greater level of food disgust sensitivity. To further increase the sense of presence in VR and to make the user experience even more immersive, it might be useful to include more sensory modalities that occur during interaction with food, such as tactile feedback and smell. With regards to food interaction measures, it might be useful to analyse them specifically in relation to the calorie content of virtual food items. In addition to the VR touch feature, other measures of food avoidance could be tested in a larger sample to draw firm conclusions regarding the link between food-related touch and food-specific disgust.

7. Conclusions

The present explorative study tested the correlations of food-specific disgust measures with eating disorder psychopathology and food interaction measures by using the VR paradigm in people with AN. The findings showed that both food-specific trait and state disgust were correlated with eating disorder severity. The correlation between food-specific trait disgust and food-related eye gaze was statistically significant in the virtual kitchen plus pet scenario, but not in the other two virtual kitchen scenarios. The inconsistent findings could be explained by the small sample size or the effect of variations in the virtual kitchen scenarios. The present study offers valuable insight into ways to modify the design of future VR studies to examine potential correlates of food-specific disgust in people with AN, which could contribute to assessment and treatment strategies.

Supplementary Materials: The following supporting information can be downloaded at the following link: https://www.mdpi.com/article/10.3390/nu15204443/s1, Table S1: List of virtual foods; Text S1: Script used in the kitchen + avatar scenario; Figure S1: Pictures of the virtual kitchen environment; Figure S2: Selection of avatars available to participants; Figure S3: SPSS output for spearman rank–order correlations; Figure S4: Scatter plot figures.

Author Contributions: Conceptualisation, V.C., L.V., J.T. and S.B.; methodology, V.C., L.V., J.T. and S.B.; formal analysis, S.B. and L.N.; investigation, S.B., K.R., J.D.P. and L.N.; resources, V.C. and L.V.; data curation, L.N., K.R., S.B. and H.M.; writing—original draft preparation, S.B.; writing—review and editing, S.B., J.T., H.H., V.C., L.N. and H.M.; visualisation, S.B.; supervision, V.C., J.T. and H.H.; project administration, V.C. and L.V.; funding acquisition, V.C. and L.V. All authors have read and agreed to the published version of the manuscript.

Funding: This research was funded by the Medical Research Council (grant number MR/V003283/1). Lucia Valmaggia, Hubertus Himmerich, and Janet Treasure received salary support from the National Institute for Health and Care Research (NIHR) Biomedical Research Centre (BRC) at the South London and Maudsley NHS Foundation Trust (SLaM) and KCL. The authors would like to thank the Ministry of National Education of Türkiye for funding Sevgi Bektas's post-graduate studentship (the funder had no role in this project).

Institutional Review Board Statement: The study was conducted in accordance with the Declaration of Helsinki and approved by the HRA (Health Research Authority) and the Northwest—Liverpool Central Research Ethics Committee (REC Ref: 18/NW/0853). The participants were reimbursed for their time.

Informed Consent Statement: Informed consent was obtained from all subjects involved in the study.

Data Availability Statement: The data that support the findings of this study are available from the corresponding author, S.B., upon reasonable request.

Acknowledgments: For the purposes of open access, the authors have applied for a Creative Commons Attribution (CC BY) license to any Accepted Author Manuscript version arising from this submission. We would like to acknowledge all of the participants involved in the present study.

Conflicts of Interest: The authors declare no conflict of interest.

References

1. Steinglass, J.E.; Sysko, R.; Mayer, L.; Berner, L.A.; Schebendach, J.; Wang, Y.; Chen, H.; Albano, A.M.; Simpson, H.B.; Walsh, B.T. Pre-Meal Anxiety and Food Intake in Anorexia Nervosa. *Appetite* **2010**, *55*, 214–218. [CrossRef]
2. Bektas, S.; Keeler, J.L.; Anderson, L.M.; Mutwalli, H.; Himmerich, H.; Treasure, J. Disgust and Self-Disgust in Eating Disorders: A Systematic Review and Meta-Analysis. *Nutrients* **2022**, *14*, 1728. [CrossRef] [PubMed]
3. Kerr-Gaffney, J.; Harrison, A.; Tchanturia, K. Eye-Tracking Research in Eating Disorders: A Systematic Review. *Int. J. Eat. Disord.* **2019**, *52*, 3–27. [CrossRef] [PubMed]
4. Stott, N.; Fox, J.R.E.; Williams, M.O. Attentional Bias in Eating disorders: A Meta-Review. *Int. J. Eat. Disord.* **2021**, *54*, 1377–1399. [CrossRef]
5. Ralph-Nearman, C.; Margaret, A.; Rachel, L.; Jennifer, L.S.; Ruth, F. A Systematic and Methodological Review of Attentional Biases in Eating Disorders: Food, Body, and Perfectionism. *Brain Behav.* **2019**, *9*, e01458. [CrossRef]
6. Knowles, K.A.; Cox, R.C.; Armstrong, T.; Olatunji, B.O. Cognitive Mechanisms of Disgust in The Development and Maintenance of Psychopathology: A Qualitative Review and Synthesis. *Clin. Psychol. Rev.* **2019**, *69*, 30–50. [CrossRef]
7. Hildebrandt, T.; Bacow, T.; Markella, M.; Loeb, K.L. Anxiety in Anorexia Nervosa and Its Management Using Family-Based Treatment. *Eur. Eat. Disord. Rev.* **2012**, *20*, e1–e16. [CrossRef]
8. Max, S.M.; Schag, K.; Giel, K.E.; Plewnia, C. Behavioural Biases in The Interaction with Food Objects in Virtual Reality and Its Clinical Implication for Binge Eating Disorder. *Eat. Weight Disord.* **2023**, *28*, 46. [CrossRef]
9. Natali, L.; Meregalli, V.; Rowlands, K.; Pietro, J.D.; Treasure, J.; Collantoni, E.; Meneguzzo, P.; Tenconi, E.; Favaro, A.; Fontanaf, F.; et al. *Virtual Food for Real Thought: An Experimental Study to Test Strategies for Inhibitory Learning of Food Anxiety in Anorexia Nervosa*, Department of General Psychology, University of Padova: Padova, Italy, 2023; manuscript to be submitted.
10. Tchanturia, K.; Dapelo, M.A.; Harrison, A.; Hambrook, D. Why Study Positive Emotions in the Context of Eating Disorders? *Curr. Psychiatry Rep.* **2015**, *17*, 537. [CrossRef]
11. Cardi, V.; Kan, C.; Roncero, M.; Harrison, A.; Lounes, N.; Tchanturia, K.; Meyer, C.; Treasure, J. Mealtime Support in Anorexia Nervosa: A Within-Subject Comparison Study of a Novel Vodcast Intervention. *Psychother. Psychosom.* **2012**, *81*, 54–55. [CrossRef]
12. Cardi, V.; Lounes, N.; Kan, C.; Treasure, J. Meal Support Using Mobile Technology in Anorexia Nervosa: Contextual Differences Between Inpatient and Outpatient Settings. *Appetite* **2013**, *60*, 33–39. [CrossRef] [PubMed]
13. Cardi, V.; Esposito, M.; Clarke, A.; Schifano, S.; Treasure, J. The Impact of Induced Positive Mood on Symptomatic Behaviour in Eating Disorders: An Experimental, AB/BA Crossover Design Testing a Multimodal Presentation During a Test-Meal. *Appetite* **2015**, *87*, 192–198. [CrossRef] [PubMed]
14. Hartmann, C.; Siegrist, M. Development and Validation of The Food Disgust Scale. *Food Qual Prefer.* **2018**, *63*, 38–50. [CrossRef]
15. Fairburn, C.G.; Beglin, S. Eating Disorder Examination Questionnaire. In *Cognitive Behavior Therapy and Eating Disorders*; Fairburn, C.G., Ed.; Guilford Press: New York, NY, USA, 2008; pp. 309–313.
16. Egolf, A.; Siegrist, M.; Hartmann, C. How People's Food Disgust Sensitivity Shapes Their Eating and Food Behaviour. *Appetite* **2018**, *127*, 28–36. [CrossRef]
17. Brown, T.A.; Menzel, J.E.; Reilly, E.E.; Luo, T.; Zickgraf, H. Exploring the Role of Disgust Sensitivity and Propensity in Selective Eating. *Appetite* **2022**, *174*, 106018. [CrossRef]
18. Thayer, K.K.; De Riso, M.; Warren, J.; Wilson, A.; Vance, S. What Therapists Need to Know About the Treatment of OCD When Disgust Is Part of the Clinical Presentation. *J. Affect. Disord. Rep.* **2021**, *6*, 100209. [CrossRef]
19. Pisetsky, E.M.; Utzinger, L.M.; Peterson, C.B. Incorporating Social Support in the Treatment of Anorexia Nervosa: Special Considerations for Older Adolescents and Young Adults. *Cogn. Behav. Pract.* **2016**, *23*, 316–328. [CrossRef]
20. Fernández-Aranda, F.; Treasure, J.; Paslakis, G.; Agüera, Z.; Giménez, M.; Granero, R.; Sánchez, I.; Serrano-Troncoso, E.; Gorwood, P.; Herpertz-Dahlmann, B.; et al. The Impact of Duration of Illness on Treatment Nonresponse and Drop-Out: Exploring the Relevance of Enduring Eating Disorder Concept. *Eur. Eat. Disord. Rev.* **2021**, *29*, 499–513. [CrossRef]
21. Meule, A. Predictors and Moderators of Treatment Outcome in Inpatients with Anorexia Nervosa. Ph.D. Thesis, Ludwig Maximilian University of Munich, Munich, Germany, 2023.
22. Petkova, H.; Simic, M.; Nicholls, D.; Ford, T.; Prina, A.M.; Stuart, R.; Livingstone, N.; Kelly, G.; Macdonald, G.; Eisler, I.; et al. Incidence of Anorexia Nervosa in Young People in the UK and Ireland: A National Surveillance Study. *BMJ Open* **2019**, *9*, e027339. [CrossRef]
23. Cheng, Z.H.; Perko, V.L.; Fuller-Marashi, L.; Gau, J.M.; Stice, E. Ethnic Differences in Eating Disorder Prevalence, Risk Factors, and Predictive Effects of Risk Factors Among Young Women. *Eat. Behav.* **2019**, *32*, 23–30. [CrossRef]
24. Strauch, J.; Agnew, A.; Meenaghan, E.; Miller, K.K.; Haines, M.S. Recruitment Strategies to Increase Racial and Ethnic Diversity in Anorexia Nervosa Clinical Research. *J. Eat. Disord.* **2023**, *11*, 118. [CrossRef] [PubMed]

nutrients

MDPI

Article

The Comparison of Disordered Eating, Body Image, Sociocultural and Coach-Related Pressures in Athletes across Age Groups and Groups of Different Weight Sensitivity in Sports

Migle Baceviciene [1,*], Rasa Jankauskiene [2] and Renata Rutkauskaite [1]

[1] Department of Physical and Social Education, Lithuanian Sports University, 44221 Kaunas, Lithuania; renata.rutkauskaite@lsu.lt
[2] Institute of Sport Science and Innovations, Lithuanian Sports University, 44221 Kaunas, Lithuania; rasa.jankauskiene@lsu.lt
* Correspondence: migle.baceviciene@lsu.lt; Tel.: +370-690-09878

Abstract: The aim of the present study was to compare disordered eating (DE), body image, and sociocultural and coach-related pressures between athletes of different age groups (adolescents and adults) and between athletes participating in weight-sensitive (WS) and less WS groups. A total of 1003 athletes participated in this study. The age range of the sample was 15 to 44 years, and the mean age was 18.9 ± 5.8 years (51.3% were female). Athletes who voluntarily agreed to participate in the study were provided with the study measures on DE, body image and sociocultural attitudes towards appearance. Vomiting, laxative misuse and excessive exercise were more prevalent in adolescent female athletes than adults, while dietary restraint was more common in adult male athletes than adolescents. Adolescent female athletes experienced higher sociocultural (family, peers) and sport-related (coach) pressures and a less positive body image compared with adult female athletes. Adult male athletes experienced higher overweight preoccupation, more DE and unhealthy eating habits, and engaged in more frequent self-weighing behaviour compared with adolescent males. When the effect of weight sensitivity in sports was tested, a higher prevalence of DE and overweight preoccupation, more frequent self-weighing, and higher body-image-related pressure from coaches were observed in female athletes participating in aesthetic weight-sensitive (WS) sports as compared with those participating in less WS sports. No differences in positive body image were observed in female WS and less WS sports. Special DE prevention and positive body image promotion programs are necessary for female competitive athletes and parents of adolescent female athletes, especially those participating in aesthetic ones. For adult male athletes, special programs aiming to promote healthy eating should be implemented to prevent DE and body image concerns. Special education about DE prevention is compulsory for coaches who train female athletes.

Keywords: eating disorders; athletes; sociocultural and sport-related pressures; body appreciation

Citation: Baceviciene, M.; Jankauskiene, R.; Rutkauskaite, R. The Comparison of Disordered Eating, Body Image, Sociocultural and Coach-Related Pressures in Athletes across Age Groups and Groups of Different Weight Sensitivity in Sports. *Nutrients* **2023**, *15*, 2724. https://doi.org/10.3390/nu15122724

Academic Editor: David C. Nieman

Received: 23 May 2023
Revised: 9 June 2023
Accepted: 11 June 2023
Published: 12 June 2023

1. Introduction

1.1. The Prevalence and Consequences of Disordered Eating in Athletes

Disordered eating (DE) and eating disorders (ED) are one of the most common mental health illnesses in athletes [1–3]. DE can have devasting effects on athlete health and performance [4]. DE and ED are classified in WHO ICD-10 and in the fifth edition of the Diagnostic and Statistical Manual of Mental Disorders (DSM-V) as anorexia nervosa, bulimia nervosa, binge eating disorder and eating disorders not otherwise specified (EDNOS). EDNOS is hereafter termed DE.

In athletes, DE might be understood in the continuum model starting with healthy dieting and exercise behaviours and the occasional use of more extreme weight loss methods, ranging to DE and ending with clinical EDs such as anorexia and bulimia nervosa [2,4–6]. Competitive athletes might move forward and back along the spectrum of the eating behaviour continuum at different points of their training cycle (i.e., during the preseason, competitive season, and off-season) and career [4]. DE is a set of disturbed eating patterns; unlike an ED, it does not conform to the clinical diagnosis of anorexia and bulimia nervosa or binge eating disorder. DE might manifest as unhealthy weight control behaviours, compensatory behaviours such as excessive exercising, binge eating, constant dieting, compulsive eating, fasting, self-induced vomiting, use of laxatives, skipped meals, use of weight loss supplements and other behaviours [2,4,5]. The prevalence of DE in competitive athletes is higher than that of ED [1,7]. Up to 45% of female athletes and up to 32.5% of male athletes are described as having DE or ED [1,8,9].

The consequences of DE attitudes and behaviours are related to the health and performance of competitive athletes [2,4,6]. DE in athletes is related to low energy availability (LEA) and the potential development of relative energy deficiency in sports (RED-S) [10]. LEA occurs when there is a mismatch between energy intake and exercise load, leaving insufficient energy to cover the body's other needs [2,11]. If DE overlaps with LEA, it negatively affects bone health, menstrual function, the endocrine, cardiovascular, gastrointestinal systems, and psychological functioning [4,11,12]. In adolescents, long-term consequences of LEA might be irreversible, since DE and LEA might have a negative effect on growth and maturity, reproductive functions, and bone mineral density [10,11]. DE also decreases athletes' psychological functioning [1,4]. Psychological consequences include depression, anxiety, social isolation, substance use, self-harm, and increased risk for suicide [1]. Further, DE might increase the risks of injury and illnesses that compromise the training regime and its quality, training adaptation, and fatigue recovery [12,13]. The training might become interrupted and less effective, and impairments might affect the results of the competition [10,11].

The analysis of factors that are associated with the development and prevalence of DE in athletes is important for DE prevention practice in athletes. The main factors predisposing athletes to DE and ED are biological, psychological, sociocultural, and gender-based [2]. A significant part of the studies that previously analysed risk factors for eating psychopathology in athletes was implemented in samples of similar age and participating in one kind of sport (i.e., gymnastics, swimming) or in one group of sports (i.e., combat sports) [14–16]. To have a deeper understanding of DE and its associated factors, it is necessary to assess them in large samples of athletes of different ages participating in various sports [14]. In the present study, we aimed to expand scientific knowledge by comparing DE, body image, and sociocultural factors in different age groups of competitive athletes and comparing athletes participating in groups of different weight sensitivity in sports. Further, most of the studies concerning DE in athletes were implemented using samples of Scandinavian and North American collegiate athletes [10,12,17]. To our best knowledge, there are no studies assessing DE and factors associated with the development of DE in a sample of athletes from Eastern European countries; therefore, one of the objectives of the present study is to provide data on this topic.

1.2. Disordered Eating, Sociocultural and Coach-Related Appearance Pressures, and Body Image in Athletes: The Effect of Age and Weight Sensitivity in Sports

The effect of age on DE, sociocultural attitudes towards appearance (general and sports-related), and body image are less studied. Research on adolescent athletes is scarce, and sport-specific risk factors for DE are less studied. Adolescence is a transition period from childhood to adulthood and a period of significant psychophysiological changes that require proper nutrition [18]. In this period, adolescents usually start specialising in a particular sport and experience major bodily changes that do not usually meet sociocultural and/or sport-related expectations. Therefore, they might experience higher general and coach-related pressures compared to adult athletes. In adolescence, DE and ED most frequently occur [19,20]. A recent systematic review concluded that age did not moderate

DE in women athletes [21]. However, the effect of age on DE in male athletes is less studied. Thus, in the present study, we aimed to extend knowledge on this issue.

Among the sport-specific factors, the most important factor associated with DE and ED in athletes is participation in weight-sensitive (WS) sports [6,8]. In WS sports, leanness and lower body weight play a major role in terms of performance, and lower body weight gives an athlete an advantage [22]. DE is more prevalent in sports with weight classes (i.e., rowing), aesthetic sports (i.e., figure skating, gymnastics), and sports where having a lower body weight is seen as advantageous (i.e., cycling) [4,8,23–25]. It is estimated that more than half of athletes participating in WS sports demonstrate DE behaviours, practising rapid weight loss (RWL) during competitive periods [16], and adolescent girls participating in aesthetic sports have significantly lower body fat compared with their peers participating in other sports [26]. Less is known about the differences in general sociocultural and coach-related pressures comparing athletes of different ages and genders participating in WS and less WS sports. Therefore, in the present study, we aimed to extend the current knowledge on this topic.

Studies show that athletes demonstrate less negative and more positive body image compared with non-athletes [21,27–30]. In male athletes, negative body image is not associated with DE [8]. However, in female athletes, dissatisfaction with body image is one of the strongest predictors of DE and ED [31]. Findings comparing the negative body image of WS and less WS sports are contradicting. Some studies found that females who participate in aesthetic sports experience a more negative body image [32]; however, others demonstrate the opposite [33,34]. Therefore, it is important to continue research on this topic. Finally, the majority of studies were focused on negative body image, and positive body image is understudied in competitive athletes [30]. Positive body image is associated with body functionality appreciation (appreciation for what one's body can do), self-esteem, self-compassion, and psychological well-being, and it is protective for DE in women of general populations [35,36]. Findings of the previous studies suggested that older women report appreciating their bodies more readily than younger women [37]. To inform DE prevention programs in athletes, it is important to understand the differences in positive body image in various age groups of athletes and in groups of weight sensitivity in sports controlling for gender. In the present study, we aimed to assess if positive body image (operating as body appreciation) is more prevalent in adult athletes compared to adolescents and in less WS sports participating athletes compared to those participating in WS sports.

The sociocultural model of Petrie and Greenleaf [38,39] posits that sociocultural (media, family, peers) and sport-specific pressures lead to DE through the internalization of appearance ideals, body dissatisfaction, and dietary restraint in athletes. Athletes are vulnerable to general pressures to attain stereotyped body ideals and also experience sport-specific pressures from sports environments (coaches, teammates, sports uniforms) [40]. More than 60% of elite athletes from leanness-focused and non-leanness-focused sports reported pressure from coaches concerning body shape [32]. Pressures from a coach are considered one of the main risk factors for body image concerns and DE in athletes [24,41,42]. However, less is known about the prevalence of sociocultural and sport-related pressures and internalization of appearance ideals in athletes of different ages. It is unclear if younger athletes experience more sociocultural and coach-related pressures compared to adult athletes. Adolescents spend a significant part of their time on social networking, which is a major source of body image concerns [43]; therefore, adolescent athletes might experience higher general pressures to attain stereotyped body ideals and to internalize these ideals more compared to adult athletes. Furthermore, adolescent athletes might perceive higher appearance pressures from coaches compared to adults. Finally, it is unclear if athletes participating in WS sports perceive higher sociocultural and coach-related pressures and internalize stereotyped appearance ideals to a greater extent compared to those participating in less WS sports. However, there is lack of empirical studies addressing these questions.

Knowledge on these issues is important for science and evidence-based practice preventing DE in athletes [14].

1.3. The Present Study

The aim of the present study was to compare DE, body image and sociocultural and coach-related pressures between athletes of different age groups (adolescents and adults) and between athletes participating in WS and less WS groups. In the present study, we hypothesized that DE, body image concerns, and general sociocultural and coach-related pressures towards appearance would be more prevalent in adolescent athletes compared with adults controlling for gender. Next, we expected that DE, body image concerns, and sociocultural pressures, including pressures from coaches, would be higher in WS sports compared to less WS sports, controlling for age and gender. Finally, we expected that positive body image would be higher in adults compared to adolescents. We developed no assumptions for positive body image differences in groups of different weight sensitivity.

2. Materials and Methods

2.1. Procedure

This cross-sectional study was approved by the Lithuanian Sports University Social Research Ethics Committee (Protocol No. SMTEK-37, 27 May 2021). The online survey was implemented by using the Survey Monkey platform from January to June of 2022. Lithuanian athletes from different sports, who had competed for at least two years in professional competitive sports and were formally included in the lists as members of sport schools and/or sport clubs or belonged to national sports teams, were invited to participate in the survey. Persons who were not involved in competitive sports were excluded from the study. Information about the organized survey was spread via emails and sent to the following organizations: National Sports Federations, sport schools or gymnasiums, and sport centres across the country. Additionally, for better survey responses from respondents, members of national teams were contacted personally via social networks. Most recruited schools and organizations agreed to disseminate the online questionnaire with the help of coaches. The email contained an invitation to participate in the study, a participant information form, and a link to the online survey. The relevant sporting body emailed this information to athletes aged over 18 years and to the parents or guardians of athletes aged <18 years with a request to forward the information and survey link to the athlete.

The inclusive criteria were being an athlete that competed for at least two years in any competitive sport (participating in sports competitions of one sport) and being an official member of a sports school, sport club, or national sports team. Exclusion criteria were not competing in any sport, competing in any sport for less than two years, not being officially listed as a member of a sports school, sport club, or national team.

The survey was completed anonymously. Before starting the survey, participants were informed that the study aimed to investigate lifestyle and health-related habits in competitive athletes, the time required to fill in the form (about 30 min), and that no identifying information will be collected. After the formalization, all the participants were asked to provide consent to participate by ticking one option "I agree to participate" or "I disagree to participate". Those who disagreed were acknowledged and the survey was terminated, while those who agreed were provided with the study measures. Participation was voluntary, and only the researchers had access to the data.

According to the Lithuanian sport centre, in 2021, 140,240 athletes participated in high-skill sports competitions, and 39,567 of them were female. Based on an a priori power calculation (small-to-medium effect size, $\alpha = 0.05$, 95% power, $n = 384$), a minimum of about 400 women and 400 men would be recruited.

2.2. Participants

One thousand three athletes ($n = 1003$) participated in this study. Of these, 488 (48.7%) were males (mean age 18.8 ± 5.7 years), and 515 (51.3%) were females (mean age 19.0 ± 5.9) from age 15 to 44 years. The mean age in both gender groups was similar: 18.8 ± 5.7 years in males and 19.0 ± 5.9 years in females. In total, 56% of all study participants were <18 years, while 44% were ≥18 years. The duration of involvement in competitive sports varied from 2 to 11 years (mean ± SD: 7.0 ± 3.2 years). Athletes reported from 2 to 21 h (mean ± SD: 11.2 ± 5.0) of exercise per week: of these, 10.8% exercised 2–5 h/week, 31.3% 6–9, 37.0% 10–15 and 20.9% 16–21 h/week.

2.3. Study Measures

The body mass index (BMI) was calculated by using self-reported weight (kg) and height (cm): weight (kg)/height (m^2). The BMI ranged from 15.2 to 37.0 kg/m^2 (mean ± SD: 21.6 ± 2.9 kg/m^2). The mean BMI in gender groups was as follows: in males 22.2 ± 3.2 kg/m^2, in females 21.0 ± 2.5 kg/m^2. Using the criteria recommended by the WHO, all the athletes were classified into underweight (4.4%), normal weight (85%) and overweight/obesity (10.6%) groups. For athletes of age <18 years, the International Obesity Task Force (IOTF) cut-offs were applied [44].

According to the proposed classification of sports [22,45], all the athletes were classified into WS aesthetic sports (for example, gymnastics, competitive dancing, figure skating, body building), other WS sports (endurance/gravitational, for example, swimming, athletics, cycling and weight class, for example, boxing, martial arts) and less WS sports (ball games, for example, basketball, handball and technical sports, for example, fencing, shooting). In the sample of athletes, 6.4% of males and 20.4% of females participated in WS aesthetic sports (n = 136), 48.2% and 32.0% in other WS sports (n = 400), and 45.5 and 47.6% in less WS sports (n = 467), respectively.

The Lithuanian version of the Eating Disorder Examination Questionnaire 6 (EDE-Q 6) was used to examine disordered eating (DE) behaviours in athletes [46]. The questionnaire consists of 28 questions and statements with the response options ranging from 0 (never) up to 6 (always). Statements 1–12 and 19–28 were used to calculate the final score, averaging the response options, where a higher score indicates more frequent DE behaviour and greater body weight and shape concerns. For this study, we used only a total score. Next, we used items 2, 14, 16, 17, and 18 to define any occurrence of DE during the last 28 days [47]. Dietary restraint was a behaviour described as going for "long periods of time (>8 h) without eating anything at all in order to influence your shape or weight" (EDE-Q 6 item 2); binge eating distinguished by loss of control (or objective binge eating) was an episode described by eating a large amount of food with the feeling of losing self-control during consumption (EDE-Q 6 item 14); self-induced vomiting was an episode described as making "yourself vomit as a means of controlling your shape or weight" (EDE-Q 6 item 16); laxative misuse was an episode described as going "to take laxatives as a means of controlling your shape or weight" (EDE-Q 6 item 17); and excessive exercising was an episode described as exercising vigorously in "a driven or compulsive way as a means of controlling your weight, shape or amount of fat, or to burn off calories" (EDE-Q 6 item 18). Psychometric characteristics of the Lithuanian version of the EDE-Q 6 were tested in our previous research [48]. For this study, Cronbach's α was 0.92 in females and 0.88 in males.

To assess negative body image, two subscales, the Overweight Preoccupation subscale and the Self-Classified Weight subscale from the Lithuanian version of the Multidimensional Self-Relations Questionnaire Appearance Scales (MBSRQ-AS), were used [49]. The Overweight Preoccupation subscale (4 items) assesses a construct reflecting fat anxiety, weight vigilance, dieting, and eating restraint, while the Self-Classified Weight subscale (2 items) reflects how one perceives and labels one's own weight, from very underweight to very overweight, irrespective of actual body fat mass. Each subscale was rated on a five-point Likert scale. The final score was calculated by averaging the response options. Psychometric properties of the Lithuanian translation of the MBSRQ-AS were tested previously in an adult sample [50]. In this study, for the Overweight Preoccupation subscale, Cronbach's α was 0.80 in females and 0.68 in males, while for the Self-Classified Weight subscale, it was 0.77 in females and 0.67 in males.

The Lithuanian version of the Body Appreciation Scale 2 (BAS-2) was employed to assess positive body image [51]. A unidimensional scale reflects a positive attitude towards one's own body and the ability to resist body beauty standards. The BAS-2 contains 10 statements with a 5-point Likert scale from never (1) to always (5). A final score is calculated by averaging the response options to demonstrate greater body appreciation. A unidimensional factor structure and good psychometric properties of the Lithuanian translation of the BAS-2 were demonstrated in our previous study [52]. In this study, for the BAS-2, Cronbach's α was 0.97 in females and 0.96 in males.

The Lithuanian version of the Drive for Muscularity Scale (DMS) was used. It reflects body image attitudes and behaviours towards body muscularity irrespective of a person's actual muscle mass [53]. The response options range in a six-point Likert scale from 1 (never) to 6 (always). A total averaged score indicates higher drive for muscularity attitudes and behaviours. The psychometric properties of the Lithuanian translation of the DMS were previously tested in a sample of male students [54]. In this study, for the DMS, Cronbach α was 0.90 in females and 0.91 in males.

The Lithuanian version of the Sociocultural Attitudes towards Appearance Questionnaire 4 (SATAQ-4) was used. SATAQ-4 is one of the most widely used measures of sociocultural factors that contribute to an acceptance of prevailing appearance ideal [55]. The SATAQ-4 consists of five subscales: internalization of a thin/low-body-fat body beauty ideal, internalization of a muscular/athletic body beauty ideal and perceived pressure on appearance from the media, family, and peers. For this study, an additional subscale was created to reflect perceived pressure on appearance from a coach. The items of coach pressure subscale were: "I feel pressure from coach to look more thinner"; "I feel pressure from coach to enhance appearance"; "I feel pressure from coach to decrease body fat"; and "My coach encourages me to decrease body weight for getting better body shape". All items are scored on a five-point Likert scale from 1 (definitely disagree) to 5 (definitely agree). The final scores are calculated by averaging the response options for each subscale and for the total scale. Previously, the psychometric properties and the original five-factor structure of the Lithuanian translation of the SATAQ-4 was confirmed in a student sample [56]. In this study, for the SATAQ-4, Cronbach's α was in the range of 0.85–0.98 in females and 0.85–0.92 in male for the subscales and 0.93 and 0.92 for the total score, respectively.

The questions about unhealthy and healthy dietary habits were taken from the national survey "Health Behavior among Lithuanian Adult Population" [57] and used in previous studies [58,59]. Unhealthy dietary habits reflect unhealthy food (sweets, chips, fast food) consumption for snacking, overeating, eating in a rush or eating while working or reading, and eating late at night less than 2 h before going to bed. Healthy dietary habits reflect regular eating regimens, having breakfast and lunch, using more healthy cooking techniques, and choosing healthy snacks (fruits, berries, vegetables). All dietary habits were assessed using a five-option Likert-type scale with the response options from 1 (never) up to 5 (always). The final averaged scores of unhealthy and healthy dietary habits represent more frequent specific dietary patterns.

Study participants were asked to indicate their weighing frequency during the past month by providing the response options from "never" (0) up to "several times a day" [6]. Similar question was used in the previous study [60].

2.4. Statistical Analysis

Descriptive statistics were calculated and the distribution of continuous variables tested for normality in a preliminary analysis. A chi-square test was used to assess the associations between pairs of categorical variables. Next, an independent-samples *t*-test was run to compare body image and DE in the two age groups with the Cohen's d calculated to represent the effect size. Effect sizes above 0.2 were considered small, and those equal to or above 0.5 were considered moderate [61]. Binary logistic regression was run to test independent multivariable adjusted effects of gender, age, and sensitivity in sports groups on any occurrence of the DE during the last week.

Finally, analysis of covariance (ANCOVA) was employed to test the DE and BI differences across three weight-sensitive sports groups where age was added as a covariate. A *Bonferroni* post hoc test was used for multiple pairwise comparisons between groups. The effect sizes, represented by eta-squared, were calculated. An effect size above 0.01 and below 0.06 was considered small, above 0.06 and below 0.12 moderate, and ≥0.12 as large [61]. A p-value less than 0.05 was considered as statistically significant. All statistical analyses were carried out with SPSS v. 29 (IBM Corp., Armonk, NY, USA).

3. Results

3.1. The Analysis of the Effect of Age on Study Variables

Table 1 presents a comparison of disordered eating (DE) behaviours during the last 28 days across age groups separately in males and females. Notably, dietary restraint behaviours were more prevalent in male athletes aged 18 years and older as compared with those <18 years. There were no significant differences when other types of DE were compared in adolescent male athletes and adults. Next, adolescent female athletes aged <18 years demonstrated more frequent self-induced vomiting, laxative misuse, and excessive exercise to reduce body weight during the last 28 days when compared with the age group ≥18 years.

Table 1. The comparison of any occurrence (%) of disordered eating behaviours across age groups during the last 28 days in male and female athletes (n = 1003).

Disordered Eating Behaviors	**Males <18 Years (n = 276)**	**Males ≥18 Years (n = 212)**	$\chi^2; p$
Dietary restraint	17.0	24.5	4.2; 0.041
Binge eating (loss of control)	38.4	46.2	3.0; 0.083
Self-induced vomiting	10.1	6.1	2.5; 0.113
Laxative misuse	9.4	5.2	3.1; 0.08
Excessive exercise	51.1	46.2	1.1; 0.287
	Females < 18 years (n = 286)	Females ≥ 18 years (n = 229)	
Dietary restraint	36.4	31.0	1.6; 0.202
Binge eating (loss of control)	55.2	47.2	3.3; 0.068
Self-induced vomiting	7.7	3.5	4.1; 0.043
Laxative misuse	8.0	3.1	5.8; 0.016
Excessive exercise	67.1	52.8	10.9; 0.001

Multivariable adjusted effects of gender (male was the reference group), age (<18 years was the reference group), and weight sensitivity in the sports group (weight-sensitive aesthetic sports group was the reference) on the occurrence of any DE behaviour during the last 28 days are presented in Table 2 with the odds ratios obtained from binary logistic regression. The models were tested on each DE separately. Females were more likely to report dietary restraint in the last 28 days, binge eating, and excessive exercise to reduce body weight by 96%, 47%, and 51%, respectively, as compared with males. Participation in less WS sports reduced the odds of dietary restraint by 49% and excessive exercise by 43% as compared with the WS aesthetic sports group. Furthermore, athletes participating in other than aesthetic WS sports were less likely to engage in excessive exercise by 44% as compared with the aesthetic sports group. Finally, older age was associated with reduced odds of vomiting, laxative misuse, and excessive exercise by 49%, 56%, and 31%, respectively, as compared with the age group <18 years.

Table 2. The effects of gender, age, and weight-sensitivity group on the odds ratios * of disordered eating behaviours during the last 28 days in the sample of athletes (n = 1003).

Dietary Restraint		**B**	**OR**	**95% CI**	***p***
Gender	Male	1.0			
	Female	0.68	1.96	1.46–2.64	<0.001
Age group	<18	1.0			
	≥18	0.11	1.12	0.84–1.49	0.448
WS group in sports	WS aesthetic	1.0			
	WS other	−0.29	0.75	0.49–1.14	0.174
	Less WS	−0.67	0.51	0.34–0.78	0.002
Binge eating					
Gender	Male	1.0			
	Female	0.38	1.47	1.13–1.89	0.003
Age group	<18	1.0			
	≥18	−0.02	0.98	0.76–1.26	0.869
WS group in sports	WS aesthetic	1.0			
	WS other	−0.09	0.91	0.61–1.36	0.648
	Less WS	0.02	1.03	0.69–1.51	0.902
Vomiting					
Gender	Male	1.0			
	Female	−0.39	0.67	0.41–1.12	0.126
Age group	<18	1.0			
	≥18	−0.67	0.51	0.30–0.87	0.012
WS group in sports	WS aesthetic	1.0			
	WS other	−0.05	0.96	0.43–2.13	0.910
	Less WS	0.07	1.08	0.49–2.35	0.853
Laxative misuse					
Gender	Male	1.0			
	Female	−0.32	0.73	0.43–1.22	0.231
Age group	<18	1.0			
	≥18	−0.81	0.44	0.25–0.78	0.004
WS group in sports	WS aesthetic	1.0			
	WS other	−0.29	0.75	0.34–1.66	0.474
	Less WS	−0.03	0.97	0.45–2.06	0.930
Excessive exercise					
Gender	Male	1.0			
	Female	0.41	1.51	1.17–1.96	0.002
Age group	<18	1.0			
	≥18	−0.38	0.69	0.53–0.89	0.004
WS group in sports	WS aesthetic	1.0			
	WS other	−0.59	0.56	0.36–0.85	0.007
	Less WS	−0.57	0.57	0.37–0.86	0.007

* Binary logistic regression; B = regression coefficient; OR = odds ratio; CI = confidence interval; WS = weight sensitive.

In Table 3, the comparison of the study measures between female athletes <18 years and adult athletes ≥18 years is presented. It was revealed that adult athletes had higher BMI, greater body appreciation, and more healthy eating habits as compared with adolescents <18 years. On the other hand, adolescent athletes demonstrated more perceived pressures regarding appearance from family, peers, and coaches and had a higher total SATAQ-4 score. All Cohen's d coefficients demonstrated a small-to-medium effect size.

Table 3. The comparison of the study measures (m ± SD) in female athletes across <18 and ≥18 years age groups (n = 515).

Study Measures	<18 Years (n = 286)	≥18 Years (n = 229)	Cohen's d	p
Body mass index, kg/m^2	20.37 ± 2.39	21.89 ± 2.35	−0.64	<0.001
EDE-Q 6	1.30 ± 1.13	1.12 ± 1.01	-	0.067
BAS-2	3.42 ± 1.07	3.76 ± 0.89	−0.38	<0.001
MBSRQ-AS: OP	2.42 ± 0.97	2.50 ± 0.98	-	0.363
MBSRQ-AS: SCW	3.04 ± 0.61	3.06 ± 0.52	-	0.731
SATAQ-4: total	2.59 ± 0.77	2.43 ± 0.73	0.22	0.015
SATAQ-4: thin	3.27 ± 1.17	3.09 ± 1.11	-	0.072
SATAQ-4: muscular	3.45 ± 0.91	3.34 ± 0.88	-	0.147
SATAQ-4: pressures media	2.54 ± 1.46	2.66 ± 1.50	-	0.361
SATAQ-4: pressures family	1.87 ± 1.01	1.64 ± 0.94	0.24	0.008
SATAQ-4: pressures peers	1.79 ± 1.12	1.60 ± 0.93	0.18	0.04
SATAQ-4: pressures coach	2.26 ± 1.21	1.88 ± 1.10	0.33	<0.001
DMS	2.36 ± 0.95	2.21 ± 0.95	-	0.075
Self-weighing frequency	2.95 ± 1.59	3.08 ± 1.68	-	0.392
Unhealthy nutrition habits	2.78 ± 0.58	2.82 ± 0.55	-	0.378
Healthy nutrition habits	3.51 ± 0.63	3.68 ± 0.57	−0.29	0.001

m = mean, SD = standard deviation; EDE-Q 6 = Eating Disorder Examination Questionnaire 6; MBSRQ-AS = Multidimensional Self-Relations Questionnaire, Appearance Scales; OP = Overweight Preoccupation; SCW = Self-Classified Weight; DMS = Drive for Muscularity Scale; SATAQ-4 = Sociocultural Attitudes towards Appearance Questionnaire 4; BAS-2 = Body Appreciation Scale 2.

In Table 4, the comparison of the study measures between adolescent male athletes <18 years and adult male athletes ≥18 years are presented. It was revealed that adult athletes had higher BMI, more frequent unhealthy dietary habits, and higher levels of DE and overweight preoccupation. Furthermore, adult male athletes indicated more frequent self-weighing. All Cohen's d coefficients demonstrated a small-to-medium effect size.

Table 4. Comparison of the study measures (m ± SD) in male athletes across <18 and ≥18 years age groups (n = 488).

Study Measures	<18 Years (n = 276)	≥18 Years (n = 212)	Cohen's d	p
Body mass index, kg/m^2	21.32 ± 3.03	23.37 ± 3.11	−0.67	<0.001
EDE-Q 6	0.51 ± 0.54	0.61 ± 0.62	−0.19	0.043
BAS-2	4.09 ± 0.82	4.05 ± 0.81	-	0.585
MBSRQ-AS: OP	1.90 ± 0.72	2.11 ± 0.82	−0.28	0.003
MBSRQ-AS: SCW	2.78 ± 0.56	2.80 ± 0.55	-	0.701
SATAQ-4: total	2.16 ± 0.66	2.10 ± 0.63	-	0.288
SATAQ-4: thin	2.30 ± 0.94	2.42 ± 0.91	-	0.157
SATAQ-4: muscular	3.32 ± 1.06	3.20 ± 0.96	-	0.213
SATAQ-4: pressures media	1.77 ± 1.03	1.69 ± 1.02	-	0.389
SATAQ-4: pressures family	1.71 ± 0.91	1.56 ± 0.81	-	0.067
SATAQ-4: pressures peers	1.64 ± 0.89	1.54 ± 0.85	-	0.207
SATAQ-4: pressures coach	1.90 ± 0.98	1.81 ± 0.95	-	0.331
DMS	3.01 ± 1.12	2.83 ± 1.12	-	0.083
Self-weighing frequency	3.16 ± 1.62	3.55 ± 1.61	−0.24	0.009
Unhealthy nutrition habits	2.63 ± 0.63	2.76 ± 0.58	−0.23	0.014
Healthy nutrition habits	3.65 ± 0.59	3.65 ± 0.56	-	0.993

m = mean, SD = standard deviation; EDE-Q 6 = Eating Disorder Examination Questionnaire 6; MBSRQ-AS = Multidimensional Self-Relations Questionnaire, Appearance Scales; OP = Overweight Preoccupation; SCW = Self-Classified Weight; DMS = Drive for Muscularity Scale; SATAQ-4 = Sociocultural Attitudes towards Appearance Questionnaire 4; BAS-2 = Body Appreciation Scale 2.

3.2. Analysis of the Effect of Weight Sensitivity on the Study Variables

Finally, DE and BI were compared across three groups of different levels of WS in sports separately in male and female athletes controlling for age (Table 5). Thus, the independent effects of weight sensitivity in sports on DE and BI could be explored. Female athletes participating in less WS sports had higher BMI and attained a higher score on DMS. By contrast, athletes involved in aesthetic WS sports perceived more pressure on appearance from their coaches, were more preoccupied with overweight, demonstrated more frequent self-weighing behaviour, and scored higher on the EDE-Q 6. Furthermore, in the aesthetic WS sports group, the score for healthy dietary habits was higher compared with the other groups.

Table 5. Comparison of the study measures m * (95% CI) across sports groups of different weight sensitivities in female athletes (n = 515).

Study Measures	**Aesthetic Weight-Sensitive Sports** n **= 105**	**Other Weight-Sensitive Sports** n **= 165**	**Less Weight-Sensitive Sports** n **= 245**	**Eta-Squared;** p
Body mass index, kg/m^2	19.68 (19.25–20.11)	20.93 [a] (20.58–21.27)	21.71 [ab] (21.43–22.00)	0.11; <0.001
EDE-Q 6	1.47 (1.26–1.68)	1.23 (1.07–1.40)	1.10 [a] (0.97–1.24)	0.02; 0.017
BAS-2	3.58 (3.39–3.78)	3.56 (3.41–3.71)	3.57 (3.44–3.70)	0.983
MBSRQ-AS: OP	2.74 (2.55–2.93)	2.41 [a] (2.26–2.56)	2.37 [a] (2.24–2.49)	0.02; 0.004
MBSRQ-AS: SCW	2.94 (2.83–3.05)	3.05 (2.96–3.13)	3.10 (3.03–3.18)	0.058
SATAQ-4: total	2.60 (2.46–2.75)	2.59 (2.48–2.71)	2.44 (2.35–2.54)	0.069
SATAQ-4: thin	3.38 (3.16–3.60)	3.25 (3.07–3.42)	3.07 (2.92–3.21)	0.051
SATAQ-4: muscular	3.34 (3.17–3.51)	3.50 (3.36–3.64)	3.36 (3.25–3.47)	0.219
SATAQ-4: pressures media	2.60 (2.31–2.89)	2.72 (2.49–2.95)	2.51 (2.32–2.70)	0.373
SATAQ-4: pressures family	1.67 (1.48–1.86)	1.78 (1.63–1.93)	1.79 (1.67–1.92)	0.578
SATAQ-4: pressures peers	1.65 (1.45–1.85)	1.72 (1.56–1.88)	1.73 (1.60–1.86)	0.789
SATAQ-4: pressures coach	2.60 (2.38–2.82)	2.19 [a] (2.02–2.36)	1.80 [ab] (1.66–1.94)	0.07; <0.001
Self-weighing frequency	3.57 (3.26–3.88)	3.04 [a] (2.79–3.29)	2.75 [a] (2.55–2.95)	0.04; <0.001
DMS	1.95 (1.77–2.14)	2.37 [a] (2.23–2.51)	2.38 [a] (2.26–2.50)	0.03; <0.001
Unhealthy nutrition habits	2.78 (2.67–2.89)	2.82 (2.73–2.90)	2.79 (2.72–2.86)	0.847
Healthy nutrition habits	3.67 (3.56–3.79)	3.70 (3.61–3.79)	3.47 [ab] (3.40–3.54)	0.03; <0.001

* = controlled by age; m = mean; CI = confidence interval. EDE-Q 6 = Eating Disorder Examination Questionnaire 6; MBSRQ-AS = Multidimensional Self-Relations Questionnaire, Appearance Scales; OP = Overweight Preoccupation; SCW = Self-Classified Weight; DMS = Drive for Muscularity Scale; SATAQ-4 = Sociocultural Attitudes towards Appearance Questionnaire 4; BAS-2 = Body Appreciation Scale 2. Etas squared are calculated only in case of significant differences between groups; [a] = $p < 0.05$ as compared with aesthetic weight-sensitive sports group; [b] = $p < 0.05$ as compared with the other weight-sensitive sports group.

The same differences across WS in sports groups in the male subsample were not apparent (Table 6). Athletes from less WS sports groups had higher BMI and attained a higher score for self-classified weight. By contrast, athletes participating in aesthetic WS sports demonstrated more frequent self-weighing behaviour than those from the less WS sports group.

Table 6. Comparison of the study measures m * (95% CI) across sports groups of different weight sensitivities in male athletes (n = 488).

Study Measures	Aesthetic Weight-Sensitive Sports n = 31	Other Weight-Sensitive Sports n = 235	Less Weight-Sensitive Sports n = 222	Eta-Squared; p
Body mass index, kg/m^2	20.69 (19.64–21.74)	22.19 [a] (21.81–22.57)	22.45 [a] (22.05–22.84)	0.02; 0.009
EDE-Q 6	0.68 (0.48–0.88)	0.56 (0.48–0.63)	0.53 (0.46–0.61)	0.403
BAS-2	3.89 (3.60–4.18)	4.11 (4.00–4.21)	4.05 (3.95–4.16)	0.372
MBSRQ-AS: OP	2.03 (1.76–2.30)	1.99 (1.90–2.09)	1.97 (1.87–2.07)	0.913
MBSRQ-AS: SCW	2.51 (2.32–2.70)	2.83 [a] (2.76–2.90)	2.77 [a] (2.70–2.85)	0.02; 0.009
SATAQ-4: total	2.28 (2.05–2.51)	2.10 (2.02–2.18)	2.15 (2.06–2.24)	0.304
SATAQ-4: thin	2.38 (2.05–2.71)	2.38 (2.26–2.50)	2.32 (2.19–2.44)	0.755
SATAQ-4: muscular	3.41 (3.05–3.77)	3.20 (3.07–3.33)	3.33 (3.19–3.46)	0.303
SATAQ-4: pressures media	2.00 (1.63–2.36)	1.72 (1.58–1.85)	1.73 (1.59–1.86)	0.351
SATAQ-4: pressures family	1.75 (1.44–2.06)	1.62 (1.50–1.73)	1.66 (1.54–1.77)	0.680
SATAQ-4: pressures peers	1.63 (1.32–1.94)	1.55 (1.44–1.67)	1.64 (1.53–1.76)	0.534
SATAQ-4: pressures coach	2.20 (1.86–2.54)	1.79 (1.67–1.91)	1.89 (1.76–2.02)	0.069
Self-weighing frequency	3.56 (3.00–4.12)	3.70 (3.50–3.90)	2.90 [b] (2.69–3.11)	0.06; <0.001
DMS	3.11 (2.72–3.51)	2.87 (2.73–3.01)	2.98 (2.83–3.12)	0.389
Unhealthy nutrition habits	2.60 (2.39–2.82)	2.71 (2.63–2.79)	2.67 (2.59–2.75)	0.599
Healthy nutrition habits	3.60 (3.39–3.80)	3.70 (3.62–3.77)	3.61 (3.53–3.68)	0.217

* = controlled by age; m = mean; CI = confidence interval. EDE-Q 6 = Eating Disorder Examination Questionnaire 6; MBSRQ-AS = Multidimensional Self-Relations Questionnaire, Appearance Scales; OP = Overweight Preoccupation; SCW = Self-Classified Weight; DMS = Drive for Muscularity Scale; SATAQ-4 = Sociocultural Attitudes towards Appearance Questionnaire 4; BAS-2 = Body Appreciation Scale 2. Etas squared are calculated only in case of significant differences between groups; [a] = $p < 0.05$ as compared with the aesthetic weight-sensitive sports group; [b] = $p < 0.05$ as compared with the other weight-sensitive sports group.

Finally, in female athletes, the correlations between the EDE-Q 6 score and body appreciation were from −0.61 to −0.63 ($p < 0.001$) and similar across different WS in sports groups. The same correlations in male athletes varied from −0.15 in the less WS group, to −0.27 in the other WS group, to −0.32 in the aesthetic WS group ($p < 0.05$). Furthermore, the correlations between overweight preoccupation and the total EDE-Q 6 score were from 0.74 to 0.79 in different WS sports groups in female and from 0.56 to 0.60 in male athletes ($p < 0.001$).

4. Discussion

4.1. Effect of Age on DE, Body Image, and Sociocultural and Coach-Related Pressures

In the present study, we aimed to compare DE, body image, and general sociocultural and coach-related pressures in groups of competitive athletes of different age (adolescents and adults). We hypothesized that DE, body image concerns, and sociocultural pressures would be more prevalent in younger athletes. The present study adds important new knowledge that age is associated with DE, body image, and sociocultural pressures for competitive female athletes. Specifically, we observed that adolescent competitive female athletes reported more frequent self-induced vomiting, laxative misuse, and excessive exercise, as well as poorer heathy eating habits, compared with adult female athletes. Further, adolescent female athletes expressed a greater desire to attain stereotypical body ideals and higher pressures to attain a stereotyped body image from family, peers, and coaches, compared with adult female athletes. This finding is novel; therefore, the comparison of the results is limited, and future studies should test it. A recent study reported that parents of adolescent athletes might use psychological violence and neglect towards their children regarding body weight, especially in adolescent girls [62]. A plethora of studies reported that coach pressures on the body image of athletes is associated with their body image concerns and DE [63]. Peers, especially teammates, are also an important source of pressure regarding appearance and body weight in adolescent athletes [64]. Adult female athletes may no longer experience appearance and body-weight-related pressures from their parents. Further, adult athletes spent more years in sport, possibly facing pressures from the sport-related environment [14]. However, they have a more stable athletic identity, higher athletic achievements, and might also have higher resilience to sociocultural pressures, including pressures from coaches. Stronger athletic identity have been associated with enhanced athletic performance, improved global self-esteem and confidence, as well as improved social relationships [65,66]. Since the study is cross-sectional, it is impossible to understand the directions of the associations. It might be that younger athletes experience more intense general sociocultural and coach-related pressures on their appearance; however, it might also be that those young female athletes, who experience high general sociocultural and coach-related pressures, have higher body image concerns and DE and withdraw from sports more frequently compared to those who experience less pressure. Future studies with experimental and longitudinal designs should be implemented to understand this topic.

Finally, the results of the present study showed that adult female athletes have a significantly more positive body image (body appreciation) compared with adolescents. Body appreciation is protective of body image in women and girls [35]. Higher body appreciation means that female athletes express respect and love for their bodies and have more resilience to the sociocultural pressures to attain stereotyped body ideals [67]. These findings suggest that special programs helping adolescent female competitive athletes to promote positive body image and resist sociocultural and sport-related pressures should be implemented. Special education for coaches who train adolescent females should be implemented to teach them how to avoid negative pressure on the body image of growing adolescent girls [40].

Importantly, no effect of age on the sociocultural pressures on body image was observed in male athletes. However, we found that higher dietary restraint, DE, self-weighing frequency, unhealthy nutrition habits and overweight preoccupation were more prevalent in adult male competitive athletes compared with adolescents. These results might be explained by the high level of competition and sport-related expectations that adult male athletes competing in elite sport possibly experience. Young adolescent athletes are in their puberty period, and it might be that they adjust body-weight-related expectations more easily compared with adult males. Since knowledge on this topic is very limited, these speculations should be tested by future studies. However, special efforts to prevent DE and body image concerns in adult male athletes are necessary. Furthermore, it is important to educate all adult male athletes about healthy nutrition and safe body weight control methods to prevent DE.

4.2. Effect of Weight Sensitivity in Sports on DE, Body Image, and Sociocultural and Coach-Related Pressures

Further, we hypothesized that DE, body image concerns, and sociocultural pressures on appearance would be more prevalent in athletes participating WS sports, controlling for age and gender. This hypothesis was partially confirmed. In females, DE was more prevalent in aesthetic sports compared with less WS sports; however, no significant differences were observed between less WS sports and other WS sports and between aesthetic WS sports and other WS sports, controlling for age. Further, we observed a significant effect of weight sensitivity for dietary restraint and excessive exercise. Specifically, we observed that female gender and participation in WS (aesthetic) sports predicted higher dietary restraint and excessive exercise in females, controlling for age. However, no significant effect of weight sensitivity in sport groups was observed for binge eating, vomiting and laxative misuse in females. These results are in accordance with previous studies which showed that disordered eating is more prevalent in female athletes participating in aesthetic sports compared with other athletes [4,6,23,68]. The present study extends previous knowledge that both adolescent and adult female athletes participating in aesthetic sports experience significantly higher DE than athletes participating in less WS sports.

Our results also add important new knowledge that, despite having the lowest body weight, females participating in aesthetic sports report the highest overweight preoccupation, which is one of the main facets of negative body image [49] and DE [69]. These findings are in line with studies reporting that females participating in aesthetic sports report a more negative body image than other athletes [32]. Further, female adults involved in aesthetic sports reported a higher self-weighing frequency compared with other weight-sensitive groups. Frequent self-weighing is related to overweight preoccupation and considered a risk factor for DE in female athletes [60,70].

The present study adds important knowledge that aesthetic female athletes do not express a more positive body image (body appreciation) than females participating in other WS sports and less WS sports. It is an important novel finding suggesting that, despite low body weight, female athletes involved in aesthetic sport do not develop a healthier body image that might protect them from disordered eating. Further, we observed that body appreciation is negatively associated with disordered eating in all three groups of female athletes. This is a novel finding that is in line with the modern DE prevention recommendations that are based on cognitive dissonance principles and the promotion of positive body image and self-compassion in female athletes [2,71,72].

Finally, the present study adds new knowledge that aesthetic sports involving females report significantly higher pressure from coaches to attain appearance ideals compared with other female athletes. Previous studies showed that pressure from coaches is one of the main risk factors for DE in athletes [24,41]. These results clearly suggest that female athletes involved in aesthetic sports experience higher body image concerns and DE compared with competitive athletes participating in less WS sports. Positive body image and self-efficacy promotion programs for female athletes participating in WS sports are of immense importance. Special education (the prevention of DE and the promotion of positive body image) for coaches of WS sports, especially aesthetic sports, must be implemented [6,73–75].

In male competitive athletes, no significant differences were observed between WS and less WS groups in DE, body image, and sociocultural and coach-related pressures, controlling for age. These results contradict previous findings that showed a higher prevalence of DE in male athletes participating in aesthetic sports compared with other athletes [8]. This might be explained by an extremely low number of male aesthetic sports participants in the present sample, which might be considered as a limitation of the present study. Therefore, the generalization of these results should be limited. However, we observed a higher self-weighing frequency and self-classified weight in a group of male athletes participating in other WS sports compared with less WS sports. Previous studies reported that self-weighing is associated with an increased risk of eating disorders in male athletes [76]. These results suggest that males participating in sports where lower body weight is advantageous and those participating in sports with weight categories also need education about safe methods of weight loss and the risks of rapid weight loss, which are prevalent in WS sports with weight categories [16]. Finally, we observed that positive body image was negatively associated with DE in less WS and other than aesthetic WS sports groups in male athletes. Thus, positive body image promotion might also be effective in DE prevention programs for men.

4.3. Limitations and Strengths of the Study

The present study has important limitations. First, it is a cross-sectional study, and the direction of the associations is unclear. Specifically, the cross-sectional design limits the ability to establish causal relationships and determine the direction of associations. Experimental or longitudinal studies are recommended to have more clarity on the associations between study variables. Second, few male athletes from aesthetic sports participated in the present research. This might have an impact on the results of the comparison of study variables between groups of weight sensitivity in sports, especially comparing aesthetic sports and less weight-sensitive sports. Next, the point of the season and years in sport might influence the results of the study [14]. Another limitation of the present study is that the sample is convenient and does not represent any population. Furthermore, in the present study, we used a modified SATAQ-4 subscale to assess coach pressures on athletes' body image. This might be considered as a limitation of the study. Future studies might additionally include other instruments that are widely accepted for assessing coach and sport-related pressures on athletes' body image [77,78].

The strengths of the study include the large sample of competitive athletes from a variety of sports, including WS and less WS sports, of different genders and ages, and the use of sound research instruments to test DE [79], body image, and sociocultural pressures to attain a stereotyped body image.

4.4. Practical Implications

Special DE prevention and positive body image promotion programs are necessary for female competitive athletes participating in weight-sensitive sports (especially aesthetic sports). Special education programs aiming to increase resistance to general sociocultural and coach-specific pressures of adolescent female athletes should be implemented. For adult male athletes participating in competitive sports, special programs that aim to prevent disordered eating and to increase healthy eating habits might be beneficial. Special education is essential for coaches who train adolescent and adult female athletes participating in weight-sensitive sports, especially aesthetic ones.

5. Conclusions

The results of the present study extend the existing knowledge about the effect of age and weight sensitivity on disordered eating in competitive athletes. When the effect of age was tested, higher DE behaviours (self-induced vomiting, laxative misuse, and excessive exercise) in adolescent female athletes were observed compared with adults. Adolescent female athletes experienced higher sociocultural pressures (family, peers) and sport-related pressures (coach) compared with adult female athletes. No differences in negative body image were observed; however, adult female athletes reported significantly higher positive body image (body appreciation) than adolescents. Adult male athletes more frequently experienced dietary restraint, disordered eating, unhealthy eating habits, self-weighing behaviour, and overweight preoccupation compared with adolescent male athletes; however, no differences were observed in sociocultural pressures and positive body image in the two age groups.

In the present study, we observed a significantly higher prevalence of disordered eating, body image concerns, self-weighing practice, and body-image-related pressures from coaches in aesthetic-sport-involved female athletes compared with athletes participating in less weight-sensitive sports. No differences in positive body image (body appreciation) were observed between weight-sensitive and less weight-sensitive female sports participants. In male athletes, no significant differences were observed between weight-sensitive and less weight-sensitive sports in disordered eating and body-image-related sociocultural pressures.

Special disordered eating prevention and positive body image promotion programs are necessary for female competitive athletes and parents of adolescent female athletes, especially those participating in aesthetic sports. For adult male athletes, special programs aiming to promote healthy eating should be implemented to prevent DE and body image concerns. Special education about DE prevention is essential for coaches who train female athletes.

Author Contributions: Conceptualization, R.J. and M.B.; methodology, R.J., M.B. and R.R.; software, M.B.; validation, R.J. and M.B.; formal analysis, M.B.; investigation, R.R.; resources, R.R.; data curation, M.B.; writing—original draft preparation, R.J., M.B. and R.R.; writing—review and editing, R.J., M.B. and R.R.; visualization, R.J., M.B. and R.R.; supervision, R.J.; project administration, R.J. All authors have read and agreed to the published version of the manuscript.

Funding: This research received no external funding.

Institutional Review Board Statement: Permission to conduct the study was received from the Lithuanian Sports University Social Research Ethics Committee (Protocol No. SMTEK-37, 27 May 2021).

Informed Consent Statement: Informed consent was obtained from all subjects involved in the study.

Data Availability Statement: The dataset generated and analysed during the current study is available from the corresponding author on reasonable request.

Conflicts of Interest: The authors declare no conflict of interest.

References

1. Reardon, C.L.; Hainline, B.; Aron, C.M.; Baron, D.; Baum, A.L.; Bindra, A.; Budgett, R.; Campriani, N.; Castaldelli-Maia, J.M.; Currie, A.; et al. Mental health in elite athletes: International Olympic Committee consensus statement (2019). *Br. J. Sports Med.* **2019**, *53*, 667–699. [CrossRef]
2. Wells, K.R.; Jeacocke, N.A.; Appaneal, R.; Smith, H.D.; Vlahovich, N.; Burke, L.M.; Hughes, D. The Australian Institute of Sport (AIS) and National Eating Disorders Collaboration (NEDC) position statement on disordered eating in high performance sport. *Br. J. Sports Med.* **2020**, *54*, 1247–1258. [CrossRef]
3. Rice, S.M.; Purcell, R.; De Silva, S.; Mawren, D.; McGorry, P.D.; Parker, A.G. The Mental Health of Elite Athletes: A Narrative Systematic Review. *Sports Med.* **2016**, *46*, 1333–1353. [CrossRef]
4. Joy, E.; Kussman, A.; Nattiv, A. 2016 update on eating disorders in athletes: A comprehensive narrative review with a focus on clinical assessment and management. *Br. J. Sports Med.* **2016**, *50*, 154–162. [CrossRef] [PubMed]
5. Sundgot-Borgen, J.; Torstveit, M.K. Aspects of disordered eating continuum in elite high-intensity sports. *Scand. J. Med. Sci. Sports* **2010**, *20*, 112–121. [CrossRef] [PubMed]
6. Sundgot-Borgen, J.; Meyer, N.L.; Lohman, T.G.; Ackland, T.; Maughan, R.J.; Stewart, A.; Müller, W. How to minimise the health risks to athletes who compete in weight-sensitive sports review and position statement on behalf of the Ad Hoc Research Working Group on Body Composition, Health and Performance, under the auspices of the IOC Medical Commission. *Br. J. Sports Med.* **2013**, *47*, 1012–1022. [CrossRef] [PubMed]
7. Sundgot-Borgen, J.; Torstveit, M.K. Prevalence of Eating Disorders in Elite Athletes Is Higher Than in the General Population. *Clin. J. Sport Med.* **2004**, *14*, 25–32. [CrossRef]
8. Karrer, Y.; Halioua, R.; Mötteli, S.; Iff, S.; Seifritz, E.; Jäger, M.; Claussen, M.C. Disordered eating and eating disorders in male elite athletes: A scoping review. *BMJ Open Sport Exerc. Med.* **2020**, *6*, e000801. [CrossRef]
9. Bratland-Sanda, S.; Sundgot-Borgen, J. Eating disorders in athletes: Overview of prevalence, risk factors and recommendations for prevention and treatment. *Eur. J. Sport Sci.* **2013**, *13*, 499–508. [CrossRef]
10. Mountjoy, M.; Sundgot-Borgen, J.; Burke, L.; Ackerman, K.E.; Blauwet, C.; Constantini, N.; Lebrun, C.; Lundy, B.; Melin, A.; Meyer, N.; et al. International Olympic Committee (IOC) Consensus Statement on Relative Energy Deficiency in Sport (RED-S): 2018 Update. *Int. J. Sport Nutr. Exerc. Metab.* **2018**, *28*, 316–331. [CrossRef]

11. Logue, D.M.; Madigan, S.M.; Melin, A.; Delahunt, E.; Heinen, M.; Mc Donnell, S.-J.; Corish, C.A. Low Energy Availability in Athletes 2020: An Updated Narrative Review of Prevalence, Risk, Within-Day Energy Balance, Knowledge, and Impact on Sports Performance. *Nutrients* **2020**, *12*, 835. [CrossRef] [PubMed]
12. Rauh, M.J.; Nichols, J.F.; Barrack, M.T. Relationships Among Injury and Disordered Eating, Menstrual Dysfunction, and Low Bone Mineral Density in High School Athletes: A Prospective Study. *J. Athl. Train.* **2010**, *45*, 243–252. [CrossRef] [PubMed]
13. Ravi, S.; Ihalainen, J.K.; Taipale-Mikkonen, R.S.; Kujala, U.M.; Waller, B.; Mierlahti, L.; Lehto, J.; Valtonen, M. Self-Reported Restrictive Eating, Eating Disorders, Menstrual Dysfunction, and Injuries in Athletes Competing at Different Levels and Sports. *Nutrients* **2021**, *13*, 3275. [CrossRef] [PubMed]
14. Stoyel, H.; Slee, A.; Meyer, C.; Serpell, L. Systematic review of risk factors for eating psychopathology in athletes: A critique of an etiological model. *Eur. Eat. Disord. Rev.* **2019**, *28*, 3–25. [CrossRef]
15. Harriger, J.A.; Witherington, D.C.; Bryan, A.D. Eating pathology in female gymnasts: Potential risk and protective factors. *Body Image* **2014**, *11*, 501–508. [CrossRef]
16. Khodaee, M.; Olewinski, L.; Shadgan, B.; Kiningham, R.R. Rapid Weight Loss in Sports with Weight Classes. *Curr. Sports Med. Rep.* **2015**, *14*, 435–441. [CrossRef]
17. Stephens, L.E.; Bowers, E.P.; Schmalz, D.L.; Duffy, L.N.; Lenhoff, J. A mixed method approach to evaluating eating-related psychopathologies in collegiate student-athletes. *J. Am. Coll. Health* 2021, *online ahead of print*. [CrossRef]
18. Desbrow, B. Youth Athlete Development and Nutrition. *Sports Med.* **2021**, *51*, 3–12. [CrossRef]
19. López-Gil, J.F.; García-Hermoso, A.; Smith, L.; Firth, J.; Trott, M.; Mesas, A.E. Global Proportion of Disordered Eating in Children and Adolescents: A Systematic Review and Meta-analysis. *JAMA Pediatr.* **2023**, *177*, 363–372. [CrossRef]
20. Schmidt, U.; Adan, R.; Böhm, I.; Campbell, I.C.; Dingemans, A.; Ehrlich, S.; Elzakkers, I.; Favaro, A.; Giel, K.; Harrison, A.; et al. Eating disorders: The big issue. *Lancet Psychiatry* **2016**, *3*, 313–315. [CrossRef]
21. Chapa, D.A.N.; Johnson, S.N.; Richson, B.N.; Bjorlie, K.; Won, Y.Q.; Nelson, S.V.; Ayres, J.; Jun, D.; Forbush, K.T.; Christensen, K.A.; et al. Eating-disorder psychopathology in female athletes and non-athletes: A meta-analysis. *Int. J. Eat Disord.* **2022**, *55*, 861–885. [CrossRef]
22. Martinsen, M.; Sundgot-Borgen, J. Higher Prevalence of Eating Disorders among Adolescent Elite Athletes than Controls. *Med. Sci. Sports Exerc.* **2013**, *45*, 1188–1197. [CrossRef]
23. Mancine, R.P.; Gusfa, D.W.; Moshrefi, A.; Kennedy, S.F. Prevalence of disordered eating in athletes categorized by emphasis on leanness and activity type—A systematic review. *J. Eat. Disord.* **2020**, *8*, 47. [CrossRef]
24. Walter, N.; Heinen, T.; Elbe, A.-M. Factors associated with disordered eating and eating disorder symptoms in adolescent elite athletes. *Sports Psychiatry* **2022**, *1*, 47–56. [CrossRef]
25. Giel, K.E.; Hermann-Werner, A.; Mayer, J.; Diehl, K.; Schneider, S.; Thiel, A.; Zipfel, S.; for the GOAL Study Group. Eating disorder pathology in elite adolescent athletes. *Int. J. Eat. Disord.* **2016**, *49*, 553–562. [CrossRef]
26. Kalnina, L.; Sauka, M.; Timpka, T.; Dahlström, Ö.; Nylander, E.; Selga, G.; Ligere, R.; Karklina, H.; Priedite, I.S.; Larins, V. Body fat in children and adolescents participating in organized sports: Descriptive epidemiological study of 6048 Latvian athletes. *Scand J. Public Health* **2015**, *43*, 615–622. [CrossRef]
27. Campbell, A.; Hausenblas, H. Effects of Exerscise Interventions on Body Image. *J. Health Psychol.* **2009**, *14*, 780–793. [CrossRef] [PubMed]
28. Hausenblas, H.A.; Downs, D.S. Comparison of Body Image between Athletes and Nonathletes: A Meta-Analytic Review. *J. Appl. Sport Psychol.* **2001**, *13*, 323–339. [CrossRef]
29. Sabiston, C.M.; Pila, E.; Vani, M.; Thogersen-Ntoumani, C. Body image, physical activity, and sport: A scoping review. *Psychol. Sport Exerc.* **2019**, *42*, 48–57. [CrossRef]
30. Soulliard, Z.A.; Kauffman, A.A.; Fitterman-Harris, H.F.; Perry, J.E.; Ross, M.J. Examining positive body image, sport confidence, flow state, and subjective performance among student athletes and non-athletes. *Body Image* **2019**, *28*, 93–100. [CrossRef]
31. Whitehead, J.; Slater, G.; Wright, H.; Martin, L.; O'connor, H.; Mitchell, L. Disordered eating behaviours in female physique athletes. *Eur. J. Sport Sci.* **2019**, *20*, 1206–1214. [CrossRef]
32. Kong, P.; Harris, L.M. The Sporting Body: Body Image and Eating Disorder Symptomatology Among Female Athletes from Leanness Focused and Nonleanness Focused Sports. *J. Psychol.* **2014**, *149*, 141–160. [CrossRef]
33. De Bruin, A.; Oudejans, R.; Bakker, F.C. Dieting and body image in aesthetic sports: A comparison of Dutch female gymnasts and non-aesthetic sport participants. *Psychol. Sport Exerc.* **2007**, *8*, 507–520. [CrossRef]
34. Kantanista, A.; Glapa, A.; Banio, A.; Firek, W.; Ingarden, A.; Malchrowicz-Mośko, E.; Markiewicz, P.; Płoszaj, K.; Ingarden, M.; Maćkowiak, Z. Body Image of Highly Trained Female Athletes Engaged in Different Types of Sport. *BioMed Res. Int.* **2018**, *2018*, 6835751. [CrossRef]
35. Linardon, J.; McClure, Z.; Tylka, T.L.; Fuller-Tyszkiewicz, M. Body appreciation and its psychological correlates: A systematic review and meta-analysis. *Body Image* **2022**, *42*, 287–296. [CrossRef] [PubMed]
36. Alleva, J.M.; Tylka, T.L. Body functionality: A review of the literature. *Body Image* **2020**, *36*, 149–171. [CrossRef]
37. Tiggemann, M.; McCourt, A. Body appreciation in adult women: Relationships with age and body satisfaction. *Body Image* **2013**, *10*, 624–627. [CrossRef] [PubMed]
38. Petrie, T.A.; Greenleaf, C.A. Eating Disorders in Sport: From Theory to Research to Intervention. In *Handbook of Sport Psychology*; Tenenbaum, G., Eklund, R.C., Eds.; Wiley: Chichester, UK, 2007; pp. 352–378.

39. Petrie, T.A. Eating Disorders in Sport. In *Handbook of Sport Psychology*; John Wiley & Sons: Hoboken, NJ, USA, 2020; pp. 694–710.
40. Reel, J.J.; Galli, N.; Miyairi, M.; Voelker, D.; Greenleaf, C. Development and validation of the intuitive exercise scale. *Eat. Behav.* **2016**, *22*, 129–132. [CrossRef] [PubMed]
41. Chatterton, J.; Petrie, T.A.; Schuler, K.L.; Ruggero, C. Bulimic Symptomatology Among Male Collegiate Athletes: A Test of an Etiological Model. *J. Sport Exerc. Psychol.* **2017**, *39*, 313–326. [CrossRef]
42. Murray, R.M.; Lucibello, K.M.; Pila, E.; Maginn, D.; Sandmeyer-Graves, A.; Sabiston, C.M. "Go after the fatty": The problematic body commentary referees hear—And experience—In adolescent girls' sport. *Sport Exerc. Perform. Psychol.* **2021**, *11*, 1–11. [CrossRef]
43. Tylka, T.L.; Rodgers, R.F.; Calogero, R.M.; Thompson, J.K.; Harriger, J.A. Integrating social media variables as predictors, mediators, and moderators within body image frameworks: Potential mechanisms of action to consider in future research. *Body Image* **2023**, *44*, 197–221. [CrossRef]
44. Cole, T.J.; Lobstein, T. Extended international (IOTF) body mass index cut-offs for thinness, overweight and obesity. *Pediatr. Obes.* **2012**, *7*, 284–294. [CrossRef]
45. Rosenvinge, J.; Sundgot-Borgen, J.; Pettersen, G.; Martinsen, M.; Stornæs, A.V.; Pensgaard, A.M. Are adolescent elite athletes less psychologically distressed than controls? A cross-sectional study of 966 Norwegian adolescents. *Open Access J. Sports Med.* **2018**, *9*, 115–123. [CrossRef]
46. Fairburn, C.G.; Beglin, S.J. Assessment of eating disorders: Interview or self-report questionnaire? *Int J Eat Disord.* **1994**, *16*, 363–370. [CrossRef]
47. Rø, O.; Reas, D.L.; Lask, B. Norms for the Eating Disorder Examination Questionnaire among female university students in Norway. *Nord. J. Psychiatry* **2010**, *64*, 428–432. [CrossRef]
48. Baceviciene, M.; Balciuniene, V.; Jankauskiene, R. Validation of the Lithuanian version of the Eating Disorder Examination Questionnaire 6.0 in a student sample. *Brain Behav.* **2020**, *10*, e01555. [CrossRef]
49. Brown, T.A.; Cash, T.F.; Mikulka, P.J. Attitudinal Body-Image Assessment: Factor Analysis of the Body-Self Relations Questionnaire. *J. Pers. Assess.* **1990**, *55*, 135–144. [PubMed]
50. Miskinyte, A.; Bagdonas, A. Jaunųjų suaugusiųjų požiūrio į savo kūną sąsajos su demografiniais rodikliais (Associations between body image and demographics variables in young adults). *Psichologija.* **2010**, *42*, 85–101.
51. Tylka, T.L.; Wood-Barcalow, N.L. The Body Appreciation Scale-2: Item refinement and psychometric evaluation. *Body Image* **2015**, *12*, 53–67. [CrossRef] [PubMed]
52. Baceviciene, M.; Jankauskiene, R. Associations between Body Appreciation and Disordered Eating in a Large Sample of Adolescents. *Nutrients* **2020**, *12*, 752. [CrossRef] [PubMed]
53. McCreary, D.R.; Sasse, D.K.; Saucier, D.M.; Dorsch, K.D. Measuring the Drive for Muscularity: Factorial Validity of the Drive for Muscularity Scale in Men and Women. *Psychol. Men Masculinities* **2004**, *5*, 49–58. [CrossRef]
54. Bacevičienė, M.; Titenytė; Balčiūnienė, V.; Jankauskienė, R. Drive for Muscularity in Lithuanian Male Students: Psychometrics and Associated Characteristics. *Balt. J. Sport Health Sci.* **2020**, *1*, 20–27. [CrossRef]
55. Schaefer, L.M.; Burke, N.L.; Thompson, J.K.; Dedrick, R.F.; Heinberg, L.J.; Calogero, R.M.; Bardone-Cone, A.M.; Higgins, M.K.; Frederick, D.A.; Kelly, M.; et al. Development and validation of the Sociocultural Attitudes Towards Appearance Questionnaire-4 (SATAQ-4). *Psychol. Assess.* **2015**, *27*, 54–67. [CrossRef]
56. Baceviciene, M.; Jankauskiene, R.; Balciuniene, V. Validation of the Lithuanian Version of the Sociocultural Attitudes towards Appearance Questionnaire-4 (SATAQ-4) in a Student Sample. *Int. J. Environ. Res. Public Health* **2020**, *17*, 932. [CrossRef]
57. Klumbiene, J.; Veryga, A.; Sakyte, E.; Petkeviciene, J.; Grabauskas, V.J.; Kriaucioniene, V. *Health Behavior among Lithuanian Adult Population, 2014*; Lithuanian University of Health Sciences: Kaunas, Lithuania, 2015.
58. Baceviciene, M.; Jankauskiene, R.; Balciuniene, V. The Role of Body Image, Disordered Eating and Lifestyle on the Quality of Life in Lithuanian University Students. *Int. J. Environ. Res. Public Health* **2020**, *17*, 1593. [CrossRef] [PubMed]
59. Baceviciene, M.; Jankauskiene, R. Changes in sociocultural attitudes towards appearance, body image, eating attitudes and behaviours, physical activity, and quality of life in students before and during COVID-19 lockdown. *Appetite* **2021**, *166*, 105452. [CrossRef] [PubMed]
60. Klos, L.A.; Esser, V.E.; Kessler, M.M. To weigh or not to weigh: The relationship between self-weighing behavior and body image among adults. *Body Image* **2012**, *9*, 551–554. [CrossRef]
61. Cohen, J. A power primer. *Psychol. Bull.* **1992**, *112*, 155–159. [CrossRef]
62. Boudreault, V.; Gagnon-Girouard, M.-P.; Carbonneau, N.; Labossière, S.; Bégin, C.; Parent, S. Extreme weight control behaviors among adolescent athletes: Links with weight-related maltreatment from parents and coaches and sport ethic norms. *Int. Rev. Sociol. Sport* **2021**, *57*, 421–439. [CrossRef]
63. Voelker, D.K.; Visek, A.J.; Learner, J.L.; DiBiasio, M. Toward understanding of coaches' role in athletes' eating pathology: A systematic review and ecological application to advance research. *Psychol. Sport Exerc.* **2021**, *58*, 102059. [CrossRef]
64. Scott, C.L.; Haycraft, E.; Plateau, C.R. Teammate influences on the eating attitudes and behaviours of athletes: A systematic review. *Psychol. Sport Exerc.* **2019**, *43*, 183–194. [CrossRef]
65. Brewer, B.W.; Van Raalte, J.L.; Linder, D.E. Athletic identity: Hercules' muscles or Achilles heel? *Int. J. Sport Psychol.* **1993**, *24*, 237–254.

66. Renton, T.; Petersen, B.; Kennedy, S. Investigating correlates of athletic identity and sport-related injury outcomes: A scoping review. *BMJ Open* **2021**, *11*, e044199. [CrossRef] [PubMed]
67. Tylka, T.L.; Wood-Barcalow, N.L. What is and what is not positive body image? Conceptual foundations and construct definition. *Body Image* **2015**, *14*, 118–129. [CrossRef] [PubMed]
68. Thiemann, P.; Legenbauer, T.; Vocks, S.; Platen, P.; Auyeung, B.; Herpertz, S. Eating Disorders and Their Putative Risk Factors Among Female German Professional Athletes. *Eur. Eat. Disord. Rev.* **2015**, *23*, 269–276. [CrossRef] [PubMed]
69. Krentz, E.M.; Warschburger, P. A longitudinal investigation of sports-related risk factors for disordered eating in aesthetic sports. *Scand. J. Med. Sci. Sports* **2011**, *23*, 303–310. [CrossRef]
70. Carrigan, K.W.; Petrie, T.A.; Anderson, C.M. To Weigh or Not to Weigh? Relation to Disordered Eating Attitudes and Behaviors Among Female Collegiate Athletes. *J. Sport Exerc. Psychol.* **2015**, *37*, 659–665. [CrossRef]
71. Menzel, J.E.; Levine, M.P. Embodying experiences and the promotion of positive body image: The example of competitive athletics. In *Self-Objectification in Women: Causes, Consequences, and Counteractions*; American Psychological Association: Worcester, MA, USA, 2011; pp. 163–186. [CrossRef]
72. Voelker, D.K.; Petrie, T.A.; Huang, Q.; Chandran, A. Bodies in Motion: An empirical evaluation of a program to support positive body image in female collegiate athletes. *Body Image* **2019**, *28*, 149–158. [CrossRef]
73. Martinsen, M.; Sherman, R.T.; Thompson, R.A.; Sundgot-Borgen, J. Coaches' knowledge and management of eating disorders: A randomized controlled trial. *Med. Sci. Sports Exerc.* **2015**, *47*, 1070–1078. [CrossRef]
74. Breslin, G.; Shannon, S.; Haughey, T.; Donnelly, P.; Leavey, G. A systematic review of interventions to increase awareness of mental health and well-being in athletes, coaches and officials. *Syst. Rev.* **2017**, *6*, 177. [CrossRef]
75. Sandgren, S.S.; Haycraft, E.; Plateau, C.R. Nature and efficacy of interventions addressing eating psychopathology in athletes: A systematic review of randomised and nonrandomised trials. *Eur. Eat. Disord. Rev.* **2020**, *28*, 105–121. [CrossRef]
76. Galli, N.; Petrie, T.; Chatterton, J. Team weigh-ins and self-weighing: Relations to body-related perceptions and disordered eating in collegiate male athletes. *Psychol. Sport Exerc.* **2017**, *29*, 51–55. [CrossRef]
77. Reel, J.J.; SooHoo, S.; Petrie, T.A.; Greenleaf, C.; Carter, J.E. Slimming Down for Sport: Developing a Weight Pressures in Sport Measure for Female Athletes. *J. Clin. Sport Psychol.* **2010**, *4*, 99–111. [CrossRef]
78. Reel, J.J.; Petrie, T.A.; SooHoo, S.; Anderson, C.M. Weight pressures in sport: Examining the factor structure and incremental validity of the weight pressures in sport—Females. *Eat. Behav.* **2013**, *14*, 137–144. [CrossRef] [PubMed]
79. Pope, Z.; Gao, Y.; Bolter, N.; Pritchard, M. Validity and reliability of eating disorder assessments used with athletes: A review. *J. Sport Health Sci.* **2015**, *4*, 211–221. [CrossRef]

nutrients

MDPI

Article

Depressive Symptoms among Bariatric Surgery Candidates: Associations with Stigmatization and Weight and Shape Concern

Alexandra Fabrig [1], Ricarda Schmidt [1], Thomas Mansfeld [2], Johannes Sander [3], Florian Seyfried [4], Stefan Kaiser [5], Christine Stroh [6], Arne Dietrich [7] and Anja Hilbert [1,*]

1 Behavioral Medicine Research Unit, Department of Psychosomatic Medicine and Psychotherapy, Integrated Research and Treatment Center AdiposityDiseases, University of Leipzig Medical Center, Stephanstrasse 9a, 04103 Leipzig, Germany; alexandra.fabrig@medizin.uni-leipzig.de (A.F.)
2 Department of General Surgery, Asklepios Clinic, 22559 Hamburg, Germany
3 Schön Klinik Hamburg Eilbek, Obesity Clinic, 22081 Hamburg, Germany
4 Department of General, Visceral, Transplant, Vascular and Pediatric Surgery, University Hospital, University of Würzburg, 97080 Würzburg, Germany
5 Department of Visceral, Pediatric and Vascular Surgery, Hospital Konstanz, 78464 Konstanz, Germany
6 Department of Surgery, SRH Wald-Klinikum, Adipositas Zentrum, 07548 Gera, Germany
7 Department of Surgery, Clinic for Visceral, Transplantation, Thoracic and Vascular Surgery, University Hospital Leipzig, 04103 Leipzig, Germany
* Correspondence: anja.hilbert@medizin.uni-leipzig.de

Abstract: Bariatric surgery candidates (BSC) are a highly vulnerable group for mental health impairments. According to the theoretical model of weight stigma, weight-related experienced stigmatization (ES) negatively influences mental health through weight bias internalization (WBI). This study tested this model among BSC and investigated whether this association depends on a negative body image in terms of weight and shape concern as a potential moderator. As part of a German multicenter study, ES, WBI, weight and shape concern, and depressive symptoms were assessed via self-report questionnaires among $n = 854$ BSC. Simple and moderated mediation analyses were applied to analyze whether WBI influences the relationship between ES and depressive symptoms, and whether this influence depends on weight and shape concern. WBI significantly mediated the relationship between ES and depressive symptoms by partially reducing the association of ES with depressive symptoms. Weight and shape concern emerged as significant moderators in the overall model and specifically for associations between WBI and depressive symptoms. The results suggest that the association between ES and depressive symptoms among BSC is stronger in those with high WBI. This association is strengthened by weight and shape concern, especially at low and mean levels. Studies evaluating longitudinal associations between weight-related stigmatization and mental health are indicated, as well as intervention studies targeting WBI in order to reduce adverse effects of ES on mental health in BSC.

Keywords: bariatric surgery candidates; weight bias internalization; weight-related experienced stigmatization; depressive symptoms; weight and shape concern

Citation: Fabrig, A.; Schmidt, R.; Mansfeld, T.; Sander, J.; Seyfried, F.; Kaiser, S.; Stroh, C.; Dietrich, A.; Hilbert, A. Depressive Symptoms among Bariatric Surgery Candidates: Associations with Stigmatization and Weight and Shape Concern. *Nutrients* **2024**, *16*, 510. https://doi.org/10.3390/nu16040510

Academic Editor: Lidia Santarpia

Received: 9 January 2024
Revised: 3 February 2024
Accepted: 7 February 2024
Published: 12 February 2024

1. Introduction

The worldwide prevalence of obesity is increasing [1–3], with 13% of adults exceeding a body mass index (BMI; kg/m^2) of 30 kg/m^2 according to the World Health Organization [4]. In 2014, 2.3% of men and 5.0% of women globally met the criteria for obesity class 2 (35.0 kg/m^2 $\leq$ BMI < 40.0 kg/m^2) and 3 (BMI $\geq$ 40.0 kg/m^2) [3]. Due to associations with physical [5] and mental health impairments including depressive symptoms [6], obesity poses a major economic challenge to healthcare systems [7]. The standard treatment of obesity, behavioral weight loss treatment, including nutritional, physical activity, and behavioral, shows overall small effects on health outcomes [8]. Although, the adjunct of the subcutaneous application of semaglutide can optimize the effects [9], bariatric surgery is the most efficacious treatment for patients with severe obesity, including obesity class 2 with physical comorbidity and class 3 leading to significant weight loss of 20–35% and long-term improvements in physical and mental comorbidities [10,11].

Among individuals with obesity, bariatric surgery candidates (BSC) comprise a notably vulnerable group to mental health impairments, given that up to 58% of these patients present with a mental disorder, especially affective and anxiety disorders, as well as eating disorders [12]. Rates of depression are significantly higher in BSC than in those seeking behavioral weight loss treatment [13]. However, so far, mechanisms related to the development of psychopathology such as depressive symptoms among BSC are not fully understood. Based on Tylka et al.'s [14] theoretical model of weight stigma, which is based on both longitudinal and cross-sectional data and supported by recent cross-sectional research [15–18], experienced stigmatization (ES) may lead to weight bias internalization (WBI) and/or body shame, thereby negatively influencing psychological well-being. While ES describes negative experiences related to one's weight [19], with weight-based teasing being the most common type of ES [20], WBI denotes individual beliefs that negative stereotypes related to one's weight are true for oneself. A variety of cross-sectional studies demonstrated that both ES and WBI increase with higher BMI [8,20–22], and are negatively associated with mental health including depressive symptoms [23–29]. This pattern is especially pronounced in BSC compared to individuals with obesity undergoing behavioral weight loss treatment [30,31], with prospective experimental evidence highlighting stronger negative effects on mental health for WBI than for ES [32]. ES and WBI were found to be associated with medium effect size, according to a systematic review with predominantly cross-sectional evidence, in adults from the population [27], and based on a cross-sectional study in BSC [21]. Notably, a systematic review revealed that ES in terms of weight-based teasing in childhood was longitudinally and cross-sectionally positively associated with depressive symptoms in both childhood and adulthood [33]. Another recent cross-sectional study among treatment-seeking adults with obesity who have experienced and internalized weight stigma found a high percentage of depressive symptoms [29].

Supporting Tylka et al.'s [14] model, a recent systematic review demonstrated that WBI may function as a mediator between ES and psychological well-being, including depression, disordered eating, and body dissatisfaction, in community-based and clinical populations [34]. Specifically, in adult patients with obesity participating in a behavioral weight loss program, ES had a direct and an indirect effect on depression through WBI [35]. Among BSC, cross-sectional evidence identified the interplay of WBI, body shame, and internalized shame as mediators in the relationship between ES and depression, though WBI as a single mediator was not significant [36]. Body shame has been highly associated with eating disorder symptoms, including weight and shape concern [37–39], and depressive symptoms in BSC. At the same time, weight and shape concern and WBI were cross-sectionally highly associated in BSC [31]. However, nothing is known about the potential impact of weight and shape concern on the relationship between ES, WBI, and mental health among BSC, specifically whether weight and shape concern strengthen the effect of WBI on the association between ES and depressive symptoms.

In this context, the aim of this cross-sectional study was to investigate, first, the mediating role of WBI on the relationship between ES and depressive symptoms among BSC and, second, the potential influence of weight and shape concern on the association between ES, WBI, and depressive symptoms. Based on Tylka et al.'s [14] theoretical model of weight stigma and related evidence, it was hypothesized that ES directly and indirectly (through WBI) will be related to depressive symptoms, and that this mediation will be moderated by BSC's weight and shape concern. An investigation of these associations was deemed to be of high clinical relevance, since both WBI and weight and shape concern may serve as potential intervention targets to improve BSC's mental health.

2. Materials and Methods

2.1. Sample

This study is part of the multicenter Psychosocial Registry for Bariatric Surgery (PRAC) study, which longitudinally assesses psychosocial aspects in a consecutive sample of patients seeking bariatric surgery in six participating study centers in Germany. Inclusion criteria for the PRAC study were a minimum age of 18 years and a planned bariatric surgery. Patients were excluded due to insufficient German language skills and inability to comply with the study protocol. Based on an eligible sample of n = 978, data on the self-reported measures of interest (see below) were missing for n = 124, leaving a total sample of n = 854 adult BSC, recruited between March 2012 and March 2023. All patients provided written informed consent before study participation. Data collection proceeded independently of clinical treatment, and all patients were informed that study data would be treated as strictly confidential and inaccessible to the surgical team.

2.2. Measures

This study used PRAC baseline data from well-established self-report questionnaires on weight and shape concern, weight-related stigmatization, and depressive symptoms, assessed prior to bariatric surgery.

2.2.1. Predictor Variable: Experienced Stigmatization

The German version of the 6-item Perception of Teasing Scale (POTS [40,41]) was used to assess how often participants had been the target of weight stigmatization by others in their childhood on a 5-point scale ranging from 0 ("never") to 4 ("very often"). The effect of teasing on the individuals, a second subscale of the POTS, was not evaluated in this study. All responses were summed up to a total score, with higher scores representing more frequently perceived teasing (Cronbach's alpha in the present study $\alpha = 0.97$).

2.2.2. Outcome Variable: Depressive Symptoms

The 9-item subscale of the German version of the Patient Health Questionnaire (PHQ-D [42–44]) was used to screen for depressive symptoms based on the Diagnostic and Statistical Manual of Mental Disorders (DSM-IV [45]). All items were rated on a 4-point Likert scale ranging from 0 ("not at all") to 3 ("nearly every day"), with higher sum scores indicating higher severity of depression ($\alpha = 0.85$).

2.2.3. Mediator Variable: Weight Bias Internalization

The German version of the 11-item Weight Bias Internalization Scale (WBIS [46,47]) was used to assess the level of weight bias internalization describing someone's belief that negative stereotypes and negative self-statements about persons with overweight or obesity apply to him- or herself. From April 2015, the WBIS was replaced by the German version of the Modified Weight Bias Internalization Scale (WBIS-M [47,48]), which assesses WBI across various weight statuses. Although the WBIS-M showed slightly better psychometric properties than the WBIS, both measures showed acceptable internal consistency as well as convergent and divergent validity [49]. In favor of good readability, only "WBIS" is referred to in tables and figures. All items were rated on a 7-point Likert scale ranging from 1 ("strongly disagree") to 7 ("strongly agree"). According to the results of psychometric analyses, item 1 was removed before computing the mean score [47]. A higher mean score indicates greater internalized weight stigma ($\alpha = 0.87$).

2.2.4. Moderator Variable: Weight and Shape Concern

The 5- and 8-item subscales of the German Eating Disorder Examination Questionnaire (EDE-Q [50,51]) on weight concern and shape concern were combined to measure a composite covering both weight and shape concern [52,53]. The items were rated on a 7-point Likert scale ranging from 0 ("no day"/"not at all") to 6 ("everyday"/"extremely") with higher mean scores indicating greater weight and shape concern ($\alpha = 0.84$).

2.2.5. Control Variables

Sociodemographic characteristics were assessed by self-report, including participants' age, sex (male, female), and education (≥10 school years, <10 school years). BMI was calculated from participants' measured weight and height using calibrated scales.

2.3. *Data Analysis*

A priori power analysis was calculated to determine the minimum sample size for detecting medium-sized effects with a statistical power of 0.80. For mediation analyses (small-sized a path, medium-sized b path, see Figure 1; percentile bootstrapping), n = 406 patients were required [54]. All statistical analyses were performed using IBM SPSS Statistics Version 29 and a two-tailed significance level of α = 0.05. Prior to conducting the main analyses, all variables were screened for plausibility and outliers. Pearson and Spearman correlation analyses were run to examine bivariate associations between all study variables and to identify relevant sociodemographic control variables.

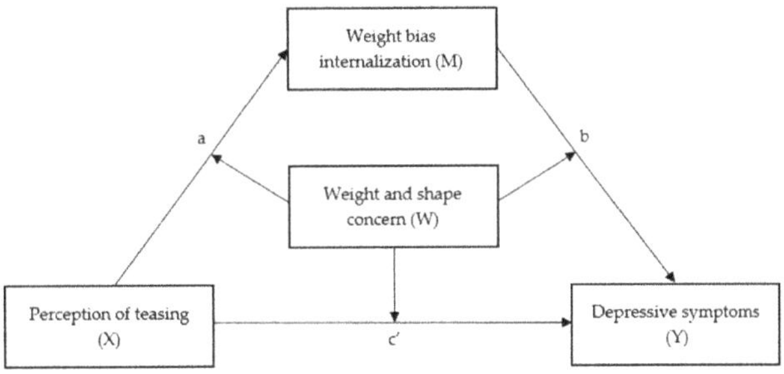

Figure 1. Moderated mediation model.

In order to examine whether there was an indirect effect of ES on depressive symptoms through WBI, a simple mediation analysis was conducted using Model 4 from the SPSS PROCESS macro 4.0 [55]. Secondly, in order to investigate the moderating influence of weight and shape concern on all paths of the mediation model, a moderated mediation analysis (Process Model 59) was conducted (see Figure 1 for the hypothesized model). Both the mediation-only and moderated mediation models were controlled for sex, age, and BMI due to significant associations of these variables with model variables. Bootstrapping was applied, which involved repeated sampling from the dataset with replacement (i.e., 10,000 bootstrap resamples), in order to achieve an approximation of the sampling distribution of the indirect effect and to generate 95% confidence intervals for these effects. For illustrative purposes, the moderating effect of weight and shape concern was calculated at three different levels of the moderator (i.e., −1 *SD*, mean, +1 *SD*).

3. Results

3.1. *Sample Characteristics*

The sample had a mean age of 46.8 ± 11.6 years, with n = 547 (67.2%) women (Table 1). Mean BMI was 48.7 ± 8.0 kg/m^2, with the majority of patients having obesity class III (n = 754, 88.3%). Most participants were married (n = 430, 51.7%) and had at least 10 years of education (n = 579, 76.3%).

Table 1. Sample characteristics.

Baseline Characteristics	*n*	*M*/*n*	*SD*/%	Min.	Max.
Sociodemographic variables					
Age	854	46.8	11.6	18	74
Sex	854				
Women		547	67.2		
Men		280	32.8		
Marital status	832				
Single		154	18.5		
Partnership		160	19.2		
Married		430	51.7		
Divorced		63	7.6		
Widowed		25	3.0		
Educational level	759				
≤10 school years		180	23.7		
>10 school years		579	76.3		
Anthropometrics					
BMI, kg/m^2	854	48.7	8.0	35.0	97.3
Obesity class 2		100	11.7		
Obesity class 3		754	88.3		
Psychological variables					
POTS, 6–30	854	16.0	7.9	6.0	30.0
WBIS, 1–7	854	4.8	1.3	1.3	7.0
PHQ-D, 0–27	854	7.8	5.2	0.0	26.0
EDE-Q WS, 0–6	854	3.7	1.1	0.0	6.0

Note: BMI = body mass index; POTS = Perception of Teasing Scale; WBIS = Weight Bias Internalization Scale; PHQ-D = Patient Health Questionnaire Depression Scale; EDE-Q WS = Eating Disorder Examination Questionnaire: composite of weight and shape concern.

Associations between all study variables can be found in Table 2. Among possible covariates, age, sex, and BMI, but not education, were significantly associated with the predictor, outcome, mediator, and moderator.

Table 2. Correlations of study variables.

	1	2	3	4	5	6	7
1. Age	–						
2. Sex	−0.11 **	–					
3. BMI	−0.14 ***	−0.03	–				
4. Education	0.02	0.11 **	−0.11 **	–			
5. POTS	−0.36 ***	0.11 **	0.29 ***	−0.01	–		
6. WBIS	−0.17 ***	0.17 ***	0.09 *	−0.03	0.38 ***	–	
7. PHQ-D	0.01	0.09 **	0.14 ***	−0.05	0.27 ***	0.56 ***	–
8. EDE-Q WS	−0.03	0.19 ***	0.09	0.01	0.33 ***	0.72 ***	0.56 ***

Note: BMI = body mass index (kg/m^2); POTS = Perception of Teasing Scale; WBIS = Weight Bias Internalization Scale; PHQ-D = Patient Health Questionnaire Depression Scale; EDE-Q WS = Eating Disorder Examination Questionnaire: composite of weight and shape concern. * $p < 0.05$. ** $p < 0.01$. *** $p < 0.001$.

3.2. Mediation

The overall prediction by the model of greater depressive symptoms by more frequent ES through the indirect effect of WBI was significant ($F(5, 848) = 89.331$, $p < 0.001$), accounting for 35% of variance (see Table 3 and Figure 2). The inclusion of WBI in the model reduced the direct effect of ES on depressive symptoms significantly ($p < 0.001$). Thus, WBI was found to be a partial mediator of the association between ES and depressive symptoms.

Table 3. Effects of simple mediation.

Path	Independent Variable	Dependent Variable	t	p	Direct Effect [95% CI]	Indirect Effect [95% CI]
a	POTS	WBIS	10.206	<0.001	0.06 [0.05, 0.07]	
b	WBIS	PHQ-D	18.018	<0.001	2.23 [2.01, 2.52]	
c	POTS	PHQ-D	7.931	<0.001	0.19 [0.15, 0.25]	
c′	POTS	PHQ-D	2.841	0.005	0.06 [0.03, 0.12]	0.13 [0.10, 0.16]

Note: POTS = Perception of Teasing Scale; WBIS = Weight Bias Internalization Scale; PHQ-D = Patient Health Questionnaire Depression Scale.

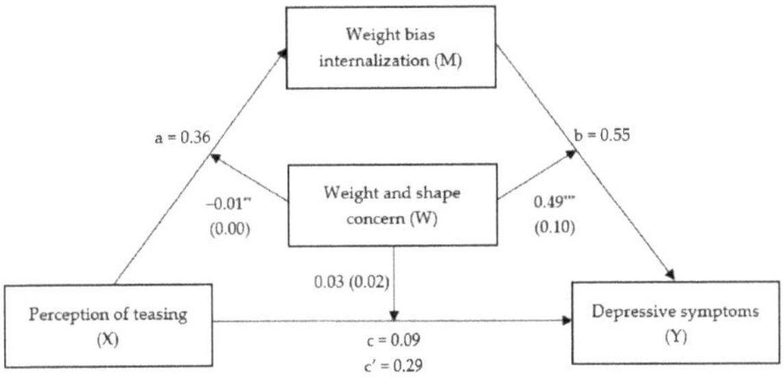

Figure 2. Moderated mediation model displayed with coefficients and standard errors. Mediation coefficients displayed are standardized. ** $p < 0.01$. *** $p < 0.001$.

3.3. Moderated Mediation

The overall prediction by the model of greater depressive symptoms by more frequent ES through WBI, while considering weight and shape concern, was significant ($F(8, 845) = 74.140$, $p < 0.001$), accounting for 41% of variance (see Table 4 and Figure 2). The interaction between the effects of ES and weight and shape concern on WBI was statistically significant, $p = 0.001$. Specifically, the conditional effect of ES on WBI was significant for low and mean values of weight and shape concern, both $p < 0.001$, while high values of weight and shape concern did not moderate the effect between ES and WBI, $p = 0.063$. The results thus indicate that the effect of ES on WBI was stronger for patients with low and mean weight and shape concern—see Figure 3.

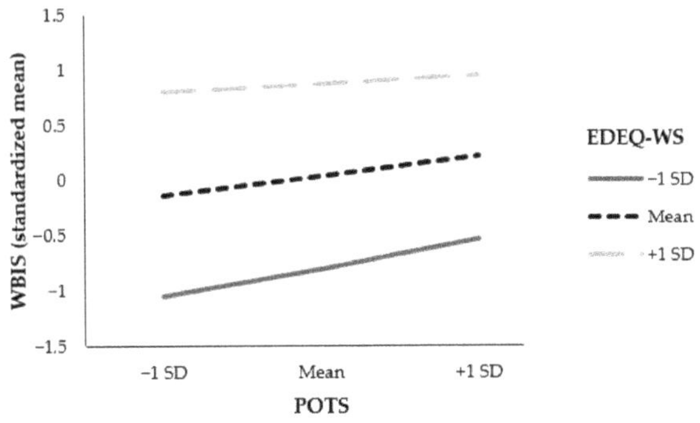

Figure 3. Conditional effect of perception of teasing on weight bias internalization at values of weight and shape concern. Note. POTS = Perception of Teasing Scale; WBIS = Weight Bias Internalization Scale; EDE-Q WS = Eating Disorder Examination Questionnaire: composite of weight and shape concern.

Table 4. Effects of moderated mediation.

	WBIS (Mediator)			PHQ-D (Outcome)		
	Effect [95% CI]	t	p	Effect [95% CI]	t	p
Path a						
POTS	0.02 [0.01, 0.03]	4.97	<0.001			
EDE-Q WS	0.74 [0.68, 0.79]	26.99	<0.001			
POTS × EDE-Q WS	−0.01 [−0.02, −0.00]	−3.08	0.002			
POTS × EDE-Q WS (−1 SD)	0.03 [0.02, 0.04]	5.37	<0.001			
POTS × EDE-Q WS (mean)	0.02 [0.01, 0.03]	4.97	<0.001			
POTS × EDE-Q WS (+1 SD)	0.01 [−0.00, 0.02]	1.86	0.063			
Age	−0.01 [−0.02, −0.01]	−4.21	<0.001			
Sex	−0.01 [−0.13, 0.13]	−0.18	0.860			
BMI	−0.01 [−0.01, 0.00]	−1.35	0.177			
Path b						
WBIS				1.55 [1.29, 1.94]	9.57	<0.001
EDE-Q WS				1.38 [0.98, 1.70]	7.77	<0.001
WBIS × EDE-Q WS				0.49 [0.32, 0.72]	5.01	<0.001
WBIS × EDE-Q WS (−1 SD)				0.99 [0.63, 1.41]	5.17	<0.001
WBIS × EDE-Q WS (mean)				1.55 [1.29, 1.94]	9.57	<0.001
WBIS × EDE-Q WS (+1 SD)				2.11 [1.80, 2.62]	10.45	<0.001
Age				0.05 [0.02, 0.08]	3.75	<0.001
Sex				−0.25 [−0.99, 0.22]	−0.84	0.399
BMI				0.05 [0.01, 0.08]	2.83	0.005
Path c′						
POTS				0.03 [−0.00, 0.08]	1.53	0.127
POTS × EDE-Q WS				0.03 [−0.01, 0.06]	1.39	0.164
Overall moderated mediation						
POTS × WBIS × EDE-Q WS (−1 SD)				0.03 [0.01, 0.05]		
POTS × WBIS × EDE-Q WS (mean)				0.03 [0.02, 0.05]		
POTS × WBIS × EDE-Q WS (+1 SD)				0.02 [−0.00, 0.04]		

Note. BMI = body mass index; POTS = Perception of Teasing Scale; WBIS = Weight Bias Internalization Scale; PHQ-D = Patient Health Questionnaire Depression Scale; EDE-Q WS = Eating Disorder Examination Questionnaire: composite of weight and shape concern.

The moderating effect of weight and shape concern on the association between WBI and depressive symptoms was significant, $p < 0.001$. Specifically, the conditional effect of WBI on depressive symptoms was significant for low, mean, and high values of weight and shape concern, all $p < 0.001$, indicating that they increased the association between WBI and depressive symptoms—see Figure 4.

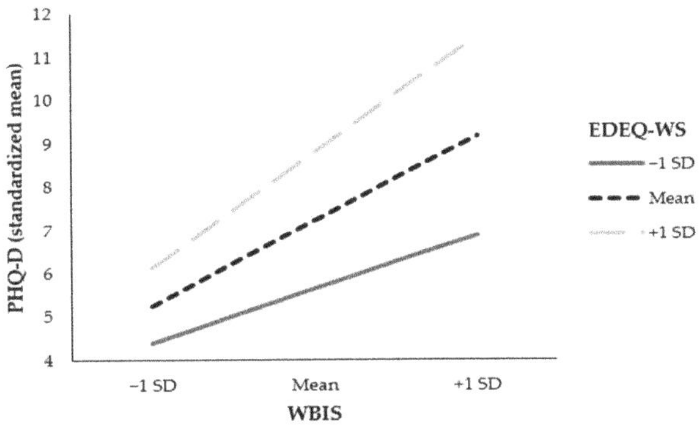

Figure 4. Conditional effect of weight bias internalization on depressive symptoms at values of weight and shape concern. Note. WBIS = Weight Bias Internalization Scale; PHQ-D = Patient Health Questionnaire Depression Scale; EDE-Q WS = Eating Disorder Examination Questionnaire: composite of weight and shape concern.

There was no moderating effect of weight and shape concern on the association between ES and depressive symptoms, $p = 0.164$, indicating that weight and shape concern did not strengthen or weaken the respective association.

The moderation of the indirect effect of ES on depressive symptoms through WBI was significant for low and mean values, but not for high values of weight and shape concern, indicating that the indirect effect of ES on depressive symptoms through WBI was stronger for patients with low and mean weight and shape concern.

4. Discussion

This cross-sectional study was the first to test the theoretical model of weight stigma [14], specifying associations between experienced stigmatization (ES), weight bias internalization (WBI), and depressive symptoms among bariatric surgery candidates (BSC), adding weight and shape concern as a potential moderator. In a large baseline sample of BSC, we found mediating effects of WBI on the relation between ES and depressive symptoms, and weight and shape concern moderated this mediation.

Notably, compared to a community sample of adults with overweight or obesity [18], the association between ES and WBI was smaller in this study of BSC, but similar to another recent study in BSC [21]. Supporting the suggested model and prior population-based research [34], WBI was here found to mediate the association between ES and depressive symptoms. The result is also in line with cross-sectional findings in individuals opting for surgical and non-surgical intervention with obesity [35], showing an indirect effect of more ES on lower mood through WBI. A recent cross-sectional study in BSC did not find a significant separate effect of WBI on ES and depressive symptoms, but only in combination with high internalized shame and body shame, and low self-compassion [36]. Based on the present results, depressive symptoms among BSC, who experienced frequent weight-based teasing in childhood, were stronger in those with a higher than lower internalized weight bias. Thus, health care professionals might pay particular attention to patients undergoing bariatric surgery, believing that negative stereotypes about weight apply to themselves, in order to improve their psychological well-being in relation to ES in childhood.

Based on Tylka et al.'s model [14], there may be a moderation effect caused by body shame on associations between ES and mental health. The present study extended this model by testing the moderating effect of weight and shape concern. As hypothesized, a significant moderating effect of weight and shape concern was found: weight and shape concern strengthened the mediating effect of WBI on the association between ES and depressive symptoms among BSC, especially in those with low and mean levels of weight and shape concern. To understand this result, it is important to evaluate the separate path connections. Weight and shape concern at all levels had a large moderating effect on the relation between WBI and depressive symptoms. This result goes in line with findings from a cluster analysis revealing that individuals with overweight or obesity and weight concern showed low levels of happiness and positivity [56]. Similarly, a cross-sectional study demonstrated that WBI was highly correlated with depressive symptoms and weight and shape concern among BSC [57]. With a significant but small effect, the composite of weight and shape concern moderated the relation between ES and WBI, indicating that the lower the patients' weight and shape concern, the stronger the positive relation between ES and WBI. Experienced teasing in childhood may thus be especially important for mental health among BSC who are less concerned about their weight and shape. In other words, the association between experienced weight teasing in childhood and current WBI is stronger for those with low or moderate levels of weight and shape concern, while for those with high weight and shape concern, WBI is generally high, whether or not the BSC reported an experience of weight teasing during childhood. This result, which is implicated in the overall result, could be due to a general negative self-image that is not necessarily weight-related.

Strikingly, the positive association between ES and depressive symptoms was not moderated by weight and shape concern, against the hypothesis. Although the association between ES and depressive symptoms was significant, as expected, it was only weak, which is congruent with the results of a community study of adults with a mean BMI of 36 kg/m^2 [18], but contrasts previous cross-sectional evidence showing strong correlations between ES and depressive symptoms in adults with BMI between 28 and 45 kg/m^2 seeking behavioral weight loss treatment [24]. The fact that the strength of association between ES and depressive symptoms was not affected by the level of weight and shape concern suggests that BSC with high weight and shape concern and a high frequency of experienced teasing do not suffer from stronger depressive symptoms than BSC with low weight and shape concern and a low frequency of experienced teasing. Given the large association of weight and shape concern and depressive symptoms, and the small association between ES and depressive symptoms, the result may suggest that weight and shape concern alone are relevant for patients' mental health, but do not serve as protective or adverse mechanisms between childhood teasing and current depressive symptoms.

The strengths of this study include the large sample size, the multicenter design, and the application of internationally well-established self-report instruments. Among the limitations, data were only quasi-longitudinal with patients reporting their current perceptions of weight teasing in childhood, thus precluding causal interpretations of the present results. Although the WBIS refers to experienced weight-based teasing in childhood, there was no objective information on patients' weight status in childhood available. With Tylka et al.'s model [14], postulating a rather longitudinal mediation, it must be taken into account that this cross-sectional study's level of evidence is lower compared to those of longitudinal studies, offering the possibility to establish causality. Further, socially desirable response behavior cannot be completely ruled out, even if the participants were informed that their answers played no role in the clinical decision to have surgery.

5. Conclusions

This study cross-sectionally confirms that WBI acts as a mediator between ES and depressive symptoms, and that weight and shape concern have moderating effects on this mediating pathway among BSC. Clinically, the results suggest that the reduction in WBI in

BSC may be a valuable target in psychological intervention. The strong correlation between WBI and depressive symptoms shown in this study, high prospective associations between presurgical and postsurgical depressive symptoms [58], and predictive effects of presurgical WBI for diminished weight loss after surgery [59] support pre-surgical assessments as well as improvements in WBI and depressive symptoms as clinical necessities in BSC [60]. A related intervention target could be weight and shape concern. Herein, patients with low weight and shape concern should not be neglected, because for these patients, the mediating effect of WBI on the association between ES and depressive symptoms was especially strong. Regarding psychotherapy in BSC, the results highlight that not the frequency of ES in the past, which cannot be changed in the present, but internal conditions, including WBI and weight and shape concern, are largely related to mental health.

In order to better understand the etiology of BSC's depressive symptoms and postsurgical outcomes, future research should use longitudinal designs prospectively assessing weight-based teasing in childhood, pre- and postsurgical WBI and outcomes. Due to the importance of WBI for mental health among the population and people with overweight or obesity [61–63], intervention studies aiming to reduce the level of WBI will be of high clinical interest. The number of recently developed studies evaluating the efficacy of lowering WBI [64,65] is small. They should be built upon in the future, focusing on the vulnerable group of BSC.

Author Contributions: Conceptualization, A.F., R.S. and A.H.; methodology, A.F., R.S. and A.H.; formal analysis, A.F.; investigation, all authors; resources, A.H.; data curation, A.F.; writing—original draft preparation, A.F.; writing—review and editing, all authors; supervision, A.H.; funding acquisition, A.H. All authors have read and agreed to the published version of the manuscript.

Funding: This research was supported by grant 01EO1501 from the German Federal Ministry of Education and Research (Dr. Hilbert). The APC was funded by the Open Access Publishing Fund of Leipzig University supported by the German Research Foundation within the program Open Access Publication Funding.

Institutional Review Board Statement: The study was conducted in accordance with the Declaration of Helsinki, and approved by the Ethics Committee of Leipzig University Medical Center (protocol code 356-11-07112011, approved on 24 February 2012).

Informed Consent Statement: Informed consent was obtained from all subjects involved in the study.

Data Availability Statement: The original contributions presented in the study are included in the article, further inquiries can be directed to the corresponding author.

Acknowledgments: The authors are very grateful to all the participants who took part in this study.

Conflicts of Interest: Dr. Hilbert reported receiving grants from the Federal Ministry of Education and Research, German Research Foundation, and Roland Ernst Foundation for Healthcare; royalties for books on the treatment of eating disorders and obesity with Hogrefe and Kohlhammer; honoraria for workshops and lectures on eating disorders and obesity and their treatment; honoraria as editor of the *International Journal of Eating Disorders* and the journal *Psychotherapeut*; honoraria as a reviewer from Mercator Research Center Ruhr, Oxford University Press, and the German Society for Nutrition; and honoraria as a consultant for WeightWatchers, Zweites Deutsches Fernsehen, and Takeda. No other competing interests were reported.

References

1. Ng, M.; Fleming, T.; Robinson, M.; Thomson, B.; Graetz, N.; Margono, C.; Mullany, E.C.; Biryukov, S.; Abbafati, C.; Abera, S.F.; et al. Global, regional, and national prevalence of overweight and obesity in children and adults during 1980–2013: A systematic analysis for the Global Burden of Disease Study 2013. *Lancet* **2014**, *384*, 766–781. [CrossRef]
2. The GBD 2015 Obesity Collaborators. Health effects of overweight and obesity in 195 countries over 25 years. *N. Engl. J. Med.* **2017**, *377*, 13–27. [CrossRef] [PubMed]
3. NCD Risk Factor Collaboration (NCD-RisC). Trends in adult body-mass index in 200 countries from 1975 to 2014: A pooled analysis of 1698 population-based measurement studies with 19.2 million participants. *Lancet* **2016**, *387*, 1377–1396. [CrossRef] [PubMed]

4. World Health Organisation. Obesity and Overweight. Available online: https://www.who.int/news-room/fact-sheets/detail/obesity-and-overweight (accessed on 2 September 2022).
5. Chu, D.-T.; Minh Nguyet, N.T.; Dinh, T.C.; Thai Lien, N.V.; Nguyen, K.-H.; Nhu Ngoc, V.T.; Tao, Y.; Le Son, H.; Le, D.-H.; Nga, V.B.; et al. An update on physical health and economic consequences of overweight and obesity. *Diabetes Metab. Syndr.* **2018**, *12*, 1095–1100. [CrossRef]
6. Avila, C.; Holloway, A.C.; Hahn, M.K.; Morrison, K.M.; Restivo, M.; Anglin, R.; Taylor, V.H. An overview of links between obesity and mental health. *Curr. Obes. Rep.* **2015**, *4*, 303–310. [CrossRef]
7. Lehnert, T.; Sonntag, D.; Konnopka, A.; Riedel-Heller, S.; König, H.-H. Economic costs of overweight and obesity. Best Practice & Research. *Clin. Endocrinol. Metab.* **2013**, *27*, 105–115. [CrossRef]
8. LeBlanc, E.S.; Patnode, C.D.; Webber, E.M.; Redmond, N.; Rushkin, M.; O'Connor, E.A. Behavioral and pharmacotherapy weight loss interventions to prevent obesity-related morbidity and mortality in adults: Updated evidence report and systematic review for the US Preventive Services Task Force. *JAMA* **2018**, *320*, 1172–1191. [CrossRef] [PubMed]
9. Wadden, T.A.; Bailey, T.S.; Billings, L.K.; Davies, M.; Frias, J.P.; Koroleva, A.; Lingvay, I.; O'Neil, P.M.; Rubino, D.M.; Skovgaard, D.; et al. Effect of subcutaneous semaglutide vs placebo as an adjunct to intensive behavioral therapy on body weight in adults with overweight or obesity: The STEP 3 randomized clinical trial. *JAMA* **2021**, *325*, 1403–1413. [CrossRef]
10. Wolfe, B.M.; Kvach, E.; Eckel, R.H. Treatment of obesity: Weight loss and bariatric surgery. *Circ. Res.* **2016**, *118*, 1844–1855. [CrossRef]
11. Hilbert, A.; Staerk, C.; Strömer, A.; Mansfeld, T.; Sander, J.; Seyfried, F.; Kaiser, S.; Dietrich, A.; Mayr, A. Nonnormative eating behaviors and eating disorders and their associations with weight loss and quality of life during 6 years following obesity surgery. *JAMA Netw. Open* **2022**, *5*, e2226244. [CrossRef]
12. Duarte-Guerra, L.S.; Coêlho, B.M.; Santo, M.A.; Wang, Y.-P. Psychiatric disorders among obese patients seeking bariatric surgery: Results of structured clinical interviews. *Obes. Surg.* **2015**, *25*, 830–837. [CrossRef]
13. Fischer, L.; Wekerle, A.-L.; Sander, J.; Nickel, F.; Billeter, A.T.; Zech, U.; Bruckner, T.; Müller-Stich, B.P. Is there a reason why obese patients choose either conservative treatment or surgery? *Obes. Surg.* **2017**, *27*, 1684–1690. [CrossRef]
14. Tylka, T.L.; Annunziato, R.A.; Burgard, D.; Daníelsdóttir, S.; Shuman, E.; Davis, C.; Calogero, R.M. The weight-inclusive versus weight-normative approach to health: Evaluating the evidence for prioritizing well-being over weight loss. *J. Obes.* **2014**, *2014*, 983495. [CrossRef] [PubMed]
15. Forbes, Y.; Donovan, C. The role of internalised weight stigma and self-compassion in the psychological well-being of overweight and obese women. *Aust. Psychol.* **2019**, *54*, 471–482. [CrossRef]
16. Pötzsch, A.; Rudolph, A.; Schmidt, R.; Hilbert, A. Two sides of weight bias in adolescent binge-eating disorder: Adolescents' perceptions and maternal attitudes. *Int. J. Eat. Disord.* **2018**, *51*, 1339–1345. [CrossRef]
17. Olson, K.L.; Mensinger, J.L. Weight-related stigma mediates the relationship between weight status and bodily pain: A conceptual model and call for further research. *Body Image* **2019**, *30*, 159–164. [CrossRef]
18. Hayward, L.E.; Vartanian, L.R.; Pinkus, R.T. Weight stigma predicts poorer psychological well-being through internalized weight bias and maladaptive coping responses. *Obesity* **2018**, *26*, 755–761. [CrossRef] [PubMed]
19. Carr, D.; Friedman, M.A. Is obesity stigmatizing? Body weight, perceived discrimination, and psychological well-being in the United States. *J. Health Soc. Behav.* **2005**, *46*, 244–259. [CrossRef] [PubMed]
20. Puhl, R.M.; Himmelstein, M.S.; Quinn, D.M. Internalizing weight stigma: Prevalence and sociodemographic considerations in US adults. *Obesity* **2018**, *26*, 167–175. [CrossRef]
21. Braun, T.D.; Gorin, A.A.; Puhl, R.M.; Stone, A.; Quinn, D.M.; Ferrand, J.; Abrantes, A.M.; Unick, J.; Tishler, D.; Papasavas, P. Shame and self-compassion as risk and protective mechanisms of the internalized weight bias and emotional eating link in individuals seeking bariatric surgery. *Obes. Surg.* **2021**, *31*, 3177–3187. [CrossRef]
22. Luck-Sikorski, C.; Bernard, M. Stigmatisierung und Diskriminierung von Patient*innen mit Adipositas. *Psychotherapeut* **2021**, *66*, 28–34. [CrossRef]
23. Alimoradi, Z.; Golboni, F.; Griffiths, M.D.; Broström, A.; Lin, C.-Y.; Pakpour, A.H. Weight-related stigma and psychological distress: A systematic review and meta-analysis. *Clin. Nutr.* **2020**, *39*, 2001–2013. [CrossRef]
24. Crockett, K.B.; Borgatti, A.; Tan, F.; Tang, Z.; Dutton, G. Weight discrimination experienced prior to enrolling in a behavioral obesity intervention is associated with treatment response among black and white adults in the Southeastern U.S. *Int. J. Behav. Med.* **2022**, *29*, 152–159. [CrossRef]
25. Duan, W.; Wang, Z. Mindfulness capability mediates the association between weight-based stigma and negative emotion symptoms. *Mindfulness* **2019**, *10*, 1074–1081. [CrossRef]
26. Emmer, C.; Bosnjak, M.; Mata, J. The association between weight stigma and mental health: A meta-analysis. *Obes. Rev. Off. J. Int. Assoc. Study Obes.* **2020**, *21*, e12935. [CrossRef] [PubMed]
27. Pearl, R.L.; Puhl, R.M. Weight bias internalization and health: A systematic review. *Obes. Rev. Off. J. Int. Assoc. Study Obes.* **2018**, *19*, 1141–1163. [CrossRef] [PubMed]
28. Spahlholz, J.; Pabst, A.; Riedel-Heller, S.G.; Luck-Sikorski, C. Coping with perceived weight discrimination: Testing a theoretical model for examining the relationship between perceived weight discrimination and depressive symptoms in a representative sample of individuals with obesity. *Int. J. Obes.* **2016**, *40*, 1915–1921. [CrossRef] [PubMed]

29. Pearl, R.L.; Hernandez, M.; Bach, C.; Groshon, L.; Wadden, T.A. Prevalence of diagnosed psychiatric disorders among adults who have experienced and internalized weight stigma. *Obes. Sci. Pract.* **2023**, *9*, 681–687. [CrossRef] [PubMed]
30. Hoffmann, K.; Paczkowska, A.; Bryl, W.; Marzec, K.; Raakow, J.; Pross, M.; Berghaus, R.; Nowakowska, E.; Kus, K.; Michalak, M. Comparison of perceived weight discrimination between Polish and German patients underwent bariatric surgery or endoscopic method versus conservative treatment for morbid obesity: An International Multicenter Study. *Nutrients* **2022**, *14*, 2775. [CrossRef] [PubMed]
31. Wagner, A.F.; Butt, M.; Rigby, A. Internalized weight bias in patients presenting for bariatric surgery. *Eat. Behav.* **2020**, *39*, 101429. [CrossRef] [PubMed]
32. Pearl, R.L.; Puhl, R.M. The distinct effects of internalizing weight bias: An experimental study. *Body Image* **2016**, *17*, 38–42. [CrossRef]
33. Szwimer, E.; Mougharbel, F.; Goldfield, G.S.; Alberga, A.S. The association between weight-based teasing from peers and family in childhood and depressive symptoms in childhood and adulthood: A systematic review. *Curr. Obes. Rep.* **2020**, *9*, 15–29. [CrossRef] [PubMed]
34. Bidstrup, H.; Brennan, L.; Kaufmann, L.; La Piedad Garcia, X.D. Internalised weight stigma as a mediator of the relationship between experienced/perceived weight stigma and biopsychosocial outcomes: A systematic review. *Int. J. Obes.* **2022**, *46*, 1–9. [CrossRef] [PubMed]
35. Magallares, A.; Bolaños-Rios, P.; Ruiz-Prieto, I.; Benito de Valle, P.; Irles, J.A.; Jáuregui-Lobera, I. The mediational effect of weight self-stigma in the relationship between blatant and subtle discrimination and depression and anxiety. *Span. J. Psychol.* **2017**, *20*, E4. [CrossRef]
36. Braun, T.D.; Quinn, D.M.; Stone, A.; Gorin, A.A.; Ferrand, J.; Puhl, R.M.; Sierra, J.; Tishler, D.; Papasavas, P. Weight bias, shame, and self-compassion: Risk/protective mechanisms of depression and anxiety in prebariatic surgery patients. *Obesity* **2020**, *28*, 1974–1983. [CrossRef]
37. Gee, A.; Troop, N.A. Shame, depressive symptoms and eating, weight and shape concerns in a non-clinical sample. *Eat. Weight Disord. EWD* **2003**, *8*, 72–75. [CrossRef]
38. Oliveira, S.; Pires, C.; Ferreira, C. Does Recall Caregiv. Eat. Messages Exacerbate Pathog. Impact Shame Eat. Weight-Relat. Difficulties? *Eat. Weight Disord. EWD* **2020**, *25*, 471–480. [CrossRef] [PubMed]
39. Troop, N.A.; Allan, S.; Serpell, L.; Treasure, J.L. Shame in women with a history of eating disorders. *Eur. Eat. Disord. Rev. J. Eat. Disord. Assoc.* **2008**, *16*, 480–488. [CrossRef] [PubMed]
40. Losekam, S.; Kraeling, S.; Goetzky, B.; Rief, W.; Hilbert, A. Evaluation of the German version of the Perception of Teasing Scale (POTS). *Unveröff. Manuskript. Univ. Marburg*, 2017; *unpublished*.
41. Thompson, J.K.; Cattarin, J.; Fowler, B.; Fisher, E. The Perception of Teasing Scale (POTS): A revision and extension of the Physical Appearance Related Teasing Scale (PARTS). *J. Personal. Assess.* **1995**, *65*, 146–157. [CrossRef]
42. Löwe, B.; Spitzer, R.L.; Zipfel, S.; Herzog, W. PHQ_D. Gesundheitsfragebogen für Patienten. Manual. Komplettversion und Kurzform: Autorisierte Deutsche Version des "Prime MD Patient Health Questionnaire (PHQ)". 2nd ed.; 2002. Available online: https://www.klinikum.uni-heidelberg.de/fileadmin/psychosomatische_klinik/download/phq_manual1.pdf (accessed on 8 January 2024).
43. Löwe, B.; Spitzer, R.L.; Zipfel, S.; Herzog, W. PHQ-D. Gesundheitsfragebogen Für Patienten. *Z. Für Med. Psychol.* **2003**, *12*, 90–95.
44. Spitzer, R.L.; Kroenke, K.; Williams, J.B. Validation and utility of a self-report version of PRIME-MD: The PHQ primary care study. Primary care evaluation of mental disorders. Patient Health Questionnaire. *JAMA* **1999**, *282*, 1737–1744. [CrossRef] [PubMed]
45. American Psychiatric Association. *Diagnostic and Statistical Manual of Mental Disorders*, 4th ed.; Text Revision (DSM-IV-TR); American Psychiatric Association: Arlington, VA, USA, 2000.
46. Durso, L.E.; Latner, J.D. Understanding self-directed stigma: Development of the weight bias internalization scale. *Obesity* **2008**, *16* (Suppl. 2), S80–S86. [CrossRef] [PubMed]
47. Hilbert, A.; Baldofski, S.; Zenger, M.; Löwe, B.; Kersting, A.; Braehler, E. Weight bias internalization scale: Psychometric properties and population norms. *PLoS ONE* **2014**, *9*, e86303. [CrossRef] [PubMed]
48. Pearl, R.L.; Puhl, R.M. Measuring internalized weight attitudes across body weight categories: Validation of the modified weight bias internalization scale. *Body Image* **2014**, *11*, 89–92. [CrossRef] [PubMed]
49. Schraven, S.; Hübner, C.; Eichler, J.; Mansfeld, T.; Sander, J.; Seyfried, F.; Kaiser, S.; Dietrich, A.; Schmidt, R.; Hilbert, A. Psychometric properties of the WBIS/-M in a representative prebariatric sample–Evidence for a 10-item version. *Obes. Facts*, 2023; *submitted*.
50. Fairburn, C.G.; Beglin, S.J. Assessment of eating disorders: Interview or self-report questionnaire? *Int. J. Eat. Disord.* **1994**, *16*, 363–370. [CrossRef] [PubMed]
51. Hilbert, A.; Tuschen-Caffier, B.; Karwautz, A.; Niederhofer, H.; Munsch, S. Eating Disorder Examination-Questionnaire. *Diagnostica* **2007**, *53*, 144–154. [CrossRef]
52. Hrabosky, J.I.; Masheb, R.M.; White, M.A.; Grilo, C.M. Overvaluation of shape and weight in binge eating disorder. *J. Consult. Clin. Psychol.* **2007**, *75*, 175–180. [CrossRef]
53. Grilo, C.M.; Hrabosky, J.I.; White, M.A.; Allison, K.C.; Stunkard, A.J.; Masheb, R.M. Overvaluation of shape and weight in binge eating disorder and overweight controls: Refinement of a diagnostic construct. *J. Abnorm. Psychol.* **2008**, *117*, 414–419. [CrossRef]
54. Fritz, M.S.; Mackinnon, D.P. Required sample size to detect the mediated effect. *Psychol. Sci.* **2007**, *18*, 233–239. [CrossRef]

55. Hayes, A.F. *Introduction to Mediation, Moderation, and Conditional Process Analysis: A Regression-Based Approach (Methodology in the Social Sciences)*; The Guilford Press: New York, NY, USA, 2022.
56. Godoy-Izquierdo, D.; Lara, R.; Ogallar, A.; Rodríguez-Tadeo, A.; Ramírez, M.J.; Navarrón, E.; Arbinaga, F. Psychosocial and diet-related lifestyle clusters in overweight and obesity. *Int. J. Environ. Res. Public Health* **2021**, *18*, 6414. [CrossRef]
57. Lawson, J.L.; Schuh, L.M.; Creel, D.B.; Blackinton, R.M.; Giambrone, S.A.; Grilo, C.M.; Ivezaj, V. Examining weight bias and loss-of-control eating among individuals seeking bariatric surgery. *Obes. Surg.* **2021**, *27*, 105–115. [CrossRef] [PubMed]
58. White, M.A.; Kalarchian, M.A.; Levine, M.D.; Masheb, R.M.; Marcus, M.D.; Grilo, C.M. Prognostic significance of depressive symptoms on weight loss and psychosocial outcomes following gastric bypass surgery: A prospective 24-month follow-up study. *Obes. Surg.* **2015**, *25*, 1909–1916. [CrossRef] [PubMed]
59. Lent, M.R.; Napolitano, M.A.; Wood, G.C.; Argyropoulos, G.; Gerhard, G.S.; Hayes, S.; Foster, G.D.; Collins, C.A.; Still, C.D. Internalized weight bias in weight-loss surgery patients: Psychosocial correlates and weight loss outcomes. *Obes. Surg.* **2014**, *24*, 195–2199. [CrossRef] [PubMed]
60. Kruseman, M.; Leimgruber, A.; Zumbach, F.; Golay, A. Dietary, weight, and psychological changes among patients with obesity, 8 years after gastric bypass. *J. Am. Diet. Assoc.* **2010**, *110*, 527–534. [CrossRef] [PubMed]
61. Curll, S.L.; Brown, P.M. Weight stigma and psychological distress: A moderated mediation model of social identification and internalised bias. *Body Image* **2020**, *35*, 207–216. [CrossRef] [PubMed]
62. Levy, M.; Nguyen, A.; Kakinami, L.; Alberga, A.S. Weight bias internalization: Relationships with mental health, physical activity, and sedentary behavior. *Stigma Health* **2021**, *8*, 453–461. [CrossRef]
63. Pearl, R.L.; Puhl, R.M.; Himmelstein, M.S.; Pinto, A.M.; Foster, G.D. Weight stigma and weight-related health: Associations of self-report measures among adults in weight management. *Ann. Behav. Med. A Publ. Soc. Behav. Med.* **2020**, *54*, 904–914. [CrossRef]
64. Pearl, R.L.; Wadden, T.A.; Bach, C.; Tronieri, J.S.; Berkowitz, R.I. Six-month follow-up from a randomized controlled trial of the Weight BIAS Program. *Obesity* **2020**, *28*, 1878–1888. [CrossRef]
65. Selensky, J.C.; Carels, R.A. Weight stigma and media: An examination of the effect of advertising campaigns on weight bias, internalized weight bias, self-esteem, body image, and affect. *Body Image* **2021**, *36*, 95–106. [CrossRef]

nutrients

Article

GFRAL Is Widely Distributed in the Brain and Peripheral Tissues of Mice

Karoline Fichtner [1,†], Hermann Kalwa [1,†], Miao-Miao Lin [1], Yuanyuan Gong [1,2], Anne Müglitz [1], Michael Kluge [3,4,‡] and Ute Krügel [1,*,‡]

1 Rudolf Boehm Institute of Pharmacology and Toxicology, Faculty of Medicine, University of Leipzig, D-04103 Leipzig, Germany
2 School of Acupuncture and Tuina, Chengdu University of Traditional Chinese Medicine, Chengdu 610075, China
3 Department of Psychiatry and Psychotherapy, University of Leipzig, D-04103 Leipzig, Germany
4 Department of Psychiatry, Rudolf-Virchow-Klinikum Glauchau, D-08371 Glauchau, Germany
* Correspondence: ute.kruegel@medizin.uni-leipzig.de
† These authors contributed equally to this work.
‡ These authors contributed equally to this work.

Abstract: In 2017, four independent publications described the glial cell-derived neurotrophic factor (GDNF) receptor alpha-like (GFRAL) as receptor for the growth differentiation factor 15 (GDF15, also MIC-1, NAG-1) with an expression exclusively in the mice brainstem area postrema (AP) and nucleus tractus solitarii (NTS) where it mediates effects of GDF15 on reduction of food intake and body weight. GDF15 is a cell stress cytokine with a widespread expression and pleiotropic effects, which both seem to be in contrast to the reported highly specialized localization of its receptor. This discrepancy prompts us to re-evaluate the expression pattern of GFRAL in the brain and peripheral tissues of mice. In this detailed immunohistochemical study, we provide evidence for a more widespread distribution of this receptor. Apart from the AP/NTS region, GFRAL-immunoreactivity was found in the prefrontal cortex, hippocampus, nucleus arcuatus and peripheral tissues including liver, small intestine, fat, kidney and muscle tissues. This widespread receptor expression, not taken into consideration so far, may explain the multiple effects of GDF-15 that are not yet assigned to GFRAL. Furthermore, our results could be relevant for the development of novel pharmacological therapies for physical and mental disorders related to body image and food intake, such as eating disorders, cachexia and obesity.

Keywords: GFRAL; brain; peripheral tissue; immunohistochemistry

Citation: Fichtner, K.; Kalwa, H.; Lin, M.-M.; Gong, Y.; Müglitz, A.; Kluge, M.; Krügel, U. GFRAL Is Widely Distributed in the Brain and Peripheral Tissues of Mice. *Nutrients* **2024**, *16*, 734. https://doi.org/10.3390/nu16050734

Academic Editor: Maria Scherma

Received: 27 January 2024
Revised: 28 February 2024
Accepted: 29 February 2024
Published: 4 March 2024

1. Introduction

Dysregulation in the control of food intake and body weight may lead to obesity or anorexia. Circulating high levels of the cell stress cytokine growth differentiation factor 15 (GDF15, also called MIC-1 and NAG-1) are linked to reduced appetite and food intake, loss of body weight and cancer-associated anorexia-cachexia [1–4]. In patients with obesity and type 2 diabetes mellitus, elevated GDF15 was found [5]. In mice models, GDF15 was effective in reducing food intake and adiposity and in counteracting metabolic dysfunction via increased energy expenditure [6,7]. GDF15 might exert its effects with multiple mechanisms, e.g., on food preference for non-fat diet, delay of gastric emptying, taste aversion or nausea and emesis [8].

In patients with anorexia nervosa, high serum GDF15 levels were found and were probably related to weight loss in this group, but can be reversed via realimentation [9–11]. Furthermore, the anti-diabetic drug metformin was shown to increase circulating GDF15 levels as a requirement for seeing beneficial effects regarding energy intake and expenditure, and weight loss, but not for lowering glucose [7].

The pleiotropic from of GDF15 is a distant member of the transforming growth factor β (TGF-β) family [12] that is widely expressed in, for example, the placenta, bladder, liver, kidney, epithelia of exocrine glands and brain [13–15]. Its synthesis is regulated by cell activation and cell stress involving p53, EGR-1 and transcription factors of the integrated stress response pathway [16]. Consequently, circulating GDF15 levels rise under 'non-homeostatic' conditions of adiposity, starvation, mitochondrial dysfunction, and insulin resistance, but also in inflammatory diseases and various cancer types [1,16].

In 2017, four research groups independently described that GDF15 binds with high affinity to the glial cell-derived neurotrophic factor (GDNF) family receptor alpha-like (GFRAL) and that this receptor has a strongly restricted expression in the brainstem [1,2,17,18].

GFRAL shares significant structural characteristics with the canonical members of the GDNF receptor family (GFR-α1-4), i.e., a highly conserved second cysteine-rich domain, a highly hydrophobic *N*-terminal signal peptide and its membrane anchorage [19]. However, in contrast to its distant homologs, a GPtdIns cleavage site at its *C*-terminal sequence is replaced by a single transmembrane region. Sequence analysis of GFRAL indicated that only 30% of its amino acid sequence is identical to that of GFRα-3 [19]. The distant relation of GFRAL from the GFR-α family is also apparent in specific ligand binding. Whereas recombinant native GDNF binds to the GFR-α1 and 2 isoforms, it does not interact with GFRAL. Otherwise, GDF15 exclusively binds to GFRAL but not to its homologs GAS-1 or GFR-α1-3 [18].

Activation of GFRAL via GDF15 recruits the tyrosine kinase rearranged during transfection (RET) as a co-receptor stimulating intracellular downstream pathways, e.g., extracellular signal-regulated kinase (ERK), protein kinase B (AKT) and phospholipase Cγ (PLCγ) [16]. The unique action of GDF15 on GFRAL at area postrema/nucleus tractus solitarii (AP/NTS) neurons is suggested to activate neurons in the parabrachial nucleus (PBN)-central amygdala circuit to mediate food and taste aversion, GDF15 driven anorexia and weight loss [1,2,20,21]. Thereby, the GDF15/GFRAL/RET signaling pathway represents a potential therapeutic target for eating disorders, metabolic diseases and cancer-related cachexia [2,22–24].

However, to date, the biological function of the GDF15/GFRAL system is poorly understood, i.e., whether elevated GFRAL signaling directed to restore homeostasis might be ineffective or even harmful in severe disease [16,25].

In fact, a further uncertainty is the widespread expression of GDF15 and its apparently pleiotropic actions [26], particularly in 'non-homoestatic' disease conditions, while its receptor GFRAL is described to be strongly restricted to the AP/NTS. An obvious explanation for this might be that there is a more extensive receptor distribution than is currently presumed at low levels, but biologically sufficient levels that evaded consideration so far.

A review of the previous literature provided some hints for such a more widespread GFRAL expression. For example, Li et al. (2005) found strong GFRAL transcripts, e.g., in the substantia nigra and the hippocampus of adult mice [19]. Apart from a high expression in the brainstem, Mullican et al. (2017) also showed low GFRAL mRNA expression in the hippocampus and cortex of mice. Further, they described strong GFRAL mRNA expression in the human testes and adipose tissues and various other human peripheral tissues also expressed GFRAL, albeit at low levels [2]. Finally, gene expression data from the GTEx portal suggest that GFRAL protein expression could be more ubiquitous distributed, although in small quantities [27].

These discrepancies to the commonly cited highly specialized receptor localization prompted us to re-evaluate the expression pattern of GFRAL in healthy mice by utilizing immunofluorescence labeling of GFRAL. For this purpose, we referred to the polyclonal sheep anti-GFRAL antigen affinity-purified antibody used by the first researchers that described its use, as well as others [1,22].

With this detailed analysis on the cellular level, in addition to the commonly accepted GFRAL expression in the AP/NTS, GFRAL-immunoreactivity (IR) was found in further brain areas and peripheral tissues. This observation is relevant for the understanding of disorders related to food intake, body image and their therapies.

2. Materials and Methods

2.1. Animals

Animal experiments were performed according to the ARRIVE guidelines, to the European (Council Directive 2010/63/EU) and to the German guidelines for the welfare of experimental animals. Experiments were approved by the local authorities (Landesdirektion, Leipzig, Germany). Male and female wild-type C57BL/6J mice (10–12 weeks old, 23–25 g, Charles River, Sulzfeld, Germany) were housed under an air-conditioned standard environment with free access to food and water and a 12 h light/dark cycle (lights on from 7:00 a.m.). A total of 4 male and 4 female mice were randomly selected; to the best of our knowledge, no gender differences in naïve animals were described. Mice were euthanized with an overdose of isoflurane and transcardially rinsed with 10 mL of heparinized saline, followed by 10% neutral buffered formalin.

2.2. Immunofluorescence Labeling and Confocal Laser Scanning Microscopy

The brain, the medial lobe of liver, visceral perigonadal white fat tissue, duodenum tissue, medulla of kidney, heart and skeletal muscle and the aorta were dissected, post-fixed in 4% paraformaldehyde (PFA) overnight, dehydrated and embedded in paraffin. Tissue sections of 6 µm thickness were cleared from paraffin and treated with citrate buffer pH 6 at 85 °C for 15 min to unmask antigenic epitopes.

Unspecific binding was blocked with 5% bovine serum albumin and 0.3% Triton in PBS for 1 h. Slices were incubated over night at 4 °C with the following primary antibodies: sheep anti-GFRAL (1:200, R&D Systems #AF5728, Lot. CDBG062305A, Minneapolis, MN, USA) and, additionally for brain, mouse anti-MAP2 (microtubule associated protein 2, a neuron specific cytoskeletal protein) (1:1000, Abcam #ab254144, Lot. GR3390046-5, Cambridge, UK).

Sections were washed with PBS followed by incubation for 2 h at room temperature with the respective secondary antibodies. For GFRAL, donkey anti-sheep IgG (H+L) cross-adsorbed secondary antibodies (1:1000, Alexa Fluor™ 680, ThermoFisher Scientific #A-21102, Waltham, MA, USA) were used. MAP2 antibodies were detected with donkey anti-mouse IgG (H+L) highly cross-adsorbed secondary antibody (1:1000, Alexa Fluor™ 488, ThermoFisher Scientific #A-21202). In negative controls, the GFRAL antibody was omitted. Sections were washed again.

In a control experiment for hippocampal expression, slices were incubated alternatively with rabbit anti-GFRAL (1:500, antibodies-online #ABIN2174394, Lot. A106214490, Limerick, PA, USA). For these slices, blocking was performed with 3% milk powder, 5% goat serum and goat anti-rabbit IgG (H+L) unconjugated (1:200, VectorLabs #AI-1000, Newark, CA, USA) in TBS was used. To counterstain GFRAL-IR against glycoproteins of the plasma membrane, wheat germ agglutinin (WGA) Alexa Fluor 488 conjugated (1:200, ThermoFisher Scientific W11261) was incubated for 2 h at room temperature followed by washing with TBS and incubation with goat anti-rabbit IgG (H+L) cross-adsorbed secondary antibody, Alexa Fluor™ 633 #A-21070. Sections were washed again.

Finally, all slices were rinsed with distilled water before the addition of Fluoromount-G™ Mounting Medium, containing DAPI to label DNA in the nuclei (ThermoFisher Scientific #00-4959-52). For imaging, a confocal live cell microscope Leica TCS SP8/DMi8 with the LAS X Software (version 3.5.7.23225, Leica Biosystems, Wetzlar, Germany) was used. Each experiment was performed in independent triplicates. Figures were arranged with the graphic software CorelDRAW 2022.

2.3. Antibody Validation with HFF-1 and HEK293 Cells

In addition to the reported proofs, the specificity of the sheep anti-GFRAL antibody was verified using two cell lines, human foreskin fibroblasts (HFF-1) and human embryonic kidney (HEK293) cells—negative and positive for GRFAL mRNA, respectively (for methods and results, see the Supplementary Materials section and Figure S1).

3. Results

3.1. GFRAL-IR Is Expressed in Various Mouse Brain Areas

In the brains of mice, immunofluorescence data confirmed the former descriptions on GFRAL in the brainstem. In addition, it was detectable in other brain areas.

In detail, strong GFRAL-IR-positive cells were found as expected in the AP/NTS region (Figure 1). A much lower GFRAL antibody specific for IR with less dense distribution could be detected in the nucleus arcuatus, a part of the hypothalamus (Figure 1). In slices from the medial prefrontal cortex, GFRAL-positive cells were observed in particular in layer 2/3 of the cingulate and prelimbic part, but also diffusely distributed in the deeper cortex (Figure 1). Very weak GFRAL-IR further appeared in the pyramidal cell layer of the CA1 region in the dorsal hippocampus (Figure 1), which could hardly be specifically assigned to a cellular structure.

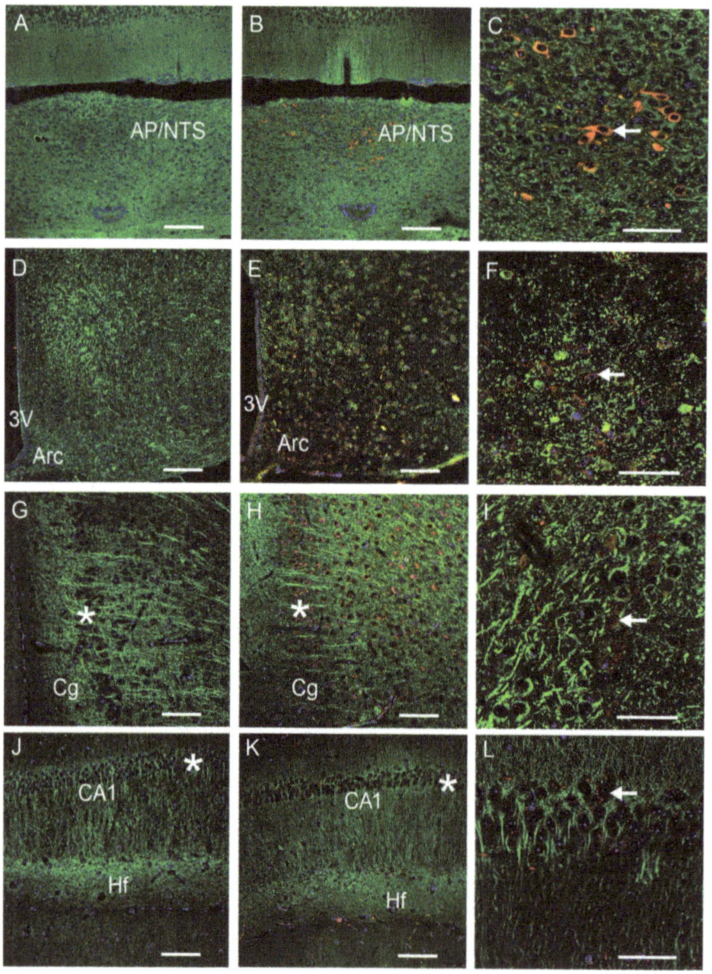

Figure 1. GFRAL-immunoreactivity (IR) in various brain areas of mice. Colors represent GFRAL-IR structures (red), nuclei (blue) and microtubule-associated protein 2 (MAP2, green). Negative controls

are shown in (**A**,**D**,**G**,**J**). Area postrema (AP) and nucleus tractus solitarii (NTS) (**A–C**); nucleus arcuatus (Arc, (**D–F**)); prefrontal cortex (**G–I**); dorsal hippocampus (CA1 region, (**J–L**)); 3V: third ventricle; Cg: cingulate cortex; Hf: hippocampal fissure. *: pyramidal cell layer, ←: GFRAL-IR-positive cell. Scale bars: 100 μm in (**A**,**B**,**D**,**E**,**G**,**H**,**J**,**K**) (20×) and 50 μm in (**C**,**F**,**I**,**L**) (63×).

Therefore, a control experiment using a rabbit anti-GFRAL antibody labeled against WGA was performed. In this approach, GFRAL-IR in the hippocampal pyramidal cell layer could conclusively be proven with an appearance compatible with a membrane-bound protein (Figure 2).

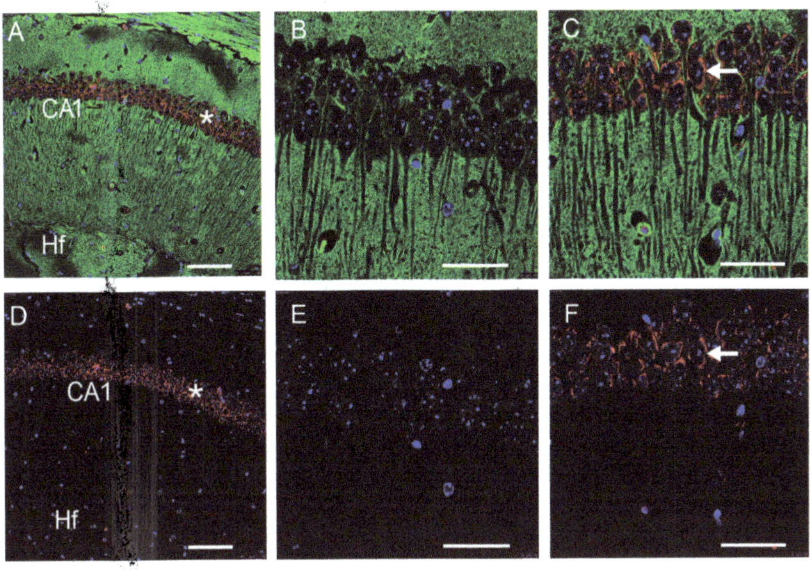

Figure 2. Verification of GFRAL-immunoreactivity (IR) in the hippocampus of mice. Colors represent GFRAL-IR structures (red), nuclei (blue) and wheat germ agglutinin (WGA, green). Negative controls are shown in (**B**,**E**). CA1: hippocampal CA1 region; Hf: hippocampal fissure; *: pyramidal cell layer; ←: GFRAL-IR-positive cell. Scale bars: 100 μm (20×) in (**A**,**D**) and 50 μm (63×) in (**B**,**C**,**E**,**F**).

3.2. Peripheral Tissues of Mice Express GFRAL-IR

GFRAL-IR was found to be widely distributed in peripheral tissues. GFRAL-positive hepatocytes were particularly observed around the central vein (Figure 3A–C). In the visceral fat tissue, GFRAL-labeling appears densely around the fat reservoir along the membrane (Figure 3D–F). GFRAL-IR was found all over the kidney, probably in the proximal tubulus cells in the parts associated with their brush border. The interior of distal tubuli and the renal corpuscles were free of GFRAL-IR (Figure 3G–I). In the intestine, cells of the mucosa (enterocytes) were also fluorescence-labelled for GFRAL, particularly the microvilli. Possibly, the IR-negative gaps at the luminal mucosal surface are attributable to goblet cells (Figure 3L).

GFRAL-positive elements were detected in the striated heart and skeletal muscle but also in the smooth muscle of the vascular wall of the aorta (Figure 4).

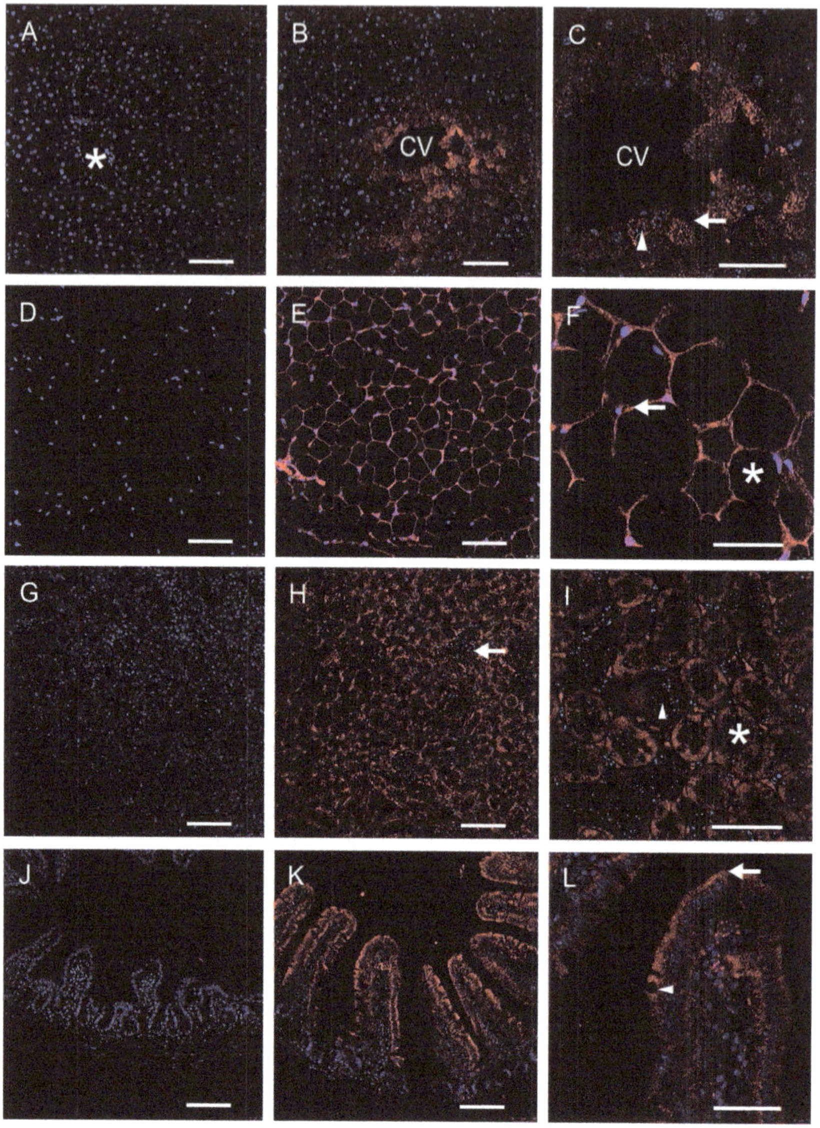

Figure 3. GFRAL-immunoreactivity (IR) in peripheral tissues of mice. In the liver (**A–C**), fat tissue (**D–F**), kidney (**G–I**) and intestine (**J–L**); GFRAL-positive structures appear in red and nuclei in blue. Negative controls are shown in (**A,D,G,J**). Liver (**A–C**): CV: central vein; *: positive hepatocytes; ▲: hepatocyte; ←: sinusoid. Visceral fat (**D–F**): *: fat reservoir; ←: cytoplasm of adipocytes. Kidney (**G–I**): ←: renal corpuscle; *: proximal tubulus; ▲: distal tubulus. Small intestine (**J–L**): ←: mucosa cells with luminal brush border; ▲: goblet cell. Scale bars: 100 µm in (**A,B,D,E,G,H,J,K**) (20×) and 50 µm in (**C,F,I,L**) (63×).

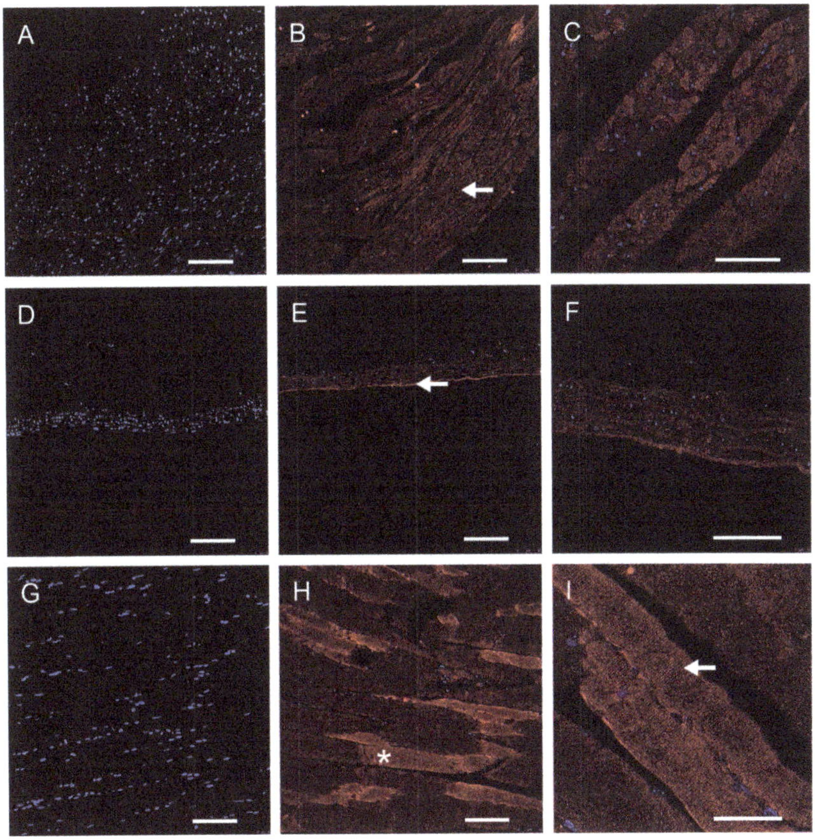

Figure 4. Immunohistochemical presentation of GFRAL-positive structures in the muscles of mice. GFRAL-positive structures (red); nuclei (blue). Negative controls are shown in (**A**,**D**,**G**). Heart muscle (**A–C**): ←:striation. cardiac muscle fibers. Aorta (**D–F**): ←: tunica. Skeletal muscle (**G–I**): * skeletal muscle fibers; ←: striation. Scale bars: 100 μm (20×) in (**A**,**B**,**D**,**E**,**G**,**H**), and 50 μm (63×) in (**C**,**F**,**I**).

4. Discussion

In summary, we showed that GFRAL is much more widely expressed in mice than is currently considered. In agreement with previous reports, high GFRAL-IR expression was found in the neurons of the AP/NTS, which is described to modulate the anorectic effects of GDF15 [1,2,17,18]. Further, this study proved that GFRAL-IR is present in the cells of the nucleus arcuatus and the pyramidal cell layers of the prefrontal cortex and hippocampus. Moreover, in selected peripheral tissues, GFRAL-IR was apparent in the liver, visceral fat tissues, kidneys, mucosa of the intestines and muscular structures.

The technique used based on the same polyclonal sheep anti-GFRAL antigen affini-ty-purified antibody originally used by the first researchers that used it, as well as others [1,22]. This antibody did not bind to GFRAL-related proteins in ELISA and failed to label GFRAL in the brainstem of GFRAL-knockout mice [1]. Here, the antibody was further validated in two cell lines proved to be negative or positive for GFRAL transcripts. HEK293 cells were specifically labeled, and no unspecific IR was detectable (see the Supplementary Materials). Li et al. (2005) described an isoform of GFRAL missing the transmembrane region that was suggested to represent a soluble form of GFRAL with a still-unknown function, which may be recognized by the used antibody as well [19]. Of note, the tissue studied here was permeabilized, a process which may interfere with an expected strong membrane-associated IR [19].

This study confirmed the intensive labeling of AP/NTS neurons, whereas the GFRAL-IR in other brain regions appeared to be specific, but weaker, and therefore may have evaded their detection so far.

For the AP/NTS region, the expression of GFRAL is conserved in rodents, monkeys and human [18]. In the other analyzed brain structures, the cellular GFRAL-IR is consistent with the GFRAL transcript distribution in adult mice, which is particularly reported for the cortex and hippocampus, although it is comparatively low [2,19]. Both groups could not detect GFRAL in peripheral tissues using RT-PCR analysis in adult mice, which might be attributable to a rare expression not recognizable via bulk PCR analysis [23]. However, in a number of human brain regions and peripheral tissues, GFRAL transcripts were detectable on a low scale, with the exception of high levels found in fat tissues, testes and mammary tissues [2,27].

The distribution of GFRAL-IR found here matches the widespread expression of its ligand, GDF15, in the brain and peripheral tissues of mice and humans [13,15,28].

Consequently, this challenges the notion that the circulating pleiotropic GDF15 exclusively activates GFRAL-expressing neurons in the AP/NTS, thereby being involved in anorexia and weight loss [4,16,23].

Rather, our immunohistochemical data point out the possibility that GFRAL is also activated in other brain areas and peripheral tissues via local or circulating GDF15. Here, GDF15/GFRAL signaling could be involved in keeping or restoring cellular homeostasis in disease or aging, as outlined below.

Similar to the AP/NTS, as a circumventricular organ with vascular permeability, neurons of the hypothalamic arcuate nucleus—a major center for appetite control—are accessible for GDF15. These neurons are open to the cerebrospinal fluid in the infundibular recess of the third ventricle [29,30]. Notably, GDF15 is also synthesized in the choroid plexus and secreted into the ventricle [14].

In the healthy mice in our study, GFRAL-IR in the arcuatus region was less intense compared to the AP/NTS, as were the GFRAL transcripts elsewhere [2]. This low expression may explain that their detection has failed using in situ hybridization analysis [18,31].

As present research mainly focuses on the exclusive role of GFRAL in the AP/NTS in anorexia, tumor-induced cachexia and obesity, the data on potential GDF15/GFRAL signaling in other brain areas are sparse. It has been consistently shown that GDF15 is synthesized in the cortex and hippocampus of mice and humans, and that it is highly regulated in the brain, for example, by ischemia, aging or Alzheimer's disease [13,32,33].

The release of GDF15 occurs downstream to cell stress related to mitochondrial and endoplasmic reticulum dysfunction and is suggested to exert anti-inflammatory effects by enhancing tissue tolerance to inflammation, as well as through the counter-regulation of inflammation-induced expression of pro-inflammatory cytokines [12,33–36]. Although the physiological functions of GDF15 are not yet fully understood, its upregulation can be considered as an attempt to compensate for cell stress [33]. However, the receptors and signaling pathways by which brain-derived GDF15 may induce such local responses are still unclear [33].

The GFRAL-IR, in comparison to its reported low mRNA outside the AP/NTS of healthy adult mice, might suggest that GFRAL-mediated signaling is repressed in normophysiological situations, e.g., for the control of appetite [24]. It should be noted that mRNA levels are not the final output of gene expression, but that they mechanistically represent the templates required for biological systems' potential highly controlled protein synthesis on demand [37]. This implies the possibility of an upregulation of GFRAL under challenging conditions, like specific developmental stages in mice and humans [19,38] or obesity. The low or even missing detection of GFRAL transcripts apart from AP/NTS may have prevented the investigation of such demanding circumstances in more detail so far.

GFRAL transcripts, even very slight, are also noticed in the peripheral tissues of humans or occasionally in mice in the current databases, but are barely further reported [27,39]. Similar to what we have found in various brain areas, GFRAL-IR was also shown to be present in the peripheral parts of the liver, visceral fat tissues, kidneys, mucosa of the intestines and muscular structures.

Our results in mice were confirmed with the detection of GFRAL-IR in human normal gastric mucosa as well as in gastric cancer cells [40]. Interestingly, in the latter GDF15, GFRAL and RET proteins are significantly and jointly elevated. Similarly, GFRAL-IR was detected in human normal pancreatic ductal epithelial and cancer cells, which was positively correlated with GDF15 expression [41]. Furthermore, in the dual immunofluorescence analysis of pancreatic cancer cells, both proteins were co-expressed. Although it was not shown whether this co-expression is also existent in normal tissues, this constellation strongly argues for carrying out feedback sensing of autocrine-released GDF15 from pancreatic cancer cells. On the other hand, pancreas carcinoma and pancreatitis are strongly associated with nausea, reduced food intake and anorexia/cachexia, and are obviously mediated by the GDF15/GFRAL brainstem pathway. The autocrine or endocrine properties of GDF15 have been discussed repeatedly, particularly regarding the adjustment of metabolic activity, and may be in line with the reported enhancement of GDF15 expression in the liver, heart and muscles after physical exercise, as reviewed by Wang (2021) [4,8,20,42,43]. However, at present, the identity of a respective receptor remains unclear, but the current data suggest that GFRAL might be a candidate.

In view of the shown GFRAL distribution in mice—results obtained with full GFRAL-KO mice or with the use of blocking antibodies should be interpreted with care. Nevertheless, full or cell-specific KO mice as well as the use of adeno-associated viruses (AAV) to assess gene function in or beyond AP/NTS are important research tools to further explore the role of GFRAL in the brain and peripheral organs under disease conditions. Although no functional investigations were performed, a more global regulatory role of GFRAL in disorders related to food intake, such as eating disorders, anorexia/cachexia and obesity, is conceivable.

In genome-wide association studies (GWAS) on variants, a GFRAL missense variant, rs12199003 (p.Arg33Cys), which obviously causes impaired binding to GDF15, was associated with BMI [8,17,44]. Further, in a transcriptome wide analysis (TWAS), GFRAL expression in human adipose tissue was negatively associated with higher levels of the orexigenic peptide ghrelin, mediating several further opposing effects to the GDF15/GFRAL signaling [45]. Interestingly, a gene variant of GFRAL was found using next-generation sequencing in a cohort of patients exhibiting diverse clinical signs of anorexia nervosa; however, its clinical relevance is unknown [46].

5. Conclusions

More insights into the genetic variants of GFRAL—together with a more comprehensive knowledge about the control of the transcription and translation of GFRAL under disease conditions, as well as further explanations of GDF15-mediated effects beyond the AP/NTS—could be relevant for the development of novel pharmacological therapies for physical and mental disorders related to body image and food intake, such as eating disorders, cachexia and obesity.

Supplementary Materials: The following supporting information can be downloaded at: https://www.mdpi.com/article/10.3390/nu16050734/s1, Figure S1: Antibody validation for GFRAL in HEK and HFF1 cells.

Author Contributions: Conceptualization, U.K. and M.K.; methodology, K.F. and A.M.; validation, H.K. and A.M.; formal analysis, A.M.; sample preparation, molecular biological techniques, and IHC, K.F., A.M., M.-M.L. and Y.G.; fluorescence microscopy, M.-M.L., Y.G. and H.K.; resources, U.K.; data curation, A.M.; writing—original draft preparation, U.K.; writing—review and editing, U.K. and

M.K.; supervision and project administration, U.K.; funding acquisition, U.K. and M.K. All authors have read and agreed to the published version of the manuscript.

Funding: This research was funded in part by the China Scholarship Council, China, grant number 201908510283 (to Y.G.). This work was supported by grants from NSFC-RSF, grant number 82261138557; NSFC, grant numbers 82274668 and 82230127; and the Sichuan Provincial Administration of Traditional Chinese Medicine, grant number 2023zd024 (to M.-M.L.).

Institutional Review Board Statement: The animal study protocol was approved by local authorities (Landesdirektion Leipzig, Regional office Saxony, ID: DD25-5131-488-13). All possible efforts were made to replace, reduce, and refine animal experiments in accordance with the ARRIVE guidelines as well as the European Council Directive 2010/63/EU and German guidelines for the welfare of experimental animals.

Data Availability Statement: The original contributions presented in this study are available upon request from the corresponding author.

Acknowledgments: The authors would like to thank Yong Tang for his scientific support and Anne-Kathrin Krause for her excellent technical support.

Conflicts of Interest: The authors declare no conflicts of interest.

References

1. Hsu, J.-Y.; Crawley, S.; Chen, M.; Ayupova, D.A.; Lindhout, D.A.; Higbee, J.; Kutach, A.; Joo, W.; Gao, Z.; Fu, D.; et al. Non-homeostatic body weight regulation through a brainstem-restricted receptor for GDF15. *Nature* **2017**, *550*, 255–259. [CrossRef] [PubMed]
2. Mullican, S.E.; Lin-Schmidt, X.; Chin, C.-N.; Chavez, J.A.; Furman, J.L.; Armstrong, A.A.; Beck, S.C.; South, V.J.; Dinh, T.Q.; Cash-Mason, T.D.; et al. GFRAL is the receptor for GDF15 and the ligand promotes weight loss in mice and nonhuman primates. *Nat. Med.* **2017**, *23*, 1150–1157. [CrossRef] [PubMed]
3. Tsai, V.W.W.; Husaini, Y.; Sainsbury, A.; Brown, D.A.; Breit, S.N. The MIC-1/GDF15-GFRAL Pathway in Energy Homeostasis: Implications for Obesity, Cachexia, and Other Associated Diseases. *Cell Metab.* **2018**, *28*, 353–368. [CrossRef] [PubMed]
4. Johann, K.; Kleinert, M.; Klaus, S. The Role of GDF15 as a Myomitokine. *Cells* **2021**, *10*, 2990. [CrossRef]
5. Dostálová, I.; Roubícek, T.; Bártlová, M.; Mráz, M.; Lacinová, Z.; Haluzíková, D.; Kaválková, P.; Matoulek, M.; Kasalicky, M.; Haluzík, M. Increased serum concentrations of macrophage inhibitory cytokine-1 in patients with obesity and type 2 diabetes mellitus: The influence of very low calorie diet. *Eur. J. Endocrinol.* **2009**, *161*, 397–404. [CrossRef]
6. Tsai, V.W.; Zhang, H.P.; Manandhar, R.; Lee-Ng, K.K.M.; Lebhar, H.; Marquis, C.P.; Husaini, Y.; Sainsbury, A.; Brown, D.A.; Breit, S.N. Treatment with the TGF-b superfamily cytokine MIC-1/GDF15 reduces the adiposity and corrects the metabolic dysfunction of mice with diet-induced obesity. *Int. J. Obes.* **2018**, *42*, 561–571. [CrossRef]
7. Coll, A.P.; Chen, M.; Taskar, P.; Rimmington, D.; Patel, S.; Tadross, J.A.; Cimino, I.; Yang, M.; Welsh, P.; Virtue, S.; et al. GDF15 mediates the effects of metformin on body weight and energy balance. *Nature* **2020**, *578*, 444–448. [CrossRef]
8. Wang, D.; Day, E.A.; Townsend, L.K.; Djordjevic, D.; Jørgensen, S.B.; Steinberg, G.R. GDF15: Emerging biology and therapeutic applications for obesity and cardiometabolic disease. *Nat. Rev. Endocrinol.* **2021**, *17*, 592–607. [CrossRef]
9. Dostálová, I.; Kaválková, P.; Papežová, H.; Domluvilová, D.; Zikán, V.; Haluzík, M. Association of macrophage inhibitory cytokine-1 with nutritional status, body composition and bone mineral density in patients with anorexia nervosa: The influence of partial realimentation. *Nutr. Metab.* **2010**, *7*, 34. [CrossRef] [PubMed]
10. Karczewska-Kupczewska, M.; Kowalska, I.; Nikolajuk, A.; Adamska, A.; Otziomek, E.; Gorska, M.; Straczkowski, M. Hyperinsulinemia acutely increases serum macrophage inhibitory cytokine-1 concentration in anorexia nervosa and obesity. *Clin. Endocrinol.* **2012**, *76*, 46–50. [CrossRef] [PubMed]
11. Plomgaard, P.; Hansen, J.S.; Townsend, L.K.; Gudiksen, A.; Secher, N.H.; Clemmesen, J.O.; Støving, R.K.; Goetze, J.P.; Wright, D.C.; Pilegaard, H. GDF15 is an exercise-induced hepatokine regulated by glucagon and insulin in humans. *Front. Endocrinol.* **2022**, *13*, 1037948. [CrossRef] [PubMed]
12. Bootcov, M.R.; Bauskin, A.R.; Valenzuela, S.M.; Moore, A.G.; Bansal, M.; He, X.Y.; Zhang, H.P.; Donnellan, M.; Mahler, S.; Pryor, K.; et al. MIC-1, a novel macrophage inhibitory cytokine, is a divergent member of the TGF-beta superfamily. *Proc. Natl. Acad. Sci. USA* **1997**, *94*, 11514–11519. [CrossRef]
13. Böttner, M.; Laaff, M.; Schechinger, B.; Rappold, G.; Unsicker, K.; Suter-Crazzolara, C. Characterization of the rat, mouse, and human genes of growth/differentiation factor-15/macrophage inhibiting cytokine-1 (GDF-15/MIC-1). *Gene* **1999**, *237*, 105–111. [CrossRef] [PubMed]
14. Strelau, J.; Sullivan, A.; Böttner, M.; Lingor, P.; Falkenstein, E.; Suter-Crazzolara, C.; Galter, D.; Jaszai, J.; Krieglstein, K.; Unsicker, K. Growth/differentiation factor-15/macrophage inhibitory cytokine-1 is a novel trophic factor for midbrain dopaminergic neurons in vivo. *J. Neurosci.* **2000**, *20*, 8597–8603. [CrossRef] [PubMed]

15. Genotype-Tissue Expression (GTEx) Portal. Available online: https://www.gtexportal.org/home/gene/GDF15 (accessed on 12 December 2023).
16. Breit, S.N.; Brown, D.A.; Tsai, V.W.-W. The GDF15-GFRAL Pathway in Health and Metabolic Disease: Friend or Foe? *Annu. Rev. Physiol.* **2021**, *83*, 127–151. [CrossRef]
17. Yang, L.; Chang, C.-C.; Sun, Z.; Madsen, D.; Zhu, H.; Padkjær, S.B.; Wu, X.; Huang, T.; Hultman, K.; Paulsen, S.J.; et al. GFRAL is the receptor for GDF15 and is required for the anti-obesity effects of the ligand. *Nat. Med.* **2017**, *23*, 1158–1166. [CrossRef]
18. Emmerson, P.J.; Wang, F.; Du, Y.; Liu, Q.; Pickard, R.T.; Gonciarz, M.D.; Coskun, T.; Hamang, M.J.; Sindelar, D.K.; Ballman, K.K.; et al. The metabolic effects of GDF15 are mediated by the orphan receptor GFRAL. *Nat. Med.* **2017**, *23*, 1215–1219. [CrossRef]
19. Li, Z.; Wang, B.; Wu, X.; Cheng, S.-Y.; Paraoan, L.; Zhou, J. Identification, expression and functional characterization of the GRAL gene. *J. Neurochem.* **2005**, *95*, 361–376. [CrossRef]
20. Wu, Q.; Clark, M.S.; Palmiter, R.D. Deciphering a neuronal circuit that mediates appetite. *Nature* **2012**, *483*, 594–597. [CrossRef]
21. Tsai, V.W.-W.; Manandhar, R.; Jørgensen, S.B.; Lee-Ng, K.K.M.; Zhang, H.P.; Marquis, C.P.; Jiang, L.; Husaini, Y.; Lin, S.; Sainsbury, A.; et al. The anorectic actions of the TGFβ cytokine MIC-1/GDF15 require an intact brainstem area postrema and nucleus of the solitary tract. *PLoS ONE* **2014**, *9*, e100370. [CrossRef]
22. Suriben, R.; Chen, M.; Higbee, J.; Oeffinger, J.; Ventura, R.; Li, B.; Mondal, K.; Gao, Z.; Ayupova, D.; Taskar, P.; et al. Antibody-mediated inhibition of GDF15-GFRAL activity reverses cancer cachexia in mice. *Nat. Med.* **2020**, *26*, 1264–1270. [CrossRef]
23. Lockhart, S.M.; Saudek, V.; O'Rahilly, S. GDF15: A Hormone Conveying Somatic Distress to the Brain. *Endocr. Rev.* **2020**, *41*, bnaa007. [CrossRef]
24. Wang, D.; Townsend, L.K.; DesOrmeaux, G.J.; Frangos, S.M.; Batchuluun, B.; Dumont, L.; Kuhre, R.E.; Ahmadi, E.; Hu, S.; Rebalka, I.A.; et al. GDF15 promotes weight loss by enhancing energy expenditure in muscle. *Nature* **2023**, *619*, 143–150. [CrossRef]
25. Iglesias, P.; Silvestre, R.A.; Díez, J.J. Growth differentiation factor 15 (GDF-15) in endocrinology. *Endocrine* **2023**, *81*, 419–431. [CrossRef]
26. Mimeault, M.; Batra, S.K. Divergent molecular mechanisms underlying the pleiotropic functions of macrophage inhibitory cytokine-1 in cancer. *J. Cell Physiol.* **2010**, *224*, 626–635. [CrossRef]
27. Genotype-Tissue Expression (GTEx) Portal. Available online: https://www.gtexportal.org/home/gene/GFRAL (accessed on 12 December 2023).
28. Fairlie, W.D.; Moore, A.G.; Bauskin, A.R.; Russell, P.K.; Zhang, H.P.; Breit, S.N. MIC-1 is a novel TGF-beta superfamily cytokine associated with macrophage activation. *J. Leukoc. Biol.* **1999**, *65*, 2–5. [CrossRef]
29. Rodríguez, E.M.; Blázquez, J.L.; Guerra, M. The design of barriers in the hypothalamus allows the median eminence and the arcuate nucleus to enjoy private milieus: The former opens to the portal blood and the latter to the cerebrospinal fluid. *Peptides* **2010**, *31*, 757–776. [CrossRef] [PubMed]
30. Korf, H.-W.; Møller, M. Arcuate nucleus, median eminence, and hypophysial pars tuberalis. *Handb. Clin. Neurol.* **2021**, *180*, 227–251. [CrossRef] [PubMed]
31. Uhlén, M.; Fagerberg, L.; Hallström, B.M.; Lindskog, C.; Oksvold, P.; Mardinoglu, A.; Sivertsson, Å.; Kampf, C.; Sjöstedt, E.; Asplund, A.; et al. Proteomics. Tissue-based map of the human proteome. *Science* **2015**, *347*, 1260419. [CrossRef] [PubMed]
32. Schindowski, K.; von Bohlen und Halbach, O.; Strelau, J.; Ridder, D.A.; Herrmann, O.; Schober, A.; Schwaninger, M.; Unsicker, K. Regulation of GDF-15, a distant TGF-β superfamily member, in a mouse model of cerebral ischemia. *Cell Tissue Res.* **2011**, *343*, 399–409. [CrossRef] [PubMed]
33. Chiariello, A.; Valente, S.; Pasquinelli, G.; Baracca, A.; Sgarbi, G.; Solaini, G.; Medici, V.; Fantini, V.; Poloni, T.E.; Tognocchi, M.; et al. The expression pattern of GDF15 in human brain changes during aging and in Alzheimer's disease. *Front. Aging Neurosci.* **2022**, *14*, 1058665. [CrossRef]
34. Conte, M.; Martucci, M.; Mosconi, G.; Chiariello, A.; Cappuccilli, M.; Totti, V.; Santoro, A.; Franceschi, C.; Salvioli, S. GDF15 Plasma Level Is Inversely Associated With Level of Physical Activity and Correlates With Markers of Inflammation and Muscle Weakness. *Front. Immunol.* **2020**, *11*, 915. [CrossRef]
35. Tanaka, T.; Biancotto, A.; Moaddel, R.; Moore, A.Z.; Gonzalez-Freire, M.; Aon, M.A.; Candia, J.; Zhang, P.; Cheung, F.; Fantoni, G.; et al. Plasma proteomic signature of age in healthy humans. *Aging Cell* **2018**, *17*, e12799. [CrossRef] [PubMed]
36. Moon, J.S.; Goeminne, L.J.E.; Kim, J.T.; Tian, J.W.; Kim, S.-H.; Nga, H.T.; Kang, S.G.; Kang, B.E.; Byun, J.-S.; Lee, Y.-S.; et al. Growth differentiation factor 15 protects against the aging-mediated systemic inflammatory response in humans and mice. *Aging Cell* **2020**, *19*, e13195. [CrossRef]
37. Buccitelli, C.; Selbach, M. mRNAs, proteins and the emerging principles of gene expression control. *Nat. Rev. Genet.* **2020**, *21*, 630–644. [CrossRef] [PubMed]
38. Jackson, T.C.; Gorse, K.; Herrmann, J.R.; Kochanek, P.M. Hippocampal and Prefrontal Cortical Brain Tissue Levels of Irisin and GDF15 Receptor Subunits in Children. *Mol. Neurobiol.* **2021**, *58*, 2145–2157. [CrossRef] [PubMed]
39. Baldarelli, R.M.; Smith, C.M.; Finger, J.H.; Hayamizu, T.F.; McCright, I.J.; Xu, J.; Shaw, D.R.; Beal, J.S.; Blodgett, O.; Campbell, J.; et al. The mouse Gene Expression Database (GXD): 2021 update. *Nucleic Acids Res.* **2021**, *49*, D924–D931. [CrossRef] [PubMed]
40. Buchholz, K.; Antosik, P.; Grzanka, D.; Gagat, M.; Smolińska, M.; Grzanka, A.; Gzil, A.; Kasperska, A.; Klimaszewska-Wiśniewska, A. Expression of the Body-Weight Signaling Players: GDF15, GFRAL and RET and their clinical relevance in Gastric Cancer. *J. Cancer* **2021**, *12*, 4698–4709. [CrossRef]

41. Zhao, Z.; Zhang, J.; Yin, L.; Yang, J.; Zheng, Y.; Zhang, M.; Ni, B.; Wang, H. Upregulated GDF-15 expression facilitates pancreatic ductal adenocarcinoma progression through orphan receptor GFRAL. *Aging* **2020**, *12*, 22564–22581. [CrossRef]
42. Laurens, C.; Parmar, A.; Murphy, E.; Carper, D.; Lair, B.; Maes, P.; Vion, J.; Boulet, N.; Fontaine, C.; Marquès, M.; et al. Growth and differentiation factor 15 is secreted by skeletal muscle during exercise and promotes lipolysis in humans. *JCI Insight* **2020**, *5*, e131870. [CrossRef]
43. Ren, Y.; Zhao, H.; Yin, C.; Lan, X.; Wu, L.; Du, X.; Griffiths, H.R.; Gao, D. Adipokines, Hepatokines and Myokines: Focus on Their Role and Molecular Mechanisms in Adipose Tissue Inflammation. *Front. Endocrinol.* **2022**, *13*, 873699. [CrossRef] [PubMed]
44. Turcot, V.; Lu, Y.; Highland, H.M.; Schurmann, C.; Justice, A.E.; Fine, R.S.; Bradfield, J.P.; Esko, T.; Giri, A.; Graff, M.; et al. Protein-altering variants associated with body mass index implicate pathways that control energy intake and expenditure in obesity. *Nat. Genet.* **2018**, *50*, 26–41. [CrossRef] [PubMed]
45. Wittekind, D.A.; Scholz, M.; Kratzsch, J.; Löffler, M.; Horn, K.; Kirsten, H.; Witte, V.; Villringer, A.; Kluge, M. Genome-wide association and transcriptome analysis suggests total serum ghrelin to be linked with GFRAL. *Eur. J. Endocrinol.* **2021**, *184*, 847–856. [CrossRef] [PubMed]
46. Donato, K.; Medori, M.C.; Macchia, A.; Cecchin, S.; Ceccarini, M.R.; Beccari, T.; Gatta, V.; Stuppia, L.; Benfatti, V.; Dalla Ragione, L.; et al. Genetic variants identified in novel candidate genes for anorexia nervosa and analysis of molecular pathways for diagnostic applications. *Eur. Rev. Med. Pharmacol. Sci.* **2023**, *27*, 77–88. [CrossRef]

nutrients

MDPI

Article

Metabolic Syndrome and Adipokines Profile in Bipolar Depression

Karolina Bilska [1], Monika Dmitrzak-Węglarz [1,*], Przemysław Osip [2], Joanna Pawlak [1], Elżbieta Paszyńska [3] and Agnieszka Permoda-Pachuta [4]

1 Department of Psychiatric Genetics, Poznan University of Medical Sciences, 61-701 Poznan, Poland
2 HCP Medical Center, 61-485 Poznan, Poland
3 Department of Integrated Dentistry, Poznan University of Medical Sciences, 61-701 Poznan, Poland
4 Department of Psychiatry, Medical University of Lublin, 20-059 Lublin, Poland
* Correspondence: mweglarz@ump.edu.pl

Abstract: Metabolic syndrome (MS) is a growing social, economic, and health problem. MS coexists with nearly half of all patients with affective disorders. This study aimed to evaluate the neurobiological parameters (clinical, anthropometric, biochemical, adipokines levels, and ultrasound of carotid arteries) and their relationship with the development of MS in patients with bipolar disorder. The study group consisted of 70 patients (50 women and 20 men) hospitalized due to episodes of depression in the course of bipolar disorders. The Hamilton Depression Rating Scale was used to assess the severity of the depression symptoms in an acute state of illness and after six weeks of treatment. The serum concentration of adipokines was determined using an ELISA method. The main finding of this study is that the following adipokines correlated with MS in the bipolar depression women group: visfatin, S100B, and leptin had a positive correlation, whereas adiponectin, leptin-receptor, and adiponectin/leptin ratio showed a negative correlation. Moreover, the adiponectin/leptin ratio showed moderate to strong negative correlation with insulin level, BMI, waist circumference, triglyceride level, treatment with metformin, and a positive moderate correlation with HDL. The adiponectin/leptin ratio may be an effective tool to assess MS in depressed female bipolar patients.

Keywords: metabolic syndrome; bipolar disorder; adipokines; metformin treatment; ADIPO/LEP ratio

Citation: Bilska, K.; Dmitrzak-Węglarz, M.; Osip, P.; Pawlak, J.; Paszyńska, E.; Permoda-Pachuta, A. Metabolic Syndrome and Adipokines Profile in Bipolar Depression. *Nutrients* **2023**, *15*, 4532. https://doi.org/10.3390/nu15214532

Academic Editors: Hubertus Himmerich and Khadijeh Mirzaei

Received: 19 September 2023
Revised: 24 October 2023
Accepted: 24 October 2023
Published: 25 October 2023

1. Introduction

Metabolic syndrome (MS) is a growing social, economic, and health problem. MS consists of abdominal obesity, hypertension, and disorders of carbohydrate and lipid metabolism [1]. The prevalence of metabolic syndrome in the Polish population is estimated at about 20%, 18% of men and 22% of women [2]. Insulin resistance and obesity (especially abdominal obesity) plays a key role in the pathogenesis of MS [1,3].

Adipose tissue is a highly specialized tissue that plays an important endocrine function through the synthesis and secretion of adipokines [3–6]. The enlarged adipocytes secrete pro-inflammatory adipokines, promoting systemic inflammation, and contributing to metabolic syndrome [1]. Among the most important of these adipokines are leptin, resistin, adiponectin, visfatin, and interleukin-6 (IL-6) [1,3,5]. Leptin regulates feeding behavior, energy homeostasis [4,7], and lipid metabolism [5,8,9]. It also controls glucose homeostasis and insulin sensitivity [3,4,9]. Moreover, leptin is known to promote a pro-inflammatory immune response, and it is suggested to be an important factor linking obesity, MS, and cardiovascular diseases [3,8,9]. Another essential role in the development of inflammation is played by resistin [1,5]. The role of resistin in the pathogenesis of insulin resistance in humans remains unclear [4,5]; however, the role in inflammation and metabolic dyslipidemia development is well-known [9]. Adiponectin plays a vital role in the regulation of glucose and lipid metabolism [1,3,5,7]. Adiponectin acts as an endogenous insulin sensitizer, increasing glucose uptake and promoting fatty acid oxidation [7]. Moreover, adiponectin plays a protective role against the development and progression of insulin resistance, MS, and cardiovascular diseases, and also inhibits pro-inflammatory factors [3,7]. Visfatin similarly affects insulin sensitivity due to its insulin-mimetic capacity [1,5,10]. It binds insulin receptors and enhances glucose uptake, transport, and lipogenesis [10]. However, unlike adiponectin, visfatin is a pro-inflammatory mediator that activates multiple inflammatory pathways (e.g., mitogen-activated protein kinase and phosphatidylinositol 3 kinase) [10]. Increased expression of IL-6 has numerous implications for the pathogenesis of obesity and its complications. IL-6, by reducing the expression of the insulin receptor, adiponectin, and inhibiting the activity of lipoprotein lipase, leads to the intensification of insulin resistance [1,11]. Moreover, IL-6, affects the functioning of the vascular endothelium by stimulating the synthesis of acute phase proteins. This leads to the formation and progression of the atherosclerotic lesions inflammation, and dysfunction [1,3].

The S100B is a calcium-binding protein responsible for transcriptional regulation and DNA repair, cell differentiation, cell growth and migration, and programmed cell death [12]. The S100B is predominantly expressed by astrocytes [13], and also by other cell types: melanocytes [14], chondrocytes, adipocytes [15,16], skeletal muscle, and a few other cell types [17]. In their research, Fujiya et al. [18] proposed that the S100B functions as an adipokine in the interaction between adipocytes and macrophages. They proved that S100B upregulated the expression of TNF-α and proinflammatory markers in macrophages, and TNF-α augmented S100B secretion from preadipocytes. Moreover, silencing of S100B in preadipocytes significantly reduced TNF-α secretion from macrophages [18]. Recent publications [17,19] suggest the involvement of S100B in obesity and diabetes mechanisms, possibly by participating in the inflammatory processes.

Bipolar disorder patients are almost twice as likely to have MS than the general population [20–22]. Moreover, patients taking antipsychotic medication have a higher risk of developing MS than antipsychotic-free patients [20,21]. Psychiatric patients are more likely to develop obesity and metabolic abnormalities than healthy people [21–23]. Factors that predispose psychiatric patients to MS include genetic, unhealthy lifestyle (e.g., smoking, excessive alcohol intake, poor sleep hygiene, physical inactivity, and unhealthy nutritional patterns), but also the use of psychotropic medication (antipsychotics, antidepressants, and mood stabilizers) [20–23].

The aim of this study was to evaluate the neurobiological parameters (clinical, anthropometric, biochemical, adipokines levels, and ultrasound of carotid arteries) and their relationship with the development of MS in patients with bipolar affective disorder. The first research hypothesis is that depending on coexisting MS, patients differ in neurobiological parameters and the level of adipokines tested. The second research hypothesis is that differences in neurobiological parameters and the level of tested adipokines depend on the state of the disease (exacerbation vs. improvement).

2. Materials and Methods

2.1. Participants

The study group included 50 women (45.0 ± 14.29 years old) and 20 men (50.9 ± 15.73 years old) (Table 1) with a diagnosis of bipolar disorder based on DSM-IV criteria [24,25]. Only patients with current depression episodes were included in the study. Patients were evaluated by psychiatrists twice, upon admission to the hospital in an acute state of illness (exacerbation) and after six weeks of treatment as recommended by the National Consultant [26], to assess depression symptoms using the 17-item version of Hamilton Depression Rating Scale (HAMD) [27].

Table 1. Characteristic of bipolar patients group with gender distinction.

Anthropometric and Biochemical Parameters	Women Mean ± SD	Men Mean ± SD
Age (years)	45.0 ± 14.29	50.9 ± 15.73
Duration of disease (years)	15.5 ± 8.88	15.9 ± 11.78
HDL (mg/dL)	51.0 ± 15.08	40.1 ± 10.43
LDL (mg/dL)	125.7 ± 39.04	125.3 ± 51.10
TG (mg/dL)	171.6 ± 82.44	196.6 ± 95.25
Insulin (mU/mL)	12.3 ± 10.90	13.3 ± 3.89
BMI (kg/m^2)	27.7 ± 6.02	28.8 ± 5.80
Waist circumference (cm)	94.9 ± 16.88	103.9 ± 10.35
Comorbidities	Women N (%)	Men N (%)
Metabolic syndrome (MS)	34 (68)	13 (65)
MS metformin treatment	16 (32)	5 (25)
Hypothyroidism	28 (56)	5 (25)
Hypertension	9 (18)	5 (25)
Hypercholesterolemia	4 (8)	1 (5)
Psychiatric drugs	Women N (%)	Men N (%)
Antidepressants	31 (62)	9 (45)
Mood stabilizers	32 (64)	16 (80)
Antipsychotics	39 (78)	12 (60)

Abbreviations: HDL—high-density lipoprotein; LDL—low-density lipoprotein; TG—triglyceride; BMI—body mass index; MS—metabolic syndrome; SD—standard deviation; N- number of patients.

The exclusion criteria were severe and unstable medical conditions, pregnancy, autoimmune diseases, severe and chronic somatic diseases (except diabetes, hypertension, and obesity), infectious diseases four weeks before and during the study period, neuropsychiatric illnesses associated with cognitive impairment, or a prior clinical diagnosis of schizophrenia or schizoaffective disorder. The study protocol did not interfere with the treatment of patients, was compliant with the indications and was supervised by the attending physicians. Patients were treated with various combinations and doses of drugs with different mechanisms of action (antidepressants, antipsychotics, and mood stabilizers). The most often prescribed drugs were: antidepressant—venlafaxine (21 women and 7 men), mood stabilizers—lithium carbonate (23 women and 10 men) and valproic acid (9 women and 11 men), antipsychotics—quetiapine (32 women and 8 men), olanzapine (13 women and 5 men) and clozapine (12 women and 3 men). More detailed information about the pharmacological therapy can be found in the Supplementary Material (Table S1). No patients were in monotherapy or with the same set of drugs in the study group.

Upon admission to the hospital, all patients underwent anthropometric measurements: height, waist circumference (with an accuracy of 1.0 cm), and body weight (with an accuracy of 0.1 kg). These data were used to estimate visceral obesity and body mass index (BMI). Following the recommendations of the World Health Organization (WHO) regarding visceral obesity [28], we adopted the following cut-off points for significantly increased risk of metabolic complications: >88 cm in women and >102 cm in men. BMI below 25 kg/m^2 was considered a normal body mass index, whereas overweight was diagnosed with BMI values between 25 and 29.9 kg/m^2, and obesity with a BMI above 30 kg/m^2 [29]. In the statistical analysis, we distinguished two groups: normal weight and obesity (BMI above 25 kg/m^2) which consists of overweight, and obesity according to

WHO. Biochemical tests determined the lipid profile (HDL, LDL, and TG [mg/dL]) and insulin concentration (mU/mL) were performed. Based on anthropometric measurements and laboratory tests, the internist diagnosed metabolic syndrome under the guidelines of the International Diabetes Federation [30] and decided which patients required metformin treatment (500 mg daily).

The intima-media complex (IMC) was measured in the common carotid arteries. Distal IMC of both carotid arteries was measured by duplex and B-mode ultrasound using SonoScape S6 ultrasound with a 6–12 Mhz linear transducer. IMC measurements were taken at several points approximately 1 cm proximal to the common carotid sinus. The IMC thickness result is presented as the average of the measurements taken. In addition, the maximum systolic velocity in the tested vessels was determined [31].

The study was approved by the Bioethics Committee of University Medical Sciences in Poznan (Resolution No. 1082/15 of 3 December 2015) [32]. All study participants provided written informed consent and were recruited in 2015–2018 in the Adult Psychiatry Clinic by the Department of Psychiatry, University of Medical Sciences in Poznan.

2.2. Biochemical Analysis

The blood samples were collected from the patients twice: upon admission to the hospital and after six weeks of treatment in order to assess the concentration of adipokines in the blood serum. In the exacerbation state, we measured concentrations of visfatin (VIS), adiponectin (ADIPO), S100B, interleukin-6 (IL-6), leptin (LEP), leptin-receptor (LEP_R), and resistin (RES). After six weeks of treatment only ADIPO, VIS, and S100B were measured. Venous blood was collected into EDTA tubes and centrifuged at 1000× g for 15 min at 4 °C to obtain serum samples, aliquoted into Eppendorf tubes, and stored at −80 °C. Commercial Enzyme-Linked Immunosorbent Assay tests (ELISA) (Table 2) were used to quantify the selected proteins in the blood serum. Optical density was read with a spectrophotometric plate reader (Asys UVM 340 Microplate Reader from Biochrom Ltd., Cambridge, UK) for a wavelength of 450 nm ± 10 nm. A four-parameter algorithm (four-parameter logistic curve) was used to assay the concentration in the tested samples. All samples and standards were run in duplicates, and the mean value of the two assays was used for statistical evaluation. All ELISA tests were performed according to the manufacturer's instructions without any modifications.

Table 2. Parameters of commercial ELISA tests used to assess concentrations of adipokines in the serum.

Adipokine Name	ELISA Kit Name and Manufacturer	Assay Range	Sensitivity	Intra-Assay Variability Coefficient (%)	Inter-Assay Variability Coefficient (%)
Visfatin	E0638h, EIAab, Wucan, China	1.56–100 ng/mL	<0.78 ng/mL	<4.9	<6.4
Adiponectin	DEE009 for the In Vitro Diagnostic (IVD) Demeditec Diagnostics GmbH, Kiel, Germany	0.27–31000 µg/mL	<0.27 µg/mL	<3.7	<8.2
S100B	Human S100B, EZHS100B-33K, EMD Millipore Corporation, Darmstadt, Germany	2.7–2000 pg/mL	2.7 pg/mL	<3.3	<5.4

Table 2. *Cont.*

Adipokine Name	ELISA Kit Name and Manufacturer	Assay Range	Sensitivity	Intra-Assay Variability Coefficient (%)	Inter-Assay Variability Coefficient (%)
Interleukin-6	Quantikine ELISA Human Leptin il-6, S6050, R&D Systems Minneapolis, Minneapolis, MN, USA	3.13–100 pg/mL	<0.70 pg/mL	<4.2	<6.4
Leptin	Quantikine ELISA Human Leptin, SLP00, R&D Systems Minneapolis, Minneapolis, MN, USA	15.6–1000 pg/mL	<7.8 pg/mL	<3.3	<5.4
Leptin receptor	Quantikine ELISA Human Leptin sR, DOBR00, R&D Systems Minneapolis, Minneapolis, MN, USA	0.313–20 ng/mL	0.020–0.128 ng/mL	<6.1	<8.6
Resistin	DRSN00 for research use only (RUO) R&D Systems Minneapolis, Minneapolis, MN, USA	0.156–10 ng/mL	0.010–0.055 ng/mL	<5.3	<9.2

2.3. Statistical Analysis

Statistical analyses were performed using the STATISTICA 13.3 (StatSoft, Krakow, Poland). The significance level $p < 0.05$ was adopted for all analyses. The distribution of variables was studied by the Shapiro–Wilk test. Variables with normal distribution were tested using parametric tests (student's *t*-test and ANOVA with Tukey's post-hock test). Variables that did not meet the normal distribution criteria were tested using non-parametric tests (Mann–Whitney U test, Wilcoxon pair order test, ANOVA rank Kruskal–Wallis test, Chi-square test, and Friedman ANOVA with Kendall Concordance). Spearman's rank correlation coefficient was applied to assess the relationship between the analyzed variables. The ROC curve was used to test the diagnostic ability of the ADIPO/LEP ratio.

3. Results

Spearman's rank correlation coefficient was applied to assess the influence of gender on studied variables. Seven of the examined variables (hypothyroidism, waist circumference, HDL, ADIPO, S100B, LEP, and LEP_R) correlated with sex; therefore, we decided to study women and men groups separately. Depression symptoms upon admission to the hospital in an acute state of illness (exacerbation) and after six weeks of treatment were assessed using the 17-item version of the Hamilton Depression Rating Scale (HAMD) [27]. Hamilton's total score was higher in exacerbation in both sexes (Table 3). Thirty-five patients (23 women and 12 men) did not achieve remission (score $\leq$ 7) [33], but they had 25–56% reduction in the total score. A total number of the uptaken antidepressant drugs correlated positively with Hamilton total score (R = 0.3633 p = 0.0195).

Because insulin resistance and obesity play a crucial role in MS pathogenesis, we conducted a statistical analysis of metabolic syndrome, and variables such as insulin levels, BMI, and metformin treatment. Women with MS had higher insulin levels (p = 0.0197) and higher BMI values (p = 0.007) than women without MS. Unfortunately, these observations were not confirmed in the group of men, probably due to the small study group. We have also tested the thickness and maximum systolic velocity of intima-media complex in the common carotid arteries, but no statistically significant results were obtained in either sex (but were correlated with age, duration of disease, and waist circumference).

Table 3. Clinical and biochemical parameters were analyzed at two-time points: upon admission to the hospital in an acute state of illness (exacerbation) and after six weeks of treatment.

Group	Women		Men	
Parameters	**Mean ± SD** **Median (Min–Max)**	***p*-Value** **(t/Z)**	**Mean ± SD** **Median (Min–Max)**	***p*-Value** **(t/Z)**
HAMD E HAMD 6	16.3 ± 1.84 17 (12–20) 7.8 ± 2.78 7 (3–15)	<0.0001 (23.34)	16. 1± 2.19 17 (12–19) 8.8 ± 2.48 9 (5–13)	<0.0001 (16.22)
VIS (ng/mL) E VIS (ng/mL) 6	16.5 ± 7.65 15.9 (2.6–33.4) 6.4 ± 4.31 5.6 (2.0–18.0)	<0.0001 (4.10)	16.2 ± 7.95 18.9 (3.7–24.9) 8.1 ± 5.08 7.4 (1.8–18.8)	0.0109 (2.58)
ADIPO (ng/mL) E ADIPO (ng/mL) 6	7913.3 ± 4840.5 7833.7 (1621.3–24,415.6) 9533.5 ± 6994.5 7498.9 (2142.1–27,335.8)	ns	4381.0 ± 2184.65 3899.8 (1912.7–8435.1) 5518.3 ±2900.62 4625.2 (2709.4–10,577.2)	0.0284 (2.19)
S100B (ng/mL) E S100B (ng/mL) 6	31.0 ± 18.71 28.25 (4.8–82.6) 27.1 ± 18.94 29.83 (4.7–77.5)	ns	29.3 ± 14.52 27.7 (9.1–55.2) 40.2 ± 14.38 40.6 (16.3–66.6)	0.038 (2.07)

Abbreviations: HAMD—Hamilton Depression Rating Scale; VIS—visfatine; ADIPO—adiponectin; E—an exacerbation; 6—after six weeks of treatment; SD—standard deviation; ns—non-significant values. Students' *t*-test was applied only to HAMD; all other parameters were calculated using the Wilcoxon pair order test, with a significance level of $p < 0.05$.

Pro-inflammatory adipokines have been tested in the context of obesity, metabolic syndrome, and metformin treatment, as well as concerning the mental state of patients. Obese women had statistically significant elevated levels of S100B and LEP and lower LEP_R levels in exacerbation (Table 4). Women with MS had statistically significant differences in the level of adipokines (VIS, ADIPO, S100B, LEP, and LEP_R) in exacerbations compared with women without MS. The VIS, S100B, and LEP levels have been increased, while ADIPO and LEP_R concentrations have been reduced. Similar results were obtained when comparing three groups: with and without MS, and MS metformin-treated. Post-hock tests showed significant differences not only between patients with and without MS, but also between MS and MS treated with metformin (Table 4). We also compared the ratio of ADIPO/LEP; women without MS (mean ± SD 1.1 ± 0.95) had significantly ($Z = 3.90$ $p = 0.0001$) higher ratio than women with MS (mean ± SD 0.4 ± 0.76). Moreover, the ADIPO/LEP ratio showed moderate to strong negative correlation with insulin level, BMI, waist circumference, TG level, metformin treatment, and positive moderate correlations with HDL concentrations. We conducted the ROC curve analysis for the ADIPO/LEP ratio regarding MS (Supplementary Table S2). The result was statistically significant ($AUC = 0.846$ $p < 0.0001$), with the cut-off point at 0.31 (Youden index 0.64). The sensitivity and specificity were 76.5% and 87.5%. In the case of our data, this meant two people without MS were classified as ill, and eight ill people were classified as healthy. We believe that 23.5% of misclassified ill people is too high a percentage, so we decided to set a better-fitting cut-off point. In relation to our data, we have selected a 0.48 cut-off point (Youden index 0.599) with 91.2% sensitivity and 68.8% specificity. In the analysis between the level of adipokines and the mental state of the patients, only the level of visfatin had changed (decreased) significantly (Table 3).

Table 4. The comparison of biochemical parameters concerns three variables: obesity, metabolic syndrome, and metformin treatment in the women group.

Group Parameters	Obesity	Metabolic Syndrome	Metformin Treatment	Metformin Treatment Post-Hock Analysis		
	p-Value (Z-Value)	*p*-Value (Z-Value)	*p*-Value (H-Value)	W vs. MSMT *p*-Value (Z-Value)	W vs. MSnT *p*-Value (Z-Value)	MSMT vs. MSnT *p*-Value (Z-Value)
VIS (ng/mL) E	ns	0.0209 (2.31)	0.0360 (6.65)	0.0187 (−2.35)	ns	ns
VIS (ng/mL) 6	ns	ns	ns	ns	ns	ns
ADIPO (ng/mL) E	ns	0.0173 (−2.38)	0.0028 (11.77)	0.0011 (3.26)	ns	0.0143 (−2.45)
ADIPO (ng/mL) 6	ns	ns	ns	0.0428 (2.02)	ns	ns
S100B (ng/mL) E	0.0012 (−3.23)	0.0022 (3.07)	0.0078 (9.71)	0.0122 (−2.51)	0.0055 (−2.78)	ns
S100B (ng/mL) 6	ns	ns	ns	ns	ns	ns
IL 6 (pg/mL)	ns	ns	ns	ns	ns	ns
LEP (pg/mL)	<0.0001 (−4.01)	0.0005 (3.48)	0.0015 (12.97)	0.0006 (−3.41)	0.0083 (−2.64)	ns
LEP_R (ng/mL)	0.0057 (2.76)	0.0254 (−2.24)	ns	ns	ns	ns
RES (ng/mL)	ns	ns	ns	ns	ns	ns

Abbreviations: VIS—visfatin; ADIPO—adiponectin; Il 6—Interleukin 6; LEP—leptin; LEP_R—receptor for leptin; RES- resistin; E—exacerbation; 6—after six weeks of treatment; W—patients without metabolic syndrome; MSMT—patients with metabolic syndrome with metformin treatment; MSnT—patients with metabolic syndrome non treated with metformin; ns—non-significant values; significance level $p < 0.05$. Comparison between normal weight and obesity (Obesity column in the table), patients with and without metabolic syndrome (Metabolic syndrome column in the table), and post-hock analysis were conducted using the Mann–Whitney U test. Comparison between three groups (patients without and with metabolic syndrome, and patients with metabolic syndrome treated with metformin) was conducted using ANOVA rank Kruskal–Wallis test (Metformin treatment column in the table).

In contrast to women, there were no changes in the level of adipokines in the group of men, in the context of obesity, metabolic syndrome, or metformin treatment (probably due to the small study group). However, the ADIPO/LEP ratio showed a strong negative correlation with BMI and waist circumference ($R = -0.6568$ $p = 0.0057$ and $R = -0.6139$ $p = 0.0088$), but no correlation with MS was observed. Interestingly, concerning the mental state of the patients, changes were observed in the level of three adipokines (VIS, ADIPO, S100B), not just one, as was the case in the group of women (Table 3). As in women group, visfatin level decreased while ADIPO and S100B levels increased after six weeks of treatment.

Spearman's rank correlation coefficient was applied to assess the influence of neurobiological parameters (clinical, anthropometric, biochemical, adipokines, and ultrasound of carotid arteries) on the development of MS in patients with bipolar affective disorder. In women, thirteen parameters were significantly correlated with MS (Table 5). A positive, strong correlation was observed with metformin treatment, whereas other parameters showed moderate (TG, insulin, BMI, waist circumference, S100B, and LEP) and weak (hypothyroidism and VIS_E) positive correlation (Table 5). A negative, weak correlation was also observed for three parameters: HDL, ADIPO_E, and LEP_R (Table 5). In the men group, only metformin treatment was significantly correlated with MS (R Spearman = 0.8819 $p < 0.0001$).

Table 5. Spearman's rank correlation coefficient and descriptive statistics for female bipolar patients with and without metabolic syndrome.

Group	Without MS	With MS		
Parameters	**Mean ± SD Median (Min–Max)**	**Mean ± SD Median (Min–Max)**	**R Spearman**	***p*-Value**
Age (years)	42.6 ± 15.90 39.5 (21.0–71.0)	46.2 ± 13.57 48.5 (18.0–68.0)	0.1174	ns
Duration of disease (years)	10.9 ± 9.51 10.0 (1.0–35.0)	13.5 ± 8.58 12.0 (1.0–39.0)	0.1661	ns
Hypothyroidism *	6	22	0.3281	0.0362
Hypertension *	2	7	0.083	ns
Hypercholesterolemia *	0	4	0.2254	ns
HDL (mg/dL)	61.2 ± 14.82 60.0 (40.0–81.0)	47.2 ± 13.56 45.0 (24.0–80.0)	−0.4049	0.0129
LDL (mg/Dl)	120.6 ± 32.64 110.0 (92.0–189.0)	141.9 ± 40.62 144.0 (85.0–230.0)	0.2317	ns
TG (mg/dL)	114.2 ± 56.62 106.0 (60.0–250.0)	195.0 ± 80.47 196.0 (59.0–344.0)	0.4657	0.0032
Insulin (mU/mL)	5.45 ± 2.20 5.85 (2.7–7.5)	16.4 ± 12.05 12.6 (3.2–44.6)	0.6161	0.0110
BMI (kg/m^2)	23.2 ± 3.35 22.5 (19.8–31.2)	29.5 ± 5.96 28.65 (21.2–44.0)	0.5391	0.0003
Waist circumference (cm)	83.5 ± 12.47 79.0 (68.0–108.0)	100.3 ± 16.11 98.0 (72.0–140.0)	0.4882	0.0009
Metformin treatment *	0	16	0.8575	<0.0001
IMC L	0.64 ± 0.26 0.57 (0.43–1.48)	0.65 ± 0.12 0.63 (0.42–0.99)	0.2093	ns
IMC R	0.80 ± 0.88 0.57 (0.35–3.83)	0.65 ± 0.16 0.67 (0.29–1.11)	0.1260	ns
V L	79.9 ± 22.86 77.0 (39.5–116.6)	75.6 ± 14.52 74.1 (52.3–110.5)	−0.0712	ns
V R	84.2 ± 25.71 93.3 (40.7–119.0)	76.1 ± 21.97 78.3 (35.7–127.5)	−0.2185	ns
HAMD E	16.7 ± 1.57 17.0 (14.0–19.0)	16.1 ± 1.95 17.0 (12.0–20.0)	−0.1529	ns
HAMD 6	7.9 ± 2.33 7.0 (5.0–13.0)	7.8 ± 3.00 8.0 (3.0–15.0)	−0.0187	ns
VIS (ng/mL) E	11.9 ± 8.44 11.07 (1.4–31.6)	17.0 ± 6.79 17.3 (2.6–33.4)	0.3350	0.0186
VIS (ng/mL) 6	5.4 ± 4.08 4.5 (2.3–14.3)	6.7 ± 4.34 5.7 (2.0–18.0)	0.2387	ns
ADIPO (ng/mL) E	11657.6 ± 6066.77 9641.0 (3227.1–24,415.6)	7709.9 ± 4782.67 6699.1 (1621.3–23,098.1)	−0.3417	0.0152
ADIPO (ng/mL) 6	12,209.6 ± 7556.85 11,491.7 (5549.0–27,335.8)	8596.9 ± 6734.39 6990.5 (2142.1–25,196.8)	−0.2930	ns
S100B (ng/mL) E	19.9 ± 9.62 17.1 (4.8–36.3)	33.0 ± 15.75 29.67 (13.3–82.8)	0.4398	0.0014

Table 5. *Cont.*

Group	Without MS	With MS		
Parameters	Mean ± SD Median (Min–Max)	Mean ± SD Median (Min–Max)	R Spearman	*p*-Value
S100B (ng/mL) 6	26.7 ± 13.69 28.2 (4.6–39.5)	30.9 ± 20.67 26.8 (8.9–77.5)	−0.0108	ns
IL 6 (pg/mL)	1.3 ± 0.97 1.2 (0.3–4.2)	2.4 ± 4.55 1.36 (0.1–26.8)	0.1278	ns
LEP (pg/mL)	18,259.1 ± 12,764.8 13,315.5 (3764.0–41,789.0)	41,818.6 ± 24,072.05 37,733.0 (5389.0–93,517.0)	0.4991	0.0002
ADIPO/LEP	1.1 ± 0.95 0.78 (0.09–3.17)	0.4 ± 0.76 0.20 (0.03–4.29)	−0.5586	<0.0001
LEP_R (ng/mL)	33.4 ± 10.04 30.8 (20.7–62.4)	27.3 ± 7.89 26.4 (13.7–47.8)	−0.3209	0.0231
RES (ng/mL)	26.5 ± 10.80 23.0 (11.4–50.7)	28.1 ± 10.65 25.9 (9.6–60.3)	0.0891	ns

Abbreviations: HDL—high-density lipoprotein; LDL—low-density lipoprotein; TG—triglyceride; BMI—body mass index; IMC—intima-media complex; L—left; R—right; V—maximum systolic velocity; HAMD—Hamilton Depression Rating Scale; VIS—visfatin; ADIPO—adiponectin; Il 6—Interleukin 6; LEP—leptin; LEP_R—leptin receptor; RES—resistin; E—exacerbation; 6—after six weeks of treatment; MS—metabolic syndrome; SD—standard deviation; significance level $p < 0.05$; * number of person instead mean ± SD and median (min-max).

4. Discussion

In our study, not all patients achieved remission, but they had shown an improvement, defined as a 20–30% reduction in the total scores of HAMD [26]. Bipolar depression is challenging to treat, and drugs should be administrated wisely to avoid mood switches in patients [34–36]. Antidepressant therapy is associated with an increased risk of mania or hypomania, so it should be taken with mood stabilizers and/or antipsychotic medications [35].

In our study, women suffering from MS had higher insulin levels, BMI, and waist circumference than patients without MS. It is unsurprising because one of the criteria of MS is overweight (mainly abdominal), whereas insulin resistance plays a crucial role in MS pathogenesis [1,3]. BP patients have a higher MS prevalence than the general population [20–22]. Most psychiatric patients have at least one metabolic disorder [22,37]. Despite other factors like genetics, physical inactivity, unhealthy diet, and addictions, medications, especially antipsychotic drugs, have well-established weight gain side effects [20,21]. In clinical practice, clozapine and olanzapine are associated with a higher risk of MS [20,21,38,39], quetiapine and risperidone cause moderate alterations [39], whereas aripiprazole has a little effect on body weight [21,38]. Unfortunately, we could not assess the effect of treatment on BMI because patients were weighed only at the time of admission to the hospital. Therefore, it is important to balance the potential benefits and damages in bipolar depression treatment, especially in long-term treatment, because high doses or multiple medications can be associated with harmful metabolic consequences [21,39]. Interestingly, there appears to be a correlation between the higher clinical effectiveness of atypical antipsychotics and the increased risk of metabolic alterations [39].

The following adipokines correlated with MS in the women group VIS, S100B, and LEP had a positive correlation, whereas ADIPO, LEP_R, and ADIPO/LEP ratio showed a negative correlation. It has been proven that leptin concentrations are significantly increased in obesity [4,5,40]. The concept of "leptin resistance" was proposed to explain this phenomenon [8]. It assesses that tissues have decreased sensitivity to leptin, so higher leptin levels are essential to correct the metabolic imbalance in obesity [3]. Leptin binds to and activates its transmembrane receptor, the LEP-R, which plays a crucial role in regulating body mass via a negative feedback mechanism between adipose tissue and the hypothalamus [8].

Similarly to our observations, higher leptin level and lowest leptin-receptor concentrations in obese patients were detected in Koch et al. study [41].

Decreased ADIPO levels in patients with obesity [40], coronary heart disease, diabetes, and hypertension demonstrate a high tendency to develop MS [3]. Moreover, patients with BP during the depression episode showed decreased levels of ADIPO [42–44]. Obesity and MS are characterized by increased leptin and decreased adiponectin concentration [6,45]. Therefore, the ADIPO/LEP ratio has been suggested as a maker of adipose tissue dysfunction [6,45]. In our study, MS patients reached significantly lower values of ADIPO/LEP ratio than patients without MS, which is consistent with the study conducted by Frühbeck et al. [45]. Moreover, the ADIPO/LEP ratio was negatively correlated with BMI and waist circumference, which is consistent with the previous study [46]. In our study, an insulin level was significantly correlated with the ADIPO/LEP ratio. In the literature, it has been claimed that the ADIPO/LEP ratio correlates with insulin resistance better than adiponectin or leptin alone [6]. In our ROC curve analysis, the statistical software proposed a 0.31 cut-off point with 76.5% sensitivity and 87.5% specificity. The 23.5% of misclassified ill people is too high a percentage, so we have chosen a better-fitting cut-off point 0.48 cut-off point with 91.2% sensitivity and 68.8% specificity. At that point, the specificity is lower, but we believe it is better to order a more detailed diagnostics for a healthy person than to miss an ill person. The ADIPO/LEP ratio has been proposed as a predictive marker for the MS with a cut-off point lower than 0.5 [45].

Visfatin is secreted by adipose and visceral adipose tissue [7], skeletal muscle, liver, and lymphocytes [40]. This cytokine acts like insulin by binding to insulin receptors, increasing glucose uptake [10,47]. The serum visfatin level correlates with the BMI, waist circumference, and insulin resistance index [47]. A meta-analysis study showed a significant increase in visfatin serum concentration in overweight/obese participants compared with normal BMI participants and also in type 2 diabetes mellitus participants compared with the control group [48]. The visfatin serum concentration was also higher in MS patients [49], consistent with our study.

The S100B, which is expressed in adipose tissue [15,16], has been associated with the pathophysiology of obesity-promoting macrophage-based inflammation [17,50]. The serum level of S100B correlates with insulin resistance, metabolic risk score, and fat cell size [17]. In studies conducted on mice [50], plasma and white adipose tissue S100B levels were increased by diet-induced obesity. Also, in a human study, serum levels of S100B positively correlate with BMI; S100B levels in obesity were significantly higher than in overweight and normal weight subjects [51]. In another study, participants with MS had a significantly higher level of S100B than the control group [19]. Moreover, serum levels of S100B were positively correlated with abdominal obesity and triglyceride serum levels [19]. In our study, BP patients with obesity and MS had a higher serum level of S100B in exacerbation than BP patients with normal BMI, and without MS. Our results show, on the one hand, the relationship between the S100B levels and obesity and MS, and on the other hand, with the mental state of patients. The meta-analysis showed elevated levels of serum S100B in patients with affective disorders (depression and mania) compared with the control group [52,53]. The same relationship was also observed in drug-naïve adolescents diagnosed with first-episode unipolar major depression [54].

5. Conclusions

We partially confirmed our first research hypothesis: 'Depending on coexisting MS, patients differ in neurobiological parameters and the level of adipokines tested'. This statement is correct for a women's group. The second research hypothesis, 'Differences in neurobiological parameters and the level of tested adipokines depend on the state of the disease (exacerbation vs. improvement)', is correct both in the groups of women and men but for a different set of proteins.

In this study, we showed that in bipolar depression, adipokines correlated with MS in the women group: VIS, S100B, and LEP had a positive correlation, whereas ADIPO,

LEP_R, and ADIPO/LEP ratio showed negative correlation. Moreover, the ADIPO/LEP ratio showed moderate to strong negative correlation with insulin level, BMI, waist circumference, TG level, treatment with metformin, and a positive correlation with HDL correlations. The ADIPO/LEP ratio has been proposed as a predictive marker for MS [6,45], and our study proved that it can be successfully applied in depressed BP women patients. It is necessary to continue research on a larger group with naturalistic treatment, taking into account different patient states: euthymia, hypomania, mania, or mixed states, to check whether the ADIPO/LEP ratio is a good predictor of MS regardless of the patient's mental state and pharmacological treatment.

6. Study Limitations

- Heterogeneous population concerning the treatment used;
- A small group of men and women subgroup;
- Only three adipokines measured twice, in an exacerbation and after six weeks of treatment;
- BMI, waist circumference, HDL, LDL, TG, and insulin level were measured only in an exacerbation state;
- Information on previous treatment was not recorded.

Supplementary Materials: The following supporting information can be downloaded at: https://www.mdpi.com/article/10.3390/nu15214532/s1, Table S1. List of psychiatric medications (active substances) patients take during six weeks of study; Table S2. The ROC curve analysis of ADIPO/LEP ratio for metabolic syndrome in bipolar depressed women group.

Author Contributions: Conceptualization, M.D.-W. and A.P.-P.; methodology, M.D.-W., A.P.-P. and P.O.; software, M.D.-W. and K.B.; validation, M.D.-W.; formal analysis, M.D.-W. and K.B.; investigation, A.P.-P., P.O. and M.D.-W.; resources, J.P. and E.P.; data curation, M.D.-W.; writing—original draft preparation, K.B.; writing—review and editing, K.B., M.D.-W. and J.P.; visualization, K.B.; supervision, M.D.-W. and A.P.-P.; project administration, M.D.-W. and A.P.-P.; funding acquisition, J.P. and E.P. All authors have read and agreed to the published version of the manuscript.

Funding: The work was supported and financed by the Poznan University of Medical Sciences from funds allocated for scientific activities for Department of Psychiatric Genetics No. 502-20-2219-6440 (JAK0000131, ID 4520) and Department of Integrated Dentistry No. JAK000111 (ID 4520).

Institutional Review Board Statement: The study was conducted in accordance with the Declaration of Helsinki, and approved by the Ethics Committee of University Medical Sciences in Poznan (Resolution No. 1082/15 of 3 December 2015).

Informed Consent Statement: Informed consent was obtained from all subjects involved in the study.

Data Availability Statement: The data that support the findings of this study are available from the second author, MDW, upon reasonable request.

Conflicts of Interest: The authors declare no conflict of interest.

References

1. Pendal-Irlik, A. Powiązania pomiędzy zespołem metabolicznym a stanem zapalnym. *Przegląd Lek.* **2018**, *75*, 308–312.
2. Kalinowski, P.; Mianowana, M. Metabolic Syndrome part II: Epidemiology of metabolic syndrome in Poland and in the World. *J. Educ. Health Sport* **2016**, *6*, 466–480. [CrossRef]
3. Fahed, G.; Aoun, L.; Bou Zerdan, M.; Allam, S.; Bou Zerdan, M.; Bouferraa, Y.; Assi, H.I. Metabolic Syndrome: Updates on Pathophysiology and Management in 2021. *Int. J. Mol. Sci.* **2022**, *23*, 786. [CrossRef]
4. Shi, J.; Fan, J.G.; Su, Q.; Yang, Z. Cytokines and Abnormal Glucose and Lipid Metabolism. *Front. Endocrinol.* **2019**, *10*, 703. [CrossRef]
5. Kojta, I.; Chacinska, M.; Blachnio-Zabielska, A. Obesity, Bioactive Lipids, and Adipose Tissue Inflammation in Insulin Resistance. *Nutrients* **2020**, *12*, 1305. [CrossRef]
6. Fruhbeck, G.; Catalan, V.; Rodriguez, A.; Gomez-Ambrosi, J. Adiponectin-leptin ratio: A promising index to estimate adipose tissue dysfunction. Relation with obesity-associated cardiometabolic risk. *Adipocyte* **2018**, *7*, 57–62. [CrossRef]

7. Reddy, P.; Lent-Schochet, D.; Ramakrishnan, N.; McLaughlin, M.; Jialal, I. Metabolic syndrome is an inflammatory disorder: A conspiracy between adipose tissue and phagocytes. *Clin. Chim. Acta Int. J. Clin. Chem.* **2019**, *496*, 35–44. [CrossRef]
8. Obradovic, M.; Sudar-Milovanovic, E.; Soskic, S.; Essack, M.; Arya, S.; Stewart, A.J.; Gojobori, T.; Isenovic, E.R. Leptin and Obesity: Role and Clinical Implication. *Front. Endocrinol.* **2021**, *12*, 585887. [CrossRef]
9. Vekic, J.; Zeljkovic, A.; Stefanovic, A.; Jelic-Ivanovic, Z.; Spasojevic-Kalimanovska, V. Obesity and dyslipidemia. *Metab. Clin. Exp.* **2019**, *92*, 71–81. [CrossRef]
10. Dakroub, A.; Nasser, S.A.; Younis, N.; Bhagani, H.; Al-Dhaheri, Y.; Pintus, G.; Eid, A.A.; El-Yazbi, A.F.; Eid, A.H. Visfatin: A Possible Role in Cardiovasculo-Metabolic Disorders. *Cells* **2020**, *9*, 2444. [CrossRef]
11. Tylutka, A.; Morawin, B.; Walas, L.; Michalek, M.; Gwara, A.; Zembron-Lacny, A. Assessment of metabolic syndrome predictors in relation to inflammation and visceral fat tissue in older adults. *Sci. Rep.* **2023**, *13*, 89. [CrossRef] [PubMed]
12. Bertheloot, D.; Latz, E. HMGB1, IL-1α, IL-33 and S100 proteins: Dual-function alarmins. *Cell Mol. Immunol.* **2017**, *14*, 43–64. [CrossRef] [PubMed]
13. Oris, C.; Kahouadji, S.; Durif, J.; Bouvier, D.; Sapin, V. S100B, Actor and Biomarker of Mild Traumatic Brain Injury. *Int. J. Mol. Sci.* **2023**, *24*, 6602. [CrossRef]
14. Carney, B.C.; Oliver, M.A.; Kurup, S.; Collins, M.; Keyloun, J.W.; Moffatt, L.T.; Shupp, J.W.; Travis, T.E. Laser-assisted drug delivery of synthetic alpha melanocyte stimulating hormone and L-tyrosine leads to increased pigmentation area and expression of melanogenesis genes in a porcine hypertrophic scar model. *Laser Surg. Med.* **2023**, *55*, 490–502. [CrossRef]
15. Zeng, X.; Ye, M.C.; Resch, J.M.; Jedrychowski, M.P.; Hu, B.; Lowell, B.B.; Ginty, D.D.; Spiegelman, B.M. Innervation of thermogenic adipose tissue via a calsyntenin 3β-S100b axis. *Nature* **2019**, *569*, 229–235, Erratum in *Nature* **2019**, *570*, E32. [CrossRef]
16. Li, P.; Song, R.J.; Du, Y.Q.; Liu, H.J.; Li, X.D. Adtrp regulates thermogenic activity of adipose tissue via mediating the secretion of S100b. *Cell Mol. Life Sci.* **2022**, *79*, 407. [CrossRef]
17. Riuzzi, F.; Chiappalupi, S.; Arcuri, C.; Giambanco, I.; Sorci, G.; Donato, R. S100 proteins in obesity: Liaisons dangereuses. *Cell Mol. Life Sci.* **2020**, *77*, 129–147. [CrossRef]
18. Fujiya, A.; Nagasaki, H.; Seino, Y.; Okawa, T.; Kato, J.; Fukami, A.; Himeno, T.; Uenishi, E.; Tsunekawa, S.; Kamiya, H.; et al. The Role of S100B in the Interaction Between Adipocytes and Macrophages. *Obesity* **2014**, *22*, 371–379. [CrossRef]
19. Kheirouri, S.; Ebrahimi, E.; Alizadeh, M. Association of S100B Serum Levels with Metabolic Syndrome and its Components. *Acta Medica Port.* **2018**, *31*, 201–206. [CrossRef]
20. Penninx, B.; Lange, S.M.M. Metabolic syndrome in psychiatric patients: Overview, mechanisms, and implications. *Dialogues Clin. Neurosci.* **2018**, *20*, 63–73. [CrossRef]
21. Mazereel, V.; Detraux, J.; Vancampfort, D.; van Winkel, R.; De Hert, M. Impact of Psychotropic Medication Effects on Obesity and the Metabolic Syndrome in People With Serious Mental Illness. *Front. Endocrinol.* **2020**, *11*, 573479. [CrossRef] [PubMed]
22. Barton, B.B.; Zagler, A.; Engl, K.; Rihs, L.; Musil, R. Prevalence of obesity, metabolic syndrome, diabetes and risk of cardiovascular disease in a psychiatric inpatient sample: Results of the Metabolism in Psychiatry (MiP) Study. *Eur. Arch. Psychiatry Clin. Neurosci.* **2020**, *270*, 597–609. [CrossRef] [PubMed]
23. Simon, M.S.; Barton, B.; Zagler, A.; Engl, K.; Rihs, L.; Glocker, C.; Musil, R. Lifestyle behaviors, metabolic disturbances, and weight gain in psychiatric inpatients treated with weight gain-associated medication. *Eur. Arch. Psychiatry Clin. Neurosci.* **2023**, *273*, 839–851. [CrossRef] [PubMed]
24. APA. *Diagnostic and Statistical Manual of Mental Disorders*, 4th ed.; Text Revision; American Psychiatric Association: Washington, DC, USA, 2000.
25. APA. *Diagnostic and Statistical Manual of Mental Disorders (DSM-IV)*, 4th ed.; American Psychiatric Association: Washington, DC, USA, 1994.
26. Galecki, P.; Blizniewska-Kowalska, K. Treatment-resistant depression—Recommendations of the National Consultant in the field of psychiatry. *Psychiatr. Pol.* **2021**, *55*, 7–21. [CrossRef]
27. Hamilton, M. A rating scale for depression. *J. Neurol. Neurosurg. Psychiatry* **1960**, *23*, 56–62. [CrossRef]
28. WHO. *Waist Circumference and Waist–Hip Ratio: Report of a WHO Expert Consultation. Geneva, 8–11 December 2008*; WHO: Geneva, Switzerland, 2008.
29. WHO. Obesity: Preventing and managing the global epidemic. In *WHO Technical Report Series*; WHO: Geneva, Switzerland, 2000; p. 894.
30. Zimmet, P.; KG, M.M.A.; Serrano Rios, M. A new international diabetes federation worldwide definition of the metabolic syndrome: The rationale and the results. *Rev. Esp. Cardiol.* **2005**, *58*, 1371–1376. [CrossRef]
31. Elwertowski, M.; Malek, G. Standards of the Polish Ultrasound Society—Update. Examination of extracranial carotid and vertebral arteries. *J. Ultrason.* **2014**, *14*, 179–191. [CrossRef]
32. Sawicka-Gutaj, N.; Gruszczyński, D.; Guzik, P.; Mostowska, A.; Walkowiak, J. Publication ethics of human studies in the light of the Declaration of Helsinki—A mini-review. *J. Med. Sci.* **2022**, *91*, e700. [CrossRef]
33. Bech, P. Rating scales in depression: Limitations and pitfalls. *Dialogues Clin. Neurosci.* **2006**, *8*, 207–215. [CrossRef]
34. McIntyre, R.S.; Berk, M.; Brietzke, E.; Goldstein, B.I.; Lopez-Jaramillo, C.; Kessing, L.V.; Malhi, G.S.; Nierenberg, A.A.; Rosenblat, J.D.; Majeed, A.; et al. Bipolar disorders. *Lancet* **2020**, *396*, 1841–1856. [CrossRef]
35. Jelen, L.A.; Young, A.H. The Treatment of Bipolar Depression: Current Status and Future Perspectives. *Curr. Behav. Neurosci. Rep.* **2020**, *7*, 1–14. [CrossRef]

36. Levenberg, K.; Cordner, Z.A. Bipolar depression: A review of treatment options. *Gen. Psychiatry* **2022**, *35*, e100760. [CrossRef] [PubMed]
37. Glodek, M.; Skibinska, M.; Suwalska, A. Diet and physical activity and metabolic disorders in patients with schizophrenia and bipolar affective disorder in the Polish population. *PeerJ* **2023**, *11*, e15617. [CrossRef] [PubMed]
38. Pillinger, T.; McCutcheon, R.A.; Vano, L.; Mizuno, Y.; Arumuham, A.; Hindley, G.; Beck, K.; Natesan, S.; Efthimiou, O.; Cipriani, A.; et al. Comparative effects of 18 antipsychotics on metabolic function in patients with schizophrenia, predictors of metabolic dysregulation, and association with psychopathology: A systematic review and network meta-analysis. *Lancet Psychiatry* **2020**, *7*, 64–77. [CrossRef] [PubMed]
39. Carli, M.; Kolachalam, S.; Longoni, B.; Pintaudi, A.; Baldini, M.; Aringhieri, S.; Fasciani, I.; Annibale, P.; Maggio, R.; Scarselli, M. Atypical Antipsychotics and Metabolic Syndrome: From Molecular Mechanisms to Clinical Differences. *Pharmaceuticals* **2021**, *14*, 238. [CrossRef]
40. Kim, J.E.; Kim, J.S.; Jo, M.J.; Cho, E.; Ahn, S.Y.; Kwon, Y.J.; Ko, G.J. The Roles and Associated Mechanisms of Adipokines in Development of Metabolic Syndrome. *Molecules* **2022**, *27*, 334. [CrossRef]
41. Koch, A.; Weiskirchen, R.; Zimmermann, H.W.; Sanson, E.; Trautwein, C.; Tacke, F. Relevance of serum leptin and leptin-receptor concentrations in critically ill patients. *Mediat. Inflamm.* **2010**, *2010*, 473540. [CrossRef] [PubMed]
42. Wu, X.; Chen, Z.; Liao, Y.; Yang, Z.; Liang, X.; Guan, N.; Gan, Z. Are serum levels of inflammatory markers associated with the severity of symptoms of bipolar disorder? *Front. Psychiatry* **2022**, *13*, 1063479. [CrossRef]
43. Misiak, B.; Stramecki, F.; Kasznia, J.; Lis, M.; Stanczykiewicz, B. Adiponectin levels in patients with bipolar disorder: A systematic review and meta-analysis. *Psychoneuroendocrinology* **2019**, *104*, 74–79. [CrossRef]
44. Platzer, M.; Fellendorf, F.T.; Bengesser, S.A.; Birner, A.; Dalkner, N.; Hamm, C.; Hartleb, R.; Queissner, R.; Pilz, R.; Rieger, A.; et al. Adiponectin is decreased in bipolar depression. *World J. Biol. Psychiatry Off. J. World Fed. Soc. Biol. Psychiatry* **2019**, *20*, 813–820. [CrossRef]
45. Fruhbeck, G.; Catalan, V.; Rodriguez, A.; Ramirez, B.; Becerril, S.; Salvador, J.; Colina, I.; Gomez-Ambrosi, J. Adiponectin-leptin Ratio is a Functional Biomarker of Adipose Tissue Inflammation. *Nutrients* **2019**, *11*, 454. [CrossRef] [PubMed]
46. Lee, J.M.; Kim, S.R.; Yoo, S.J.; Hong, O.K.; Son, H.S.; Chang, S.A. The relationship between adipokines, metabolic parameters and insulin resistance in patients with metabolic syndrome and type 2 diabetes. *J. Int. Med. Res.* **2009**, *37*, 1803–1812. [CrossRef]
47. Ragino, Y.I.; Stakhneva, E.M.; Polonskaya, Y.V.; Kashtanova, E.V. The Role of Secretory Activity Molecules of Visceral Adipocytes in Abdominal Obesity in the Development of Cardiovascular Disease: A Review. *Biomolecules* **2020**, *10*, 374. [CrossRef] [PubMed]
48. Chang, Y.H.; Chang, D.M.; Lin, K.C.; Shin, S.J.; Lee, Y.J. Visfatin in overweight/obesity, type 2 diabetes mellitus, insulin resistance, metabolic syndrome and cardiovascular diseases: A meta-analysis and systemic review. *Diabetes Metab. Res. Rev.* **2011**, *27*, 515–527. [CrossRef] [PubMed]
49. Ugur, K.; Erman, F.; Turkoglu, S.; Aydin, Y.; Aksoy, A.; Lale, A.; Karagoz, Z.K.; Ugur, I.; Akkoc, R.F.; Yalniz, M. Asprosin, visfatin and subfatin as new biomarkers of obesity and metabolic syndrome. *Eur. Rev. Med. Pharmacol. Sci.* **2022**, *26*, 2124–2133. [CrossRef]
50. Buckman, L.B.; Anderson-Baucum, E.K.; Hasty, A.H.; Ellacott, K.L.J. Regulation of S100B in white adipose tissue by obesity in mice. *Adipocyte* **2014**, *3*, 215–220. [CrossRef]
51. Steiner, J.; Schiltz, K.; Walter, M.; Wunderlich, M.T.; Keilhoff, G.; Brisch, R.; Bielau, H.; Bernstein, H.G.; Bogerts, B.; Schroeter, M.L.; et al. S100B serum levels are closely correlated with body mass index: An important caveat in neuropsychiatric research. *Psychoneuroendocrinology* **2010**, *35*, 321–324. [CrossRef]
52. Schroeter, M.L.; Sacher, J.; Steiner, J.; Schoenknecht, P.; Mueller, K. Serum S100B represents a new biomarker for mood disorders. *Curr. Drug Targets* **2013**, *14*, 1237–1248. [CrossRef]
53. Schroeter, M.L.; Steiner, J.; Schonknecht, P.; Mueller, K. Further evidence for a role of S100B in mood disorders: A human gene expression mega-analysis. *J. Psychiatr. Res.* **2014**, *53*, 84–86. [CrossRef]
54. Bilginer, C.; Yaman, H.; Karadeniz, S.; Hizarci Bulut, S.; Yaman, S.O.; Aydogdu, S. Oxidative Stress and Serum S100B Levels in Adolescents with First-Episode Drug-Naive Unipolar Depression. *Psychiatr. Danub.* **2021**, *33*, 158–164. [CrossRef]

nutrients MDPI

Review

The Impact of Anorexia Nervosa and the Basis for Non-Pharmacological Interventions

Vicente Javier Clemente-Suárez [1], Maria Isabel Ramírez-Goerke [1], Laura Redondo-Flórez [2], Ana Isabel Beltrán-Velasco [3], Alexandra Martín-Rodríguez [1,*], Domingo Jesús Ramos-Campo [4,*], Eduardo Navarro-Jiménez [5], Rodrigo Yáñez-Sepúlveda [6] and José Francisco Tornero-Aguilera [1]

[1] Faculty of Sports Sciences, Universidad Europea de Madrid, Tajo Street, s/n, 28670 Madrid, Spain; vctxente@yahoo.es (V.J.C.-S.); mariaisabel.ramirez@universidadeuropea.es (M.I.R.-G.); josefrancisco.tornero@universidadeuropea.es (J.F.T.-A.)
[2] Department of Health Sciences, Faculty of Biomedical and Health Sciences, Universidad Europea de Madrid, C/Tajo s/n, Villaviciosa de Odón, 28670 Madrid, Spain; lauraredondo_1@hotmail.com
[3] Psychology Department, Facultad de Ciencias de la Vida y la Naturaleza, Universidad Antonio de Nebrija, 28240 Madrid, Spain; abeltranv@nebrija.es
[4] LFE Research Group, Department of Health and Human Performance, Faculty of Physical Activity and Sport Science-INEF, Universidad Politécnica de Madrid, 28040 Madrid, Spain
[5] Facultad de Ciencias de la Salud, Universidad Simón Bolívar, Barranquilla 080002, Colombia; eduardo.navarro@unisimon.edu.co
[6] Faculty of Education and Social Sciences, Universidad Andres Bello, Viña del Mar 2520000, Chile; rodrigo.yanez.s@unab.cl
* Correspondence: sandra.martin.rodriguez8@gmail.com (A.M.-R.); domingojesusramos@gmail.com (D.J.R.-C.)

Abstract: Anorexia nervosa is a psychiatric disorder with an unknown etiology that is characterized by an individual's preoccupation with their weight and body structure while denying the severity of their low body weight. Due to the fact that anorexia nervosa is multifaceted and may indicate the coexistence of genetic, social, hormonal, and psychiatric disorders, a description of non-pharmacological interventions can be used to ameliorate or reduce the symptoms of this condition. Consequently, the purpose of the present narrative review is to describe the profile's context in the anorexic person as well as the support they would require from their family and environment. In addition, it is aimed at examining preventative and non-pharmacological interventions, such as nutritional interventions, physical activity interventions, psychological interventions, psychosocial interventions, and physical therapy interventions. To reach the narrative review aims, a critical review was conducted utilizing both primary sources, such as scientific publications, and secondary sources, such as bibliographic indexes, web pages, and databases. Nutritional interventions include nutritional education and an individualized treatment for each patient, physical activity interventions include allowing patients to perform controlled physical activity, psychological interventions include family therapy and evaluation of the existence of other psychological disorders, psychosocial interventions include management of the relationship between the patient and social media and physical therapy interventions include relaxation massages and exercises to relieve pain. All these non-pharmacological interventions need to be individualized based on each patient's needs.

Keywords: anorexia nervosa; interventions; non-pharmacological; individual profile; therapy; eating disorder; microbiota; physical exercise; nutrition; psychology

Citation: Clemente-Suárez, V.J.; Ramírez-Goerke, M.I.; Redondo-Flórez, L.; Beltrán-Velasco, A.I.; Martín-Rodríguez, A.; Ramos-Campo, D.J.; Navarro-Jiménez, E.; Yáñez-Sepúlveda, R.; Tornero-Aguilera, J.F. The Impact of Anorexia Nervosa and the Basis for Non-Pharmacological Interventions. *Nutrients* **2023**, *15*, 2594. https://doi.org/10.3390/nu15112594

Academic Editor: Palmiero Monteleone

Received: 10 May 2023
Revised: 29 May 2023
Accepted: 30 May 2023
Published: 1 June 2023

1. Introduction

Eating disorders are diseases whose main characteristics are distorted eating behavior and an extreme concern for self-image and body weight. The two main eating disorders are anorexia nervosa (AN) and bulimia nervosa (BN), referring in this review exclusively to AN. According to the diagnostic and statistical manual for mental disorders (5th edition) (DSM-5), individuals with AN are disturbed by their weight and body shape, without recognizing the graveness of their low weight. The prevalence of AN is 0.4% among women and 0.1% among men at any point in time, worldwide [1]. However, recent reviews suggest that the lifetime prevalence rates of AN might be up to 4% among females and 0.3% among males. Regarding BM, up to 3% of females and more than 1% of males suffer from this disorder during their lifetime. According to epidemiological studies, the trend is for a significant percentage increase in AN, this being greater in women than in men [2].

New fashion trends and new standards in physical aspects and eating patterns are indicated as possible triggers for the increase in the frequency of AN. Yet, as with many psychiatric illnesses, the etiology of AN is not yet known with certainty, but is thought to be multifactorial and may include components of genetic, social, metabolic, personality, hormonal, sexual, the way of expressing emotions, learning, abuse history, mistreatment, perfectionism, and the coexistence of other psychiatric disorders (such as a depressive component, obsessive–compulsive, anxiety disorders and loss of impulse control) [3]. The initial consequence of AN is decreased caloric and nutrient intake. With this deficit and malnutrition, disorders appear in different organ systems, including cardiovascular, gastrointestinal, endocrine, neurological and skeletal [4]. These organic dysfunctions in the different systems are accompanied by dysfunctions in the mental health of patients with AN such as mood and anxiety disorders. If AN is considered as a pathology of a psychiatric nature, it is one with the highest mortality rate compared to other mental disorders [4], due to increased suicidal thoughts (up to 10% more). Indeed, one in five deaths associated with AN has been due to suicide [5].

A greater number of resources are needed, including more healthcare professionals to deal with the increased patient-related workloads. In the case of AN, the treatments are highly expensive given the frequency of relapses and the high disease burden [6]. Given the prevalence of the disease and the epidemiological curve, special focus should be made since it can increase the global healthcare costs. Furthermore, from an economic point of view, it is not only the associated health costs, but also the decreased work capacity of subjects with AN, which affect the world economy indirectly, due to a decrease in workplace productivity produced by the physical and mental health implications of AN [7].

Interventions can be of two types: pharmacological treatments and non-pharmacological interventions. The interventions that concern this review are of a non-pharmacological nature, and among most traditional interventions we can find [6]:

- Family therapy (changing and solving family problems to cure AN)
- Family based treatment (parents' involvement in adolescents' food consumption)
- Joint family therapy (collaborative work among adolescents, the entire family and a therapist with monitoring of the family's emotional issues)
- Behavioral family system therapy (three-step behavioral weight gain program with family involvement)
- Cognitive-behavioral therapy (modifications of irrational beliefs and problematic eating behavior)
- Specialist supportive clinical management (Education with supportive therapy)

The aforementioned refers to psychological treatments, however, like any pathology, prevention and treatment from a holistic point of view is essential. Especially since current treatment efficacy remains limited; around 40% of AN patients after 10 years of medical care still show prolonged symptoms and disabilities [8]. This highlights the importance of developing new alternative therapies against AN. Therefore, the inclusion of other areas such as nutrition, physical exercise, and psychosocial aspects among others, are essential and will be discussed throughout this review. Therefore, the present narrative review focuses on the analysis from a holistic perspective of the multifactorial etiology of AN, addressing the different medical–scientific disciplines considered non-pharmacological interventions, to give a broader view of the pathology of AN, generating holistic practical applications that can improve the patient intervention processes.

reinforce the idea that established social and cultural body ideals must be met as proof of acceptance [45,46].

Furthermore, previous authors suggested that adolescents with better parent–adolescent relationships presented a lower probability of suffering from body dissatisfaction [47]. Body disappointment is a complex risk factor which may predispose to anorexia development, maintaining and lengthening illness [48,49]. Previous researchers have pointed out how anorexia patients can be inclined to overvalue their body size, more specifically controversial parts such as hips, buttocks, abdomen and arms [50]. As a consequence, anorexia patients suffer different emotions such as anxiety, disapprobation, guilt and social withdrawal, increasing their discomfort [48]. Thus, it should be considered how powerfully parental influence may affect anorexia progress.

Previous authors supported the idea that neuropsychological deficits, such as cognitive inflexibility and decision-making, could be also observed in first-degree relatives of the patients. Regarding decision-making, it may be that anorexia patients present a dysfunction at this level, since previous research has shown how decision-making was significantly diminished in eating disorder patients, leading to restrictive eating behaviors [51].

Finally, family-based interventions have been established as a useful tool especially in children and adolescents, showing encouraging results in anorexia management and recovery [52–54]. Thus, recent authors developed a specific model treatment, which consists of an intervention constituting four phases, which may have a positive effect in families supporting anorexia patients. These phases include commitment and growth of the therapeutic agreement, aiding families to relieve the symptoms of an eating disorder, discover individual and family development issues and finally, to complete treatment and discuss future plans [55]. Naturally, parental influence in this case plays a beneficial role in improving anorexia management, as patients may feel supported by their family members, promoting better strategies in order to decrease the severity and overall health impact of anorexia or even overcoming the disease.

3.3. Social Context and Anorexia

The sociocultural context is recognized as an important risk factor for the development and for the specificity of eating disorders [56].

Social performance generally refers to an individual's ability to interact successfully with their environment (including work, school, social activities, and relationships with partners and family). This is possible by the development of a variety of social skills including verbal and nonverbal gestures, social cognition, and interpersonal performance. Although not a diagnostic criterion, some studies suggest that eating disorders are associated with atypical social and emotional behaviors [57–59]. At the same time, other studies highlight the important role of social support [59], social behavior and social inclusion [60] in the successful recovery of adults with ED.

Currently, the media promotes pre-set standards of physical appeal based on thinness. Media coverage interaction with body dissatisfaction, and personality traits could intensify specific behaviors in women that should aid them in achieving an ideal body image, e.g., excessive focus on body image, weight control, increased physical activity. However, the intensification of these behaviors may develop anorexia readiness syndrome (ARS) in women [61]. Such behaviors can be attributed to comparing oneself to unrealistic ideals while setting aside one's own attractiveness. These behaviors include: increased interest in nutrition, calorie counting, dieting, weight control methods, increased physical activity, excessive focus on body image, emotional lability related to eating and body perception, the desire to control one's own body dimensions and weight, high competitiveness and perfectionism, as well as the need for control [61].

Several studies have confirmed the significance of sociocultural messages, in highly industrialized countries, in reinforcing the perception that 'thin female bodies' are more attractive. Levine et al. [62] found in a group of 10–14-year-old girls that most reported receiving a clear message from fashion magazines and peers or family members that slimness is important and can be achieved through dieting and other methods. Two strong influences on the urge for thinness and altered eating patterns were reading magazines containing information and ideas about attractive body shapes and weight control, and receiving weight/body shape-related criticism from family [62]. In a follow-up study of 428 boys and girls, Bearman et al. [47] found that body dissatisfaction was higher among girls. For both sexes, lack of parental support, negative affectivity, self-reported dietary restraint, body mass index, and eating pathologies, showed significant associations with future increases in body dissatisfaction.

Factors such as body image dissatisfaction, the restrictive pursuit of thinness, the adoption of a perfectionist attitude towards the body, and the development of bulimic tendencies are often indicated in scientific research as predictors of eating disorders. Peer pressure is common to both sexes: A study conducted among adolescent schoolchildren indicated that, while girls talk more about appearance, boys had a higher perception of appearance pressure and teasing. Boys also admitted that they talked to their friends about muscle development more often than girls did about diet. Most researchers, however, focus on the selected risk factor [63].

Additionally, while peers are a major influence in a teenager's life, another strong social influence for teenagers is social media, media advertisements, and the internet in general. Social media is a part of most people's lives, especially Facebook, Twitter, Tiktok, and Instagram. These accounts contain dangerous information about body image, eating habits, and physical appearance. Social media is a way for some people to inform themselves and has become a career for others in the form of brand ambassadors or 'influencers'. However, most research on social media and social networking is somewhat outdated for today's teenagers, as technology changes so rapidly. Although there is little research on social networking sites, advertisements are a great way to reach different types of people, whether through social media or other media. The internet in general is a tool that many young adults use to define themselves and they can find themselves on sites that are considered "pro-anorexia", where adolescent girls can find like-minded people [64].

Pro-anorexia websites are virtual spaces where adolescents can exchange ideas about their body image and physical appearance. Uncontrolled use of these websites is a common practice among adolescents, particularly among young women, and is a factor related to eating disorders. According to different studies, looking at images of underweight celebrities is associated with an ideal body image and aspiration to lose weight, and these conditions may promote eating disorders [65,66]. The COVID-19 pandemic influenced media exposure to adolescents due to increased overall media consumption, increased exposure to harmful content related to diet and physical appearance, particularly on social networking sites, and increased use of video conferencing and subsequent self-image exposure while working and/or studying from home. All of these may be associated with an increase in eating-disordered behaviors [67].

Knowledge of the relationship between the use of social media and self-esteem related to eating disorders is necessary to promote prevention of the use of these electronic media, generating skills and providing sufficient information to avoid being negatively influenced by messages broadcast on social media about the 'Western ideal' of body perfection [68,69].

3.4. Bone and Muscular Implications of Anorexia

Eating Disorders (EDs), specifically AN, are associated with multiple neuroendocrine disruptions and low bone mineral density (BMD) [70,71]. Increased calorie restriction may impede reaching optimum bone mass, especially during adolescence, which can lead to long-term skeletal disorders [71,72]. Dual energy X-ray absorptiometry (DXA) measurements have shown that adolescents with AN have lower BMD and lower rates of bone accretion than normal-weight adolescents of comparable age and maturity [70,71]. Specifically, Z-scores of less than -1 are present in up to 50% of adolescent girls with anorexia nervosa, while Z-scores of less than -2 are present in 11% of these individuals. Despite the fact that there are fewer data on AN in males than in girls, the research that is available is even more alarming, with 70% of boys having Z-scores of less than −1 at age [71]. Thus, osteopenia, lower than normal bone mass and lower BMD, is a frequent and often chronic consequence of AN that causes clinical fractures and an elevated lifetime risk of fracture (Figure 1). Another study reported that only 15% of women with AN in their mid-20 s had a normal BMD across all bone regions investigated, while 55% had osteopenia and 35% had osteoporosis [73]. Moreover, studies in adults have shown lower markers of bone formation and higher markers of bone resorption, which has led to the theory that osteopenia is brought on by low osteoblast activity and high osteoclast activity. Conversely, AN results in a broad decrease in bone turnover indicators during adolescence, a period of significant bone turnover [74,75].

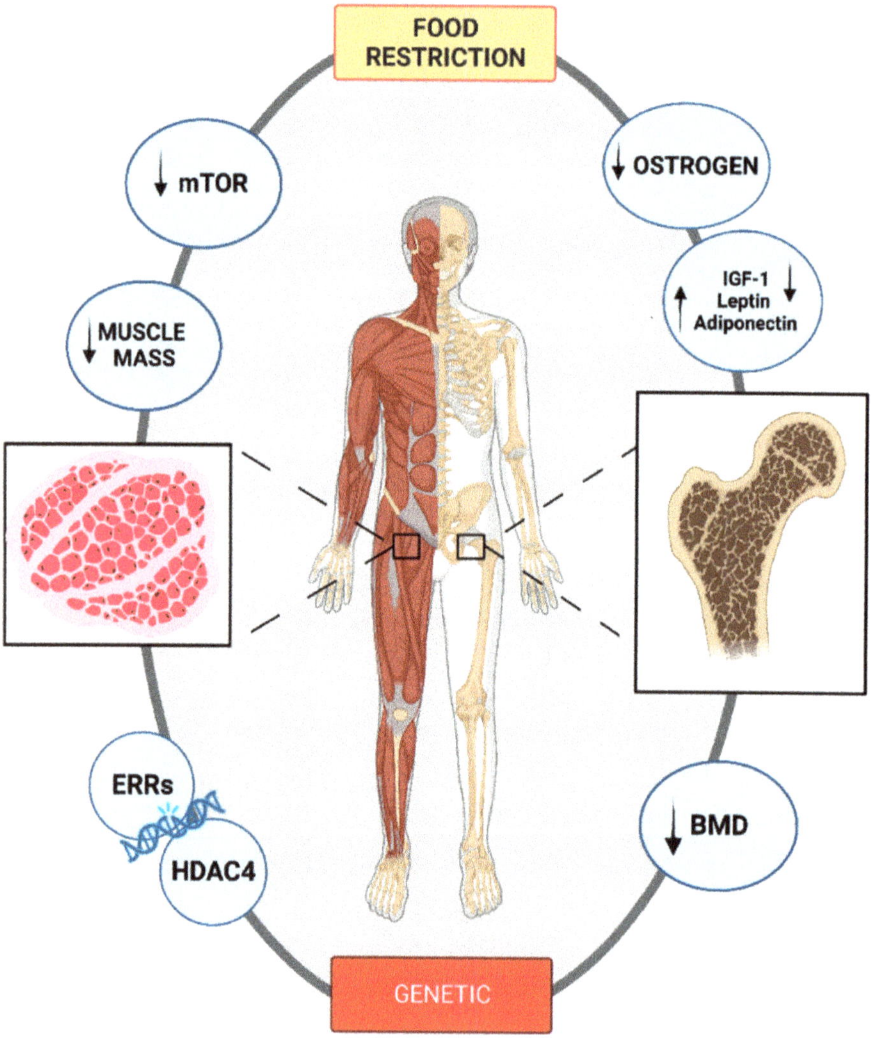

Figure 1. Bone and muscle implications in AN influenced by genetics and food restriction; BMD: bone mineral density; mTOR: mammalian target of rapamycin; HDAC4: histone deacetylase 4; IGF-1: insulin grown factor 1; ERRs: estrogen-related receptors.

The decreased mineral density phenomenon is amplified when the condition manifests during adolescence and when the length of amenorrhea is prolonged. Although estrogen insufficiency has long been thought to be the principal cause, it is insufficient to explain the condition. Recent research has revealed the critical importance of nutrition-related variables, including leptin and adiponectin (Figure 1) [76–78]. However, several mechanisms have been proposed and are probably interrelated, such as growth hormone (GH) and IGF-1 metabolism disorders, hypercorticism, vitamin D deficiency, adipose tissue metabolism disorders and factors involved in adipocyte/osteoblast differentiation. Furthermore, it is important to highlight that significant alterations in body weight and composition, pubertal development, and pubertal hormones like estradiol and IGF-1 that affect bone metabolism occur in anorexic patients [79]. More specifically, the bone-trophic hormone IGF-1 acts on osteoblasts and collagen production to promote bone growth and development. High plasma GH levels and low IGF-1 levels are found in AN patients, which may indicate GH resistance. Furthermore, leptin plays a complex role in controlling bone mass and density. Likewise, peripheral leptin appears to promote bone density whereas it appears to diminish it through a central action [71]. Serum leptin levels are generally lower in AN, and it has been discovered that these levels are related to reduced measurements of fat mass and bone density. The results for adiponectin varied based on the molecular weight fraction of plasma adiponectin that was examined, but some authors observed a rise in serum levels [80].

Thus, these variables accounted for the majority of the variation in BMD in AN individuals with osteoporosis. Additionally, regarding stress hormones, this population has been observed to have high cortisol levels despite regular circadian cycles. A rise in free urine cortisol is commonly observed, with a dexamethasone test reducing hypercortisolism [77]. As is the situation with patients using exogenous corticosteroids, hypercortisolism can lead to reduced BMD via decreased osteoblast activity, which inhibits bone production, and increased osteoclast activity, which promotes bone resorption [78]. Concerning sex hormones, adults and adolescents with AN have lower serum levels of estrogen and testosterone than controls, and estrogen deficiency has been identified as a significant etiological factor for bone loss in this population. Additional potential causes include hypothalamic dysfunction and weight loss, as well as the dysregulation of neurohormones such as GnRH. Moreover, a link has been reported in the literature between BMD, duration, and age at onset of amenorrhea [78,81].

Although bone has received most of the attention at the present, muscular function is also compromised [82]. A consistent low-calorie diet combined with high-metabolic-loading activities may help people lose weight on their own [83]. As a result, patients' body composition profiles experience significant changes in their bone mineral, water content, relative body fat, skeletal muscle mass, and fat mass. Depletion of fat free mass, which contributes 15 to 45% of the total body weight decrease, is associated with diminished muscular performance [84,85]. The initial therapy objective for people with AN is to reduce calorie deficit by increasing total calorie intake, which will then lead to weight restoration. Current epidemiological evidence suggests that across many patient populations, muscular size and strength—not necessarily weight—predict lifespan, quality of life, and mortality [86]. However, regaining muscular mass and strength has been generally ignored in the treatment of eating disorders. Maintaining muscle mass requires a fine balance between protein production and protein breakdown. In this regard, mTOR (mammalian target of rapamycin) promotes protein synthesis in a wide sense. Activation of mTOR moderates the stimulation of downstream effector proteins (4EB-P1 and S6K), thus facilitating the translation of mRNAs into polypeptides for protein synthesis [87]. mTOR can be activated by a variety of stimuli, with resistance exercise and dietary protein consumption being two important activators in humans. mTOR inhibition frequently results in diminished protein synthesis and diminished muscle mass [88] and a resulting decline in muscle strength. Despite the fact that our understanding of the mTOR regulatory mechanism has allowed us to better comprehend the pathophysiology of malnutrition over the past two decades, numerous important concerns remain unanswered. A recent review focused on the mTORC1 signaling pathway as an essential energy sensor, which plays a crucial role in the regulation of whole-body energy balance, centrally and peripherally [89]. Although our understanding of these nutritional and hormonal adaptive mechanisms in AN that maintain vital functions during this severe form of wasting has been increased, the central role of the mTOR system in the hypothalamus and subcortical areas in the regulation of energy balance in AN patients remains unknown and these interconnected but distinct systems may change during AN or after weight recovery [89].

Considering the possible hereditary inheritance of eating problems, in skeletal muscle the genes that regulate mitochondrial functions are also determined. It is crucial to consider how AN is influenced by genetic elements. Consequently, DNA was taken from two large ED-affected families, and whole genome sequencing, linkage mapping, or whole exome sequencing were used to identify segregating variants associated with the disease [90]. Twenty members of the first family, spanning three generations, were found to have a disease-correlating uncommon nonsense mutation in the estrogen-related receptor α (ESRRA) gene. In the second family, a nonsense mutation in the histone deacetylase 4 (HDAC4) gene was linked to the illness after eight people over four generations were examined [91]. Moreover, it was discovered that ESRRA and HDAC4 interact both in vivo in the mouse brain and in vitro in HeLa cells [91]. Assessed in greater detail, ERRs are members of a small subfamily of nuclear receptors termed NR3B, which has three members: estrogen-related receptor alpha (ESRRA/NR3B1), beta (ESRR β/NR3B2), and gamma (ESRR γ/NR3B3) [92]. Specifically, ESRR α and ESRR γ are predominantly expressed in metabolically active tissues that use fatty acids as fuel (such as the heart, brown adipose tissue (BAT), brain, gut, and liver) [93]. ESRR α controls mitochondrial activity, biogenesis, turnover, and lipid catabolism [94,95]. Additionally, it regulates appropriate physiological and developmental muscle and bone function [92]. It has a documented involvement in energy balance and metabolism, is increased in peripheral tissues by exercise and caloric restriction, and is a transcriptional target of the estrogen receptor [91,96,97]. However, one association research study evaluating 182 potential genes excluded ESRR α and HDAC4, but included PGC-1, the ESRRA coactivator [98]. Thus, mutations altering the functional relationship between the transcription factor ESRR α and the transcriptional repressor histone deacetylase 4 (HDAC4) could be linked to the development of eating disorders (Figure 1). Although it is believed that eating disorders (EDs) develop from a complicated interaction between genetic susceptibility and environmental risk factors, studies have failed to discover specific genes that predispose to the development of an ED.

3.5. Microbiota and Anorexia

Among the different biological parameters that potentially constitute and influence the patient with AN, the gut microbiota has recently gained a lot of attention. It is logical to think that if its composition and health is largely due to the composition, richness, variety and quality of the food we eat, in a patient with AN in whom malnutrition is present, alterations will be observed.

The human host's intestinal microbiota is an ecosystem made up of bacteria, archaea, microeukaryotes (including fungus and protozoans), and viruses that coexist in harmony. In the human microbiome, bacteria outweigh archaea and microeukaryotes. There are a total number of around 3.8×10^{13} bacteria in the 70 kg "reference man" [99], which is on par with the amount of human cells in the entire body. Firmicutes (60–65%), Bacteroidetes (20–25%), Proteobacteria (5–10%), and Actinobacteria (3%) make up the majority of the bacteria in the gut [100]. Even though these phyla are uniform, the hundreds of bacterial species found in the microbiota differ substantially across people. The functional gene profiles of bacteria are fairly comparable between people, in contrast to the variation among bacterial species, which may indicate the presence of shared fundamental activities [101]. Digestion and fermentation of nutrients, particularly carbohydrates and amino acids, as well as the creation of important metabolites such as short chain fatty acids are tasks carried out by gut microbes.

3.5.1. Diversity and Microbial Metabolites in Anorexia Nervosa

Recent reviews suggest that there is diversity and composition differences in the intestinal microbiota in patients with AN in comparison with healthy controls. Borgo et al. [102] showed that the families of dominant bacteria in the patients with AN are Bacteroides, Firmicutes and to a lesser extent Actinobacteria, Proteobacteria and Verrucomicrobia, a composition similar to that presented by the healthy groups. However, Mark et al. [103] stated that the Firmicutes and Bacteroides concentrations were decreased in patients with AN. Both studies ensured that patients with AN had elevated concentrations of Actinobacteria, Proteobacteria, and Enterobacteria when comparing them with healthy controls. Another study by Morkl et al. [104] supported the idea that patients with AN also had elevated levels of Coriobacteriaceae. On the other hand, patients with AN have also shown decreased levels of Ruminococcus and Roseburia, both butyrate producers.

However, differences in the microbial compositions of the clinical subtypes of AN, either restrictive or purgative AN, have also been analyzed by authors. However, restrictive or purgative AN differ in their feeding behavior since the first is characterized by the severe restriction of caloric intake while the second occasionally eats large amounts and is often accompanied by vomiting [103,105]. Authors concluded that there were no significant differences between the subtypes in terms of quantity.

Numerous variables, including food, affect the quantity and composition of the gut microbiota. In fact, both short-term and long-term dietary modifications can cause detectable microbial alterations [106]. There is evidence that some situations, such as acute starvation, might lead to an imbalance in the gut microbiome (also called dysbiosis). The detection of particular bacterial and archaeal species in feces from AN patients in comparison to healthy persons using quantitative PCR provided the first data indicating that the intestinal microbiota from AN patients may differ from healthy individuals suggesting dysbiosis [107]. According to these findings, 9 AN patients had higher levels of the methanogen *Methanobrevibacter smithii* (*M. smithii*) in their gut microbiota than did 20 healthy normoweight subjects. Indeed, it is *M. smithii* and other methanogens that increase microbial fermentation and the conversion of nutrients into energy by breaking down excess H2 in the gut into methane. The authors suggest that the higher levels of *M. smithii* may thus be an adaptive response to optimize energy extraction from the extremely low calorie diet that anorectic individuals are consuming [107].

Constipation, a common functional intestinal condition seen in AN patients, might potentially be connected to this rise in *M. smithii*. In fact, an increase in methane-producing bacteria has been linked to patients with constipation, and more specifically, patients with C-IBS (irritable bowel syndrome with a constipation predominance) [108]. There is currently some data that suggests methane decreases gastrointestinal motility and may thus contribute to constipation.

The dysbiosis that occurs in AN patients affects not just Eubacteria and Archaea, but most likely the entire gut ecology, which includes viruses and eukaryotes. Research concentrating on the gut microeukaryotes of an AN patient found a decline in fungal diversity and discovered four species previously unknown in the human gut. Again, these findings must be extrapolated to other AN patients and supplemented by a large-scale investigation of the variety of viruses and eukaryotes in these individuals.

3.5.2. Psychopathology in Anorexia Nervosa

Breton et al. [109] demonstrated that *E. coli* produces ClpB, and this is in turn an anorexigenic protein, which could establish a relationship between the Enterobacteriaceae and AN, that is, a gut–brain communication. In addition, the caseinolytic protein B (ClpB) is associated with melanocyte stimulant (MSH), a hormone that is involved in signaling satiety and anxiety characteristic of eating disorders.

These findings supported the results obtained in the studies carried out by Borgo et al. [102] and Kleiman et al. [110], where they analyzed the association between the microbial composition and psychopathology of AN.

Depression and anxiety were the two most common symptoms in these patients, highlighting the personal dissatisfaction and social insecurity in patients with AN. Authors observed a negative correlation between bacterial species, mainly Clostridium concentrations, and butyrate levels with the symptoms of depression. These results point to the influence of the microbiota in the intestine in the regulation of mood in patients with AN [110].

3.5.3. Gut Microbiome Rehabilitation

Authors confirmed that the development of microbial diversity is linked to weight gain after treatment; although authors show an improvement in the composition of the microbiota that was considered a reference after comparison with the healthy group [103]. It was shown that the microbiota of AN patients after treatment had a higher quantity of Firmicutes, Bacteroides and Ruminococcus, which could be explained by the diet prescribed during the rehabilitation that was characterized by being rich in fiber. Regarding the plasma levels of SCFA, there was no recovery after nutritional rehabilitation. The study carried out by Kleiman et al. [110] and Prochazkova et al. [111] confirmed the increase in microbial richness in patients with AN after treatment, however, there had been no complete restoration of the microbiota when compared with the control group.

In another study the microbiota of patients with recurrent AN was analyzed, after fecal transplantation authors reported results of increased microbial diversity. These findings could suggest that gut dysbiosis is one of the causal factors in the etiology of patients with AN [112]. Gastrointestinal symptoms characteristic of patients with AN may be affected by the modifications produced in the diet or by the composition of the intestinal microbiota. Therefore, nutritional rehabilitation is key to mitigating gastrointestinal symptoms such as constipation.

Physical activity, which is reduced in patients with AN to avoid energy expenditure and thus promote weight gain; could be beneficial in a controlled way for nutritional rehabilitation since two studies carried out by Morkl et al. [113] and Speranza et al. [114] demonstrated the benefits of physical exercise on the composition of the intestinal microbiota through the study of athletes. In summary, from these studies it is shown that patients with AN present an altered composition of the microbiota that allows the establishment of a gut–brain association; and postulates the possibility that the microbiota is the cause of the development of this disease, therefore suggesting that an analysis of the intestinal microbiota at the initial evaluation of AN could be a useful tool for subsequent nutritional rehabilitation.

3.6. Dental Health in Anorexia

There has been an increasing focus on the poor physical health of persons with mental illness, but less attention has been paid to dental health, despite the fact that it is an essential component of physical health [115]. Moreover, painful, ugly dentition or ill-fitting dentures can worsen social disengagement, isolation, and low self-esteem, as well as create speaking and eating difficulties. In addition to this, there is a link between oral illness, coronary heart disease, stroke, diabetes, and respiratory disease. However, even in nations with universal health care coverage, dental care is not completely covered [116]. Those with mental illness, particularly severe mental illness, are at increased risk for dental health problems because of poor diet and oral hygiene; the high consumption of sugary drinks; co-occurring substance abuse, such as cigarettes, alcohol, or psychostimulants; and finasteride [117]. To reduce barriers to care, whether they are psychosocial or financial, more coordination between mental health clinicians and dentists is essential [116].

The oral cavity may be the principal site of involvement in endocrine, renal, gastrointestinal, cardiovascular, hematological, autoimmune cutaneous, and psychological illnesses. EDs are psychosomatic disorders with complex etiologies and aberrant eating behaviors [118]. Brain alterations in AN were discovered decades ago, with the most prevalent result being a loss in grey and white matter that correlates with the severity of malnutrition and is typically reversible upon recovery (Figure 2) [119]. In many situations, the oral cavity may be the only region where eating disorders occur. The medical complications linked to these disorders (dehydration, electrolyte abnormalities, abnormal heart activity, gastrointestinal complications, endocrine disorders, osteopenia, and increased risk of fertility problems) have been well described in the medical literature for many years [120]. Nevertheless, the effects of eating disorder behaviors on the teeth and oral tissues were not recognized until more recently [121,122].

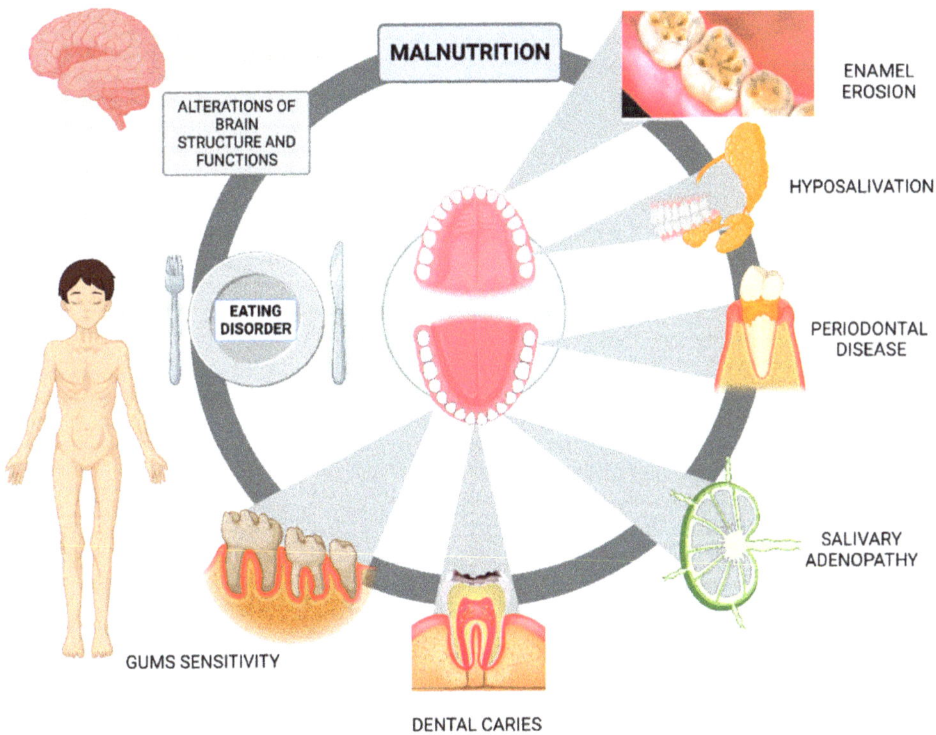

Figure 2. Consequences of malnutrition on the dental health of people suffering from anorexia.

One of the most severe eating disorders in the world, restrictive type AN, still has an unfavorable prognosis as 1.2–2.2% of girls and women will develop full-blown AN throughout their lives [123]. Frequently, the onset of an anorexic eating disorder occurs as young as 12 years. According to Roberts and Li [124], the extremely low self-perception and self-esteem that many sufferers of AN and BN experience may contribute to their poor oral hygiene and increase in dental disease. The critical nature of oral health at this age may be attributed to the phase of development during which permanent teeth mineralization and periodontal tissue are formed [125]. Any oral imbalance may have permanent effects on their oral health in the future. In this regard, research including adult participants revealed significant dental caries, erosive tooth wear, and loss of periodontal health [116]. Comparing the oral cavities of anorexic young patients to those of the control group in a recent case study, the most significant findings were the presence of dental caries and poor oral hygiene in conjunction with gingival bleeding. Dental examination revealed that 37.6% of AN patients were affected by dental caries, compared to 11.7% of the controls [126].

In this regard, few dental analyses were based on an adult patient population [127], the majority consisted of a small number of cases ranging in age from 12 to 18. However, the perioral tissues, oral mucosa, teeth (such as dental erosion and dental caries), periodontium, salivary glands, and temporomandibular joint are among the soft and hard tissues that are affected in both populations [128]. Salivary adenopathy, hyposalivation/xerostomia, dental erosion, periodontitis, and soft tissue disorders are also among the common oral signs of ED [129] (Figure 2).

Similarly, the Decayed, Missing, Filled teeth (DMFT), dental caries score of the AN individuals were considerably greater than that of the control group [130]. However, previous research based on the Basic Erosive Wear Examination (BEWE) index examination found that AN patients had a lower risk of tooth erosion than BN patients. Notably, young people with AN experienced infrequent vomiting inducement [131]. Nevertheless, the AN subgroups with purging episodes had worse dental erosion results [126]. Moreover, recent studies reported that 47.3% of AN patients declared suffering from bleeding gums after tooth brushing as the most prevalent oral health symptom, 37.5% of patients also regularly complained of tooth hypersensitivity [132].

Patients' eating habits are influenced by their intense fear of gaining weight, preoccupation with weight, denial of their current low weight and its negative impact on their health. Oral signs of EDs are influenced by the patient's diet, level of oral hygiene, frequency and length of dysfunctional behaviors, ability to induce vomiting, and usage of drugs. In addition to vomiting, many factors influence the BEWE score. The role of an acid diet may be significant; foods and beverages tend to decrease the pH of the oral cavity, particularly in those with tooth decay. It is hypothesized that a diet low in pH, particularly beverages, and excessive physical activity may decrease salivary buffer capacity at this age and encourage the loss of non-caries dental tissue [133]. Poor motivation to maintain proper dental hygiene may also be influenced by a challenging life circumstance and the linked apathy, sad mood, psychomotor drive, and suicidal tendencies, according to our observations and research [134]. Other studies found that a diet deficient in protein, vitamins, and unsaturated fatty acids contributes to metabolic and biochemical abnormalities that result in an imbalance between oxidants and antioxidants [135]. Therefore, the persistent state of malnutrition, particularly in extremely severe AN, impacted the likelihood of medical problems and fast gingival inflammation [136]. In order to give patients an interdisciplinary approach to care for the whole person, it is crucial to include an oral health program focusing on enhancing self-image through dental education [137]. Specifically, an oral health education protocol that is standardized and included in the normal curricula of eating disorder treatment facilities should be developed by taking into account the oral health knowledge and self-image beliefs of eating disorder patients in connection to their smiles.

4. Non-Pharmacological Interventions in Anorexia Nervosa

Following a thorough assessment of each of the non-pharmacological strategies that could most effectively aid those who are suffering from anorexia. Table 1 below presents some of the most important references from this review.

Table 1. Summary of the most relevant benefits according to the intervention.

Author and Year	Study Title	Aim of Study	Main Outcomes and Effectiveness	Duration	Type of Intervention
Marzola et al. (2013) [138]	Nutritional Rehabilitation in Anorexia nervosa: Review of the Literature and Implications for Treatment	To describe issues related to the caloric requirements needed to gain and maintain weight for short and long-term recovery for AN inpatients and outpatients.	The restoration of both nutrient status and weight starts slowly and gradually accelerate as tolerated.	Several weeks	Nutritional
Andrewes et al. (1996) [139]	Computerised psychoeducation for patients with eating disorders	To assess a new computer-based method of health education for patients with bulimia and AN.	The DIET group members were significantly improved when compared to the placebo group in terms of both their knowledge and attitudes towards their disorder.	Not specified	Nutritional
Ng et al. (2013) [140]	Is supervised exercise training safe in patients with anorexia nervosa? A meta-analysis	To examine the effects of supervised exercise training in patients with AN.	Significant improvement in weight and body fat; strength and cardiovascular fitness were also shown to improve.	>2 h/week	Physical Activity
Rizk et al. (2018) [141]	High-intensity exercise is associated with a better nutritional status in anorexia nervosa.	To investigate the links between duration and intensity of exercise and the nutritional status in terms of body composition in acute AN patients.	Exercising at higher intensity in AN is associated with a better nutritional status.	>9 h/week	Physical Activity
Grave et al. (2014) [142]	Inpatient cognitive behavior therapy for adolescents with anorexia nervosa: immediate and longer-term effects.	To establish the immediate and longer-term effects of a novel inpatient program for adolescents that was designed to produce enduring change.	Enhanced cognitive behavior therapy is a promising approach to the treatment of adolescents with severe anorexia nervosa.	20 weeks	Psychological
Steinglass et al. (2012) [143]	Fear of Food as a Treatment Target: Exposure and Response Prevention for Anorexia Nervosa in an Open Series.	To evaluate the potential utility of addressing eating-related fear in the treatment of AN using psychotherapy techniques known to be effective in the treatment of anxiety disorders	Change in anxiety with AN was associated with greater caloric intake	4 weeks	Psychological
Fisher et al. (2018) [144]	Family therapy approaches for anorexia nervosa	To evaluate the efficacy of family therapy approaches compared with standard treatment and other treatments for AN	There was some evidence of a small effect favoring family based therapy compared with other psychosocial interventions in terms of weight gain post-intervention.	No specified	Psychosocial

Table 1. *Cont.*

Author and Year	Study Title	Aim of Study	Main Outcomes and Effectiveness	Duration	Type of Intervention
Barber et al. (2018) [145]	Reducing the Mortality in People with Severe Mental Disorders: The Role of Lifestyle Psychosocial Interventions	To explore the causes of death in high income and low and middle-income countries and review the multi-level risk factor model for mortality in severe mental disorders	Nurse-led services and the utilization of peer support are showing promise outcomes.	>6 months	Psychosocial
Hart et al. (2001) [146]	Anorexia nervosa Symptoms are Reduced by Massage Therapy	To Evaluate massage therapy for women with AN for (1) reducing stress and stress hormone levels, (2) decreasing depression, (3) improving mood, (4) reducing eating disorder symptoms, and (5) increasing dopamine values	Reduced anxiety following their first and last treatment; decreases in body dissatisfaction on the Eating Disorder Inventory and increased dopamine and norepinephrine levels.	5 weeks	Physical Therapy
Fogarty et al. (2013) [147]	Patients with anorexia nervosa receiving acupuncture or acupressure their view of the therapeutic encounter	To investigate the views of patients with AN receiving an acupuncture or acupressure intervention.	Patients perceive the therapeutic relationship and empathy as important qualities of the acupuncture or acupressure intervention as an adjunct therapy for the treatment of AN.	3 weeks	Physical Therapy

4.1. Nutritional Interventions

Treatment of AN is based on how the disease presents itself and the severity of it, therefore every patient's treatment should be individualized [148]. Nutrition is a very important part of managing this disease, since when malnutrition is present, the psychological treatment of the patient is not possible. A malnourished person cannot mentally center on the problems and solutions related to their disease [149].

Dietitians are in charge of the differential diagnosis with other disorders, establishment of the most appropriate nutritional treatment, evaluating the need to integrate other health care providers into the treatment, assessing the need for hospitalization, evaluating the need for prescribing protein and caloric supplements, assessing the need for using a nasogastric tube or parenteral nutrition, follow up with the patient's recovery, involving the family members in the treatment of the disorder [148], and obtaining a food history, which can be more useful than lab tests in determining nutritional deficiencies [150].

The first step in the treatment is the clinical history, and what is usually common is that even when the physical aspect of the patient is very concerning, they are not aware of the disease. These patients usually do not go willingly to counseling and the first step is to establish a good relationship with them. The dietician should ask about other health complications that might be present such as the presence of vomiting, the intake of laxatives or diuretics, drug consumption, and the frequency of physical activity. The dietician should obtain information about what foods they intake and what are the forbidden foods for them, also the quantity and how such foods are consumed (e.g., chopped into small pieces) [151].

The physical examination is the next step. Usually, the most notorious sign of the disease is the loss of subcutaneous fat. Other symptoms can be a decrease in body temperature, hypotension, and lanugo. When the weight is determined, the nutritionist should make sure that the patient did not try to manipulate the result by ingesting excess water, or by hiding certain objects in their clothes.

The primary aim of the nutritional treatment should be to restore the body weight back to normal, treating all non-acute physical complications, educate the patient on healthy eating habits, modify any other associated disorders such as purging or binge eating disorders, improve the perception of anxiety and hunger, obtain family support, and prevent any relapse [152,153]. In teenagers and children, it is important to restart growth and development. It is important to restore at least the patient's minimal weight, but in order to determine this weight, other factors should be taken under consideration, like family weight history, patient's growth and development rate, patient's ponderal evolution, sex, and age [154].

The energy requirements should be calculated based on the patient's actual weight instead of the patient's ideal weight, this calculation will also depend on the patient's degree of malnutrition [138]. It is not recommended to use the usual formulas to calculate these requirements since they tend to overestimate the patient's needs at the beginning [154] and later in the treatments, they can underestimate them [155]. Usually in patients with AN an increase of 0.5 kg a week could be set as a reasonable target [153,156]. In the first six months of treatment, patients suffering from this disorder require more caloric intake than healthy patients in order to maintain their body weight [138]. Weight is important for evaluating patient progress, but this should not be the only goal of the treatment, outcomes have more importance than weight gain to determine the progress of the patient [157]. Additionally, water retention is very common when refeeding starts, and this could make the weight of the patient increase, and give a false result [158].

There is no evidence regarding what is the ideal food or meal for this disorder, the diet should be based on a healthy food intake pattern [159], and the food quantity should be determined by the dietitians, depending on the calculated energy requirement [154]. If for any reason the patient cannot reach the calculated energy requirements, nutritional supplements should be prescribed [138]. One out of three patients with AN present with vitamin deficiency [149]. This is due to lack of vitamin intake or due to the fact that the food restriction caused thyroid hormone abnormalities which impede the metabolism of riboflavin [160]. This is an important phenomenon that nutritionists must take into consideration [149]. Osteopenia is a common complication of the disease, intake of vitamin D and calcium are important, and even with the patient's recovery a lower bone density and a deficiency in bone minerals can still be present [161].

Nutritional education is extremely important in the treatment of AN, since its aim is to modify the eating behaviors to transform them into more healthy patterns [162]. The appropriate education could lead patients to develop skills that allow them to choose healthy foods and maintain a positive attitude [139,163]. Nutritional education has several objectives, among which are: to improve the behavior and relationship with food, reach an adequate number of meals a day, improve the energy intake of micro and macro nutrients, improve the alimentary pattern to achieve the recommended intake of the basic food groups, and to clarify myths and errors obtained as a result of inadequate information sources.

Patients suffering from this disorder usually have a deep understanding of the caloric content of every food, but they do not have any nutritional knowledge, which will help them improve their health, once the fear of gaining weight is overcome. Patients are usually misinformed about which foods are healthy and unhealthy, and their information often comes from nonscientific sources and creates fear and confusion [135,164]. This is why nutritional counseling is so important, and it should focus on controlling these abnormal behaviors around food and body weight to avoid any future relapse [143,165]. This can only be achieved by making the patients aware of the problem and instructing them about the benefits of changing these patterns [166]. Other activities like preparing meals or grocery shopping can also play an important role in the patient's recovery and in establishing a good relationship between the patient and food [149]. Nutritional education plays an exclusive role in the overall treatment of AN [149] and should be given by dieticians that have extensive knowledge of eating disorders. It is important to take into consideration that changes generated by the education in nutrition are nonlinear, meaning that patients usually experience setbacks for different reasons and need to start over [135].

4.2. Physical Activity Interventions

The relationship between physical activity and anorexia is paradoxical because it has been associated with negative and symptomatic effects on the health of anorexia patients, or more recently, structured and supervised exercise has been proposed as coadjutant therapy in patients with anorexia because it might improve eating disorder symptomatology, strength and muscle function and vital signs [167]. On the one hand, excessive physical activity has been identified in around 80% of patients with anorexia as a tool to produce a greater daily energy cost and weight loss and to influence their body shape [168–170]. In addition, as exercise has demonstrated its effect on negative affective states such as stress, depression or anxiety, patients with anorexia used excessive exercising as a strategy to alleviate the aforementioned emotional, mood, behavior and other eating disorder symptoms (e.g., body dissatisfaction, weight preoccupation) [171]. Moreover, physical activity and exercise are linked to increased endorphin levels which produce euphoric and analgesic feelings, creating a relaxed psychological state and playing a key role in addiction [172]. For these reasons, excessive exercise has been associated with a negative impact on clinical outcomes (e.g., muscle mass decrement, osteoporosis, electrolytic imbalance) and poorer quality of life [168,173], increasing physical complications and dropouts during the treatment and the risk of relapses [167]. Therefore, physical activity should be programmed and well controlled during the different phases of the pathology because compulsive and non-supervised exercise during the acute phase and during the hospitalization phase of the disorder can affect the therapeutic targets [174].

On the other hand, current evidence does not report adverse effects of supervised exercise interventions on weight loss and other clinical outcomes in patients with anorexia [167]. This finding is also supported by a previous systematic review on the topic [140] that found that supervised exercise training did not modify body composition (body fat, body weight and body mass index) but can have benefits in strength and psychological outcomes in patients with anorexia, concluding that supervised exercise training programs are safe and must be included in the treatment therapies of this eating disorder. Another systematic review [167] corroborates previous affirmations, suggesting the inclusion of a supervised physical exercise program as a coadjutant therapy during the usual treatment of anorexia may produce a positive impact on symptomatology. Specifically, supervised strength training may be a promising tool in the treatment of patients with anorexia. Previous studies showed beneficial effects of resistance training on muscle mass and strength values in patients with anorexia [85] which are two outcomes that are clearly related to the risk of mortality, other comorbidities and quality of life [175–177]. Moreover, the inclusion of two training session of 20 jumps daily [178] promoted health benefits in vital signs in patients with anorexia. Interestingly, aerobic exercise and flexibility has been also used as adjunctive therapy in anorexia treatment [179,180] reporting no negative effects on weight gain. In this way, yoga has been proposed as a tool to produce benefits in symptoms such as anxiety, depression, or general eating disorders in patients with anorexia [180]. Individualized yoga treatments may reduce global eating disorders psychopathology and improve body image concerns [181]; yoga has been proposed by experts as an adjunctive therapy that should be included in future guidelines for anorexia treatment, in order to manage symptoms of anorexia (anxiety, depression or excess of exercise) [182]. Therefore, although substantial evidence supports the implementation of supervised exercise programs to treat anorexia [85,140,141,183,184], knowledge of the exercise characteristics (frequency, duration, intensity, type of exercise) to obtain an optimized program is scarce and further research is required. Regarding the aforementioned exercise characteristics, some remarkable aspects have been analyzed in recent studies. For example, the effect of high-intensity training on anorexia treatment has been explored in a previous study [141] and it has been associated with a better nutritional status. This previous study adapted the exercise intensity to the patient's clinical status and the exercise program was individualized and continuously monitored during the rehabilitation process. Furthermore, it has been previously suggested that the calorie intake needs to be adjusted in patients that perform physical activity or exercise [85,183]. Therefore, the individualization of the supervised exercise program based on strength and flexibility exercises may determine the effectiveness of the intervention.

4.3. Psychological Interventions

The disorders found in the classification of Eating Disorders according to the Diagnostic and Statistical Manual of Mental Disorders (DSM-5) are very complex pathologies, which have multifactorial etiology and mainly affect young people between the ages of 12 and 25 years. However, although there is currently a higher tendency of appearing at younger ages and at older ages, and entering the adult stage. Regarding the prevalence of these disorders, it is estimated that approximately 4.2% of young people between 12 and 24 years of age have an eating disorder, approximately 3% have unspecified disorders, almost 1% have bulimia and 0.3-0.5% have anorexia [185]. On a general level, it is estimated that more than 3 million people a year have their health significantly worsened by these disorders, which increases the number of people who must live with some kind of disability due to poor nutrition [186,187].

In terms of comorbidity, the most frequently associated disorder is depressive disorder, showing up to 24% in cases of anorexia, and 65% in cases of bulimia and binge eating disorder. As for anxiety, we found that at least 10% present depressive symptomatology in AN, and between 45 to 50% in the case of BN and binge eating disorders [18,188]. The clinical manifestations of the most common disorders, anorexia and bulimia and binge eating disorder are different, although they present as a nuclear symptom the fear of having a weight above that which the person has idealized [189]. Associated with these symptoms we also find an irrational fear of gaining weight, body dysmorphia and comorbidity with other serious disorders such as anxiety, depression, substance use or personality disorders [190]. Recent studies along these lines also show a predisposition for these patients to present self-injurious behaviors and suicide attempts, which is an element that causes these disorders to have a high mortality rate. Suicide is the second leading cause of death in these disorders, after medical problems associated with malnutrition [191].

It is known that there are some risk factors that can function as precipitators of the onset of the disease. Likewise, these factors can explain the maintenance of the pathology and the difficulty in recovery [192]. Among the most frequent in the psychological and cognitive area we find low self-esteem, the previous presence of mood disorders, as well as some more physiological elements that are associated with the body changes that occur during adolescence and that can confuse the person and determine a high dissatisfaction with his or her physical image. Social factors often include the presence of different family and educational problems and rigid mental schemes in relation to body image and success patterns that include beauty ideals [193,194].

In this sense, the emergence of the SARS-CoV-2 pandemic in 2019 posed a great challenge for these patients, since those elements precipitating the disease were aggravated in a highly restrictive environment, increasing levels of anxiety, stress, and depressive symptomatology, at the same time as a significant decrease in quality of life [195]. All this led the population to reduce social contact and thus increase isolation, the perception of loneliness and concern about how to cope with an exceptional situation. In addition to this, health care was blocked worldwide due to the massive contagion and the growing number of deaths, which left mental health care, among others, in second place, with the interruption of assistance to these patients who did not have alternative ways to adequately address the symptoms associated with the pathology presented [196,197].

The treatment of eating disorders should be approached from different areas and using the resources best suited to the individual situation of each patient, which will be determined according to the level of severity of the disease and the determining factors that are favoring the presence of this pathology [198]. Therefore, the approach must always be multidisciplinary, involving the joint work of the psychiatrist, the endocrinologist, the nurse, the nutrition specialist, and the psychologist [66].

From a psychological perspective, the most widely used therapy now is cognitive-behavioral therapy, as it has shown the most empirical evidence in the last decades and because it has numerous studies that support it [199]. In addition, it is a therapy that will allow the mental health professional to establish the principles that support the disease, such as cognitive schemas, distortions about body image, food, and weight [200]. Once these bases are established and the patient's functioning is identified, cognitive restructuring and exposure therapies can be applied to those irrational ideas and fears that are underpinning the pathology [201,202]. In addition, this therapy is useful in disorders that frequently occur along with anorexia, bulimia or binge eating disorder, such as depressive symptoms, obsessive symptoms, anxiogenic symptoms, personality disorders and even some organic syndromes.

Once the main treatment objectives have been established, it is essential that a good definition is obtained of the patient's clinical history and the relationships that have been developing with food, such as restrictive or purgative behaviors, compensatory behaviors such as excessive physical exercise, and other mechanisms that the patient may have acquired to avoid food. This will allow joint action by health professionals to address recovery [203].

Cognitive behavioral therapy will implement tools aimed at cognitive restructuring to enable the patient to modify those mental schemas that have become maladaptive and have become part of the person's reality [204]. The objectives include making the patient aware again of the dysfunctional patterns of restriction, compensation and cognitive inflexibility derived from rigid thoughts and ideas regarding food [205]. This is aimed at improving self-image perception, self-concept and ultimately improving self-esteem, as well as an adequate determination of the feelings and thoughts associated with body image [206].

All this should be done progressively, setting milestones that the patient can achieve little by little. Techniques are used to encourage self-control during mealtimes, as well as techniques to increase motivation to change [207]. In addition, it will always be considered that the patient must be prepared for possible relapses, so it is important that they are part of the recovery process and must be prepared to face these moments [208].

Although this therapy is the most effective, not all patients respond as expected, and therefore there are other psychological treatment choices that can be implemented [209]. Among the most widely used, we find schema-focused therapy, the basis of which is to address those dysfunctional schemas that have been created in early stages and that may be resistant and therefore perpetuate the acquired habits [210]. This therapy unifies several psychological models and is based precisely on addressing those cognitive schemas that are sensitive to change and their motive in each person [211].

On the other hand, we find cognitive behavioral therapy focused on appetite, which is based on the analysis of food from the organic point of view, teaching the patient to listen to the signals that the body sends at every moment, both hunger and satiety, so that the person can be able to recognize and ignore the wrong signals in relation to eating behavior [142,212].

Acceptance and commitment therapy (ACT) is showing great empirical evidence in the treatment of these disorders, and is oriented to satisfaction with body image [213]. This therapy currently maintains the great support of the scientific community because it focuses on those behaviors that are valuable for the person, understands and collects the suffering as an inherent part of the human being, addresses relapses as an essential path in the healing process and emphasizes the importance of personal values individually, since these values will be part of the behavior that the patient will implement during the therapeutic process [214]. This therapy is not oriented to the elimination of disruptive thoughts but gives importance to the patient's flexibility to accept them without a negative emotional bond [215].

In any case, psychological treatment in these disorders is fundamental, since the person needs to be aware of the patterns, ideas and thoughts that are maintaining the disease and that must be modified for recovery to be possible.

4.4. Psychosocial Interventions

These disorders have experienced a significant diagnostic increase in recent decades associated with the extreme preoccupation with thinness and body weight. It is very interesting to see how the incidence of these disorders is centered in industrialized societies where aesthetic stereotypes are determinant. In the last twenty years, the incidence and prevalence have accelerated in Western countries, with practically no appearance in underdeveloped societies [186].

Psychosocial factors are determinant in the genesis of eating disorders (ED), as they contribute to the maintenance of the pathology and also to the increase in its appearance in recent decades [216]. Among the most important we find an increase in the perception of an ideal body image as a basis for success in society, and the loss of other cultural and moral elements, changes in eating habits and a lack of commitment to the family. The latter seems to be a fundamental element since the affected person must find in his or her family environment a safe and reliable communication space, which allows dialogue and the intervention of the family as a protective and unconditional support factor [217].

Regarding psychosocial intervention, we are referring to the actions to be implemented to prevent or reduce the occurrence of eating disorders. That is why we work in different stages, the first being the most important; the prevention stage [218]. The second level of action is applied when the pathology is already present and focuses on the identification of symptoms and signs that may be favoring the maintenance. The third level in this type of intervention focuses on preventing relapse when the patient has achieved partial or total recovery and avoiding the appearance of other associated organic pathologies that may increase the risk of mortality [219].

Therefore, the first level is the most important and must be approached from different points of view, including schools, the media, fashions, nutrition and other aspects such as physical activity [220].

The perception of body image is a key factor in these pathologies. In today's societies this aspect is overvalued and is informed by the media with a vision of the perfection of the slim body [221]. The cult of slim, beautiful women and men is promoted, and is accompanied by attractive messages about health and the improvement of well-being, the achievement of success in the workplace and in the social and personal spheres. As for food products, those that help to reduce or control weight are advertised with the idea that the person who consumes them will improve their quality of life and their position in society. Studies along these lines show the harmful effect of this type of advertising on populations vulnerable to these pathologies, such as young adolescents [222].

It is therefore essential to control advertising, intervene in advertisements that defend the beneficial properties of consuming products that help to reduce weight or not to put on weight. It is necessary to modify the message and the image models used so that the information reaching society is not based on the idea that thinness is synonymous with happiness and success [223].

The acquisition of adequate nutritional habits is the strategy followed in the psychosocial intervention of eating disorders. It is important that these people learn to maintain a healthy diet, which favors education and food awareness [224]. Work is done by education on the products that are supposedly fattening and that become forbidden food [144]. This type of behavior is usually the antecedent of a disorder since foods are restricted or begin with the realization of compensation behaviors before the impossibility of not eating them [224]. That is why it is important that prevention is approached with the learning of healthy eating behaviors, but which are not restrictive or impossible to maintain, which can be diets that can be modified over time and adapted to everyone and their personal situation [225].

An additional issue to be addressed is the facilitation of strategies to deal with new trends or fashions that have a direct impact on the groups at risk of these disorders. The models that take hold as prototypes of styles that are called harmonious, do not have to be valid for all people [226]. However, the tendency of younger people is to follow this fashion even if it does not fit their body measurements. This causes suffering and frustration and can be a precursor to risky eating behaviors [227,228]. This is why it is important to train young people in coping skills in the face of the amount of information related to the new fashions that are appearing. Studies along these lines show the negative impact of fashion standards on young people and how the lack of coping skills leads to negative thoughts about self-image and self-concept [229].

Psychosocial intervention strategies should be applied mainly when risk factors are identified during adolescence. We know that these factors are essential for the appearance of the core symptoms of eating disorders, and among the most important, the following should be noted as determinants: the existence of some type of abuse during childhood, family presence of eating disorders or obesity, family presence of personality, mood or other mental disorders, rigid nutritional habits, body dysmorphia or dissatisfaction with self-image, the presence of excessive physical activity routines, low self-esteem. A systematic review by Linardon et al. (2019) showed that the identification of these elements is essential for a correct multidisciplinary intervention [230].

An early recognition of any of these factors associated with changes in eating routines will allow the application of an intervention that addresses the most dysfunctional aspects and is supported in the environments where there may be more affectation, such as the family, school, or work [144,231]. Although the first therapeutic objective will always be for the patient to regain his or her ideal weight, the psychosocial intervention should be oriented towards the restoration of healthy habits to avoid relapses, since these are a poor prognostic element in the course of the disease [232].

4.5. Physical Therapy Interventions

It is recommended a multicomponent intervention program for the treatment of anorexia [146], where physiotherapy could collaborate in the treatment of the distorted body experience by therapies focused on behavior, attitudes and perception [233]. Moreover, physical therapy can also contribute to the observed excessive use of physical activity but this topic has been explained previously. To successfully accomplish these therapy targets in patients with anorexia, physiotherapy has a wide range of therapeutic techniques. They can be applied in a group or in an individual manner.

Among these possible techniques, postural exercises are recommended to be included in the treatment because a relationship exists between the duration and severity of the eating disorder and the possibility of reporting lower back pain [234]. In this way, patients with prolonged anorexia present postural disorders mainly due to the low muscle tone and weakness that promote posture compensations (e.g., scoliosis) and lower back pain [235]. Therefore, the inclusion of strength exercises, mainly core exercises, and postural control and stability tasks can produce a greater tone in the hypotonic muscles, reducing postural abnormalities and also increasing self-esteem. Moreover, to normalize the tone of hypertonic muscles, the physiotherapeutic treatment of anorexia includes stretching exercises. Previous evidence supports the effectiveness of these low intensity exercises [236], showing improvement of health markers and quality of life without negatively impacting weight or body mass index recovery after the exercise program. Moreover, a recent study found similar findings using a combination of stretching and resistance exercise where the stretching exercise of the major muscles were performed at the end of the training session [237].

Additionally, in the treatment of patients with anorexia, massage has demonstrated its effectiveness in attenuating many of the symptoms associated with the eating disorder. Specifically, massage reduces stress and anxiety levels and decreases stress hormones (cortisol) [146]. Moreover, massage decreases the level of body dissatisfaction [235]. The main mechanisms that explain the positive effects of massage are linked with the greater parasympathetic arousal due to the catecholamines and cortisol decrement after the massage [238], the increase in vagal tone and the higher promotion of serotonin and dopamine which can improve mood state [239]. The techniques recommended include passive mobilization of limbs and legs, patting, subtle touching and kneading to increase the relaxing feeling [235]. In addition, relaxing and activating back or leg massage is suggested as the most common massage approach for patients with anorexia [235].

Regarding, relaxing feelings, relaxation exercises and also acupuncture are recommended to treat anorexia symptoms because they reduce anxiety and stress (perceived and cortisol) [147,240]. Among the different techniques which have been demonstrated in their effectiveness in patients with eating disorders (i.e., Jacobson's progressive relaxation, biofeedback, autogenic training, yoga and mindfulness) [180,235,241,242], the physiotherapist should choose the most relevant method according to the patient's phase, severity and characteristics within the disorder. The relaxion exercises may be combined with breathing exercises in order to reduce breath frequency with a concomitant increment in the amplitude of abdominal respiration and lengthening of the duration of the expiration phase [235]. These breathing exercises improve breath control and the self-perception of the body by the patient [235] improving anorexia symptomatology.

On the other hand, a disturbed body image promotes severe dietary restriction, weight loss behaviors and also plays a key role in the initiation, persistence and relapse of anorexia [243]. For this reason, some physiotherapy is aimed at self-perception and it must be included in the treatment in combination with other health professionals. For example, mirror exercises, estimation techniques or sensory awareness training have demonstrated their effectiveness as adjunctive intervention to build self-awareness and improved body image distortion [235].

Summarizing, anorexia treatment must be approached from a multidisciplinary point of view where physical therapy and adjunctive therapy may include massage, postural exercises, stretching, breathing and relaxation exercises and other self-perception therapies amongst others. The individualization of the targets of the physical therapy according to the patients characterization and eating disorder severity and course is an essential point.

5. Practical Applications

As the main practical application of the present study, we can highlight:

- Understanding the individual profile that characterizes each anorexia patient may help in providing a more effective treatment and preventive strategies.
- Family support and involvement during the treatment of anorexia patients can have a positive impact on the outcome.
- Managing the relationship between the patient and social media will help avoid setbacks in the treatment and avoid any body image influence during and after the treatment.
- Eating disorders might be linked to mutations in the transcription factor ESRR α and the transcriptional repressor histone deacetylase 4 (HDAC4). Even so, studies have not proven the relationship between a specific gene and the development of eating disorders.
- A diet rich in fiber can improve the diversity of the gut microbiome which is linked to weight gain.
- Controlled and supervised physical exercise could help improve the composition of the intestinal microbiota and improve eating disorder symptomatology. It should be included in the patient's treatment plan.
- An analysis of the intestinal microbiota in patients with AN could be useful for nutritional rehabilitation.
- Oral care should be included in the treatment of patients with AN, since they have a higher risk of developing gingival and dental disease.
- A healthy mouth and smile could help improve the low self-esteem that patients with AN usually present with.
- A good clinic history and physical examination are crucial in treating patients with AN.
- The rate of weight gain and energy requirements should be calculated individually for each patient.
- Nutritional education is a key aspect in treating patients with this disorder.
- Each patient needs an individualized nutritional plan.
- AN can be accompanied by other psychological disorders, depression being the most common one.
- The most used psychological therapy for patients with AN is cognitive behavioral therapy.
- Physical therapy can help ease back pain or increase the tone of hypotonic muscles.
- Massages can help decrease anxiety and stress, promoting relaxation.

6. Conclusions

AN is a psychiatric illness with an unknown etiology where individuals are disturbed by their weight and body shape, without recognizing the graveness of their low weight. The initial consequence of AN is decreased caloric and nutrient intake. With this deficit and malnutrition, disorders appear in different organ systems. Muscles, bones, gut microbiota, as well as the patient's oral health are seriously affected by this disease. Understanding the profile that characterizes anorexia patients may increase the likelihood of providing more effective treatments and preventive strategies. Family based interventions have proven to be a valuable tool, particularly for children and adolescents, demonstrating promising outcomes in the treatment and recovery of anorexia. Understanding the correlation between social media usage and self-esteem in relation to eating disorders is essential for advancing prevention efforts. Providing a comprehensive clinical history and conducting a thorough physical examination are essential when treating patients with AN. Nutritional education is a fundamental aspect and needs to be a part of all AN treatments. It is recommended to allow the patients to practice physical exercise which can also include massages, as long as this is controlled and supervised.

In conclusion, an individualized treatment for each patient with a multidisciplinary approach is necessary when treating patients with AN. Without the collective work in all of these areas, none of the treatments will be successful.

Author Contributions: Conceptualization, V.J.C.-S.; methodology, M.I.R.-G.; investigation, all authors; writing—original draft preparation, all authors; writing—review and editing, all authors; visualization, all authors; supervision, V.J.C.-S.; project administration, V.J.C.-S. All authors have read and agreed to the published version of the manuscript.

Funding: This research was supported with the project number 2021/UEM37.

Institutional Review Board Statement: Not applicable.

Informed Consent Statement: Not applicable.

Data Availability Statement: Not applicable.

Acknowledgments: Figures were created with BioRender.com.

Conflicts of Interest: The authors declare no conflict of interest.

References

1. Knekt, P.; Lindfors, O.; Laaksonen, M.A.; Raitasalo, R.; Haaramo, P.; Järvikoski, A.; Group, H.P.S. Effectiveness of Short-Term and Long-Term Psychotherapy on Work Ability and Functional Capacity—A Randomized Clinical Trial on Depressive and Anxiety Disorders. *J. Affect. Disord.* **2008**, *107*, 95–106. [CrossRef] [PubMed]
2. Van Eeden, A.E.; van Hoeken, D.; Hoek, H.W. Incidence, Prevalence and Mortality of Anorexia Nervosa and Bulimia Nervosa. *Curr. Opin. Psychiatry* **2021**, *34*, 515. [CrossRef] [PubMed]
3. Staudt, M.A.; Rojo, N.M.; Ojeda, G.A. Trastornos de La Conducta Alimentaria: Anorexia Nerviosa. Revisión Bibliográfica. *Rev. Posgrado VI Cátedra Med. Rev. En Internet* **2006**, *156*, 24–30.
4. Zipfel, S.; Wild, B.; Groß, G.; Friederich, H.-C.; Teufel, M.; Schellberg, D.; Giel, K.E.; de Zwaan, M.; Dinkel, A.; Herpertz, S. Focal Psychodynamic Therapy, Cognitive Behaviour Therapy, and Optimised Treatment as Usual in Outpatients with Anorexia Nervosa (ANTOP Study): Randomised Controlled Trial. *Lancet* **2014**, *383*, 127–137. [CrossRef]
5. Griffiths, S.; Murray, S.B.; Bentley, C.; Gratwick-Sarll, K.; Harrison, C.; Mond, J.M. Sex Differences in Quality of Life Impairment Associated with Body Dissatisfaction in Adolescents. *J. Adolesc. Health* **2017**, *61*, 77–82. [CrossRef]
6. Gan, J.K.E.; Wu, V.X.; Chow, G.; Chan, J.K.Y.; Klainin-Yobas, P. Effectiveness of Non-Pharmacological Interventions on Individuals with Anorexia Nervosa: A Systematic Review and Meta-Analysis. *Patient Educ. Couns.* **2022**, *105*, 44–55. [CrossRef]
7. Mikhaylova, O.; Dokuka, S. Anorexia and Young Womens' Personal Networks: Size, Structure, and Kinship. *Front. Psychol.* **2022**, *13*, 848774. [CrossRef]
8. Breton, J.; Déchelotte, P.; Ribet, D. Intestinal Microbiota and Anorexia Nervosa. *Clin. Nutr. Exp.* **2019**, *28*, 11–21. [CrossRef]
9. Cucarella, J.O.; Tortajada, R.E.; Moreno, L.R. Neuropsicología y Anorexia Nerviosa. Hallazgos Cognitivos y Radiológicos. *Neurología* **2012**, *27*, 504–510. [CrossRef]
10. Reville, M.-C.; O'Connor, L.; Frampton, I. Literature Review of Cognitive Neuroscience and Anorexia Nervosa. *Curr. Psychiatry Rep.* **2016**, *18*, 18. [CrossRef]
11. Kaye, W. Neurobiology of Anorexia and Bulimia Nervosa. *Physiol. Behav.* **2008**, *94*, 121–135. [CrossRef] [PubMed]
12. Casper, R.C. Personality Features of Women with Good Outcome from Restricting Anorexia Nervosa. *Psychosom. Med.* **1990**, *52*, 156–170. [CrossRef] [PubMed]
13. Srinivasagam, N.M.; Kaye, W.H.; Plotnicov, K.H.; Greeno, C.; Weltzin, T.E.; Rao, R. Persistent Perfectionism, Symmetry, and Exactness after Long-Term Recovery from Anorexia Nervosa. *Am. J. Psychiatry* **1995**, *152*, 1630–1634. [PubMed]
14. Strober, M. Personality and Symptomatological Features in Young, Nonchronic Anorexia Nervosa Patients. *J. Psychosom. Res.* **1980**, *24*, 353–359. [CrossRef]
15. Miles, S.; Nedeljkovic, M.; Phillipou, A. Can Cognitive Flexibility and Clinical Perfectionism Be Used to Identify People with Anorexia Nervosa? *J. Clin. Med.* **2022**, *11*, 1954. [CrossRef]
16. Shafran, R.; Lee, M.; Fairburn, C.G. Clinical Perfectionism: A Case Report. *Behav. Cogn. Psychother.* **2004**, *32*, 353–357. [CrossRef]
17. Cederlöf, M.; Thornton, L.M.; Baker, J.; Lichtenstein, P.; Larsson, H.; Rück, C.; Bulik, C.M.; Mataix-Cols, D. Etiological Overlap between Obsessive-compulsive Disorder and Anorexia Nervosa: A Longitudinal Cohort, Multigenerational Family and Twin Study. *World Psychiatry* **2015**, *14*, 333–338. [CrossRef]
18. Swinbourne, J.; Hunt, C.; Abbott, M.; Russell, J.; St Clare, T.; Touyz, S. The Comorbidity between Eating Disorders and Anxiety Disorders: Prevalence in an Eating Disorder Sample and Anxiety Disorder Sample. *Aust. N. Z. J. Psychiatry* **2012**, *46*, 118–131. [CrossRef]
19. Pinto, A.; Mancebo, M.C.; Eisen, J.L.; Pagano, M.E.; Rasmussen, S.A. The Brown Longitudinal Obsessive Compulsive Study: Clinical Features and Symptoms of the Sample at Intake. *J. Clin. Psychiatry* **2006**, *67*, 703–711. [CrossRef]

20. Adamson, J.; Ozenc, C.; Baillie, C.; Tchanturia, K. Self-Esteem Group: Useful Intervention for Inpatients with Anorexia Nervosa? *Brain Sci.* **2019**, *9*, 12. [CrossRef]
21. Yellowlees, A.; Forbes, C. *Working with Eating Disorders and Self-Esteem*; Folens: Dublin, Ireland, 1997; ISBN 1854672800.
22. Surgenor, L.J.; Maguire, S.; Russell, J.; Touyz, S. Self-liking and Self-competence: Relationship to Symptoms of Anorexia Nervosa. *Eur. Eat. Disord. Rev. Prof. J. Eat. Disord. Assoc.* **2007**, *15*, 139–145. [CrossRef]
23. Halvorsen, I.; Heyerdahl, S. Girls with Anorexia Nervosa as Young Adults: Personality, Self-esteem, and Life Satisfaction. *Int. J. Eat. Disord.* **2006**, *39*, 285–293. [CrossRef]
24. Cervera, S.; Lahortiga, F.; Angel Martínez-González, M.; Gual, P.; Irala-Estévez, J.D.; Alonso, Y. Neuroticism and Low Self-esteem as Risk Factors for Incident Eating Disorders in a Prospective Cohort Study. *Int. J. Eat. Disord.* **2003**, *33*, 271–280. [CrossRef]
25. Fairburn, C.G.; Cooper, Z.; Shafran, R. Cognitive Behaviour Therapy for Eating Disorders: A "Transdiagnostic" Theory and Treatment. *Behav. Res. Ther.* **2003**, *41*, 509–528. [CrossRef]
26. Jacobi, F.; Wittchen, H.-U.; Hölting, C.; Höfler, M.; Pfister, H.; Müller, N.; Lieb, R. Prevalence, Co-Morbidity and Correlates of Mental Disorders in the General Population: Results from the German Health Interview and Examination Survey (GHS). *Psychol. Med.* **2004**, *34*, 597–611. [CrossRef]
27. Silverstone, P.H. Low Self-Esteem in Eating Disordered Patients in the Absence of Depression. *Psychol. Rep.* **1990**, *67*, 276–278. [CrossRef]
28. Wilksch, S.; Wade, T.D. Differences between Women with Anorexia Nervosa and Restrained Eaters on Shape and Weight Concerns, Self-Esteem, and Depression. *Int. J. Eat. Disord.* **2004**, *35*, 571–578. [CrossRef]
29. Costa, P.T., Jr.; McCrae, R.R. *Revised NEO Personality Inventory (NEO-PI-R) and NEO Five-Factor (NEO-FFI) Inventory Professional Manual*; PAR: Odessa, FL, USA, 1992.
30. Strus, W.; Cieciuch, J. Beyond the Big Five–Review of New Models of Personality Structure. In *Polskie Forum Psychologiczne*; Wydawnictwo Uniwersytetu Kazimierza Wielkiego w Bydgoszczy: Bydgoszcz, Poland, 2014; Volume 1, no. XIX; pp. 17–49.
31. MacNeill, L.P.; Best, L.A.; Davis, L.L. The Role of Personality in Body Image Dissatisfaction and Disordered Eating: Discrepancies between Men and Women. *J. Eat. Disord.* **2017**, *5*, 44. [CrossRef]
32. Swami, V.; Taylor, R.; Carvalho, C. Body Dissatisfaction Assessed by the Photographic Figure Rating Scale Is Associated with Sociocultural, Personality, and Media Influences. *Scand. J. Psychol.* **2011**, *52*, 57–63. [CrossRef]
33. Miller, J.L.; Schmidt, L.A.; Vaillancourt, T.; McDougall, P.; Laliberte, M. Neuroticism and Introversion: A Risky Combination for Disordered Eating among a Non-Clinical Sample of Undergraduate Women. *Eat. Behav.* **2006**, *7*, 69–78. [CrossRef]
34. Sternheim, L.C.; van Passel, B.; Dingemans, A.; Cath, D.; Danner, U.N. Cognitive and Experienced Flexibility in Patients with Anorexia Nervosa and Obsessive Compulsive Disorder. *Front. Psychiatry* **2022**, *13*, 868921. [CrossRef] [PubMed]
35. Murray, D.; Lesser, M.; Lawson, W. Attention, Monotropism and the Diagnostic Criteria for Autism. *Autism* **2005**, *9*, 139–156. [CrossRef] [PubMed]
36. Abbate-Daga, G.; Buzzichelli, S.; Amianto, F.; Rocca, G.; Marzola, E.; McClintock, S.M.; Fassino, S. Cognitive Flexibility in Verbal and Nonverbal Domains and Decision Making in Anorexia Nervosa Patients: A Pilot Study. *BMC Psychiatry* **2011**, *11*, 162. [CrossRef]
37. Jáuregui-Lobera, I. Neuropsychology of Eating Disorders: 1995–2012. *Neuropsychiatr. Dis. Treat.* **2013**, *9*, 415–430. [CrossRef] [PubMed]
38. Chamberlain, S.R.; Fineberg, N.A.; Menzies, L.A.; Blackwell, A.D.; Bullmore, E.T.; Robbins, T.W.; Sahakian, B.J. Impaired Cognitive Flexibility and Motor Inhibition in Unaffected First-Degree Relatives of Patients with Obsessive-Compulsive Disorder. *Am. J. Psychiatry* **2007**, *164*, 335–338. [CrossRef] [PubMed]
39. Meiran, N.; Diamond, G.M.; Toder, D.; Nemets, B. Cognitive Rigidity in Unipolar Depression and Obsessive Compulsive Disorder: Examination of Task Switching, Stroop, Working Memory Updating and Post-Conflict Adaptation. *Psychiatry Res.* **2011**, *185*, 149–156. [CrossRef]
40. Fineberg, N.A.; Day, G.A.; de Koenigswarter, N.; Reghunandanan, S.; Kolli, S.; Jefferies-Sewell, K.; Hranov, G.; Laws, K.R. The Neuropsychology of Obsessive-Compulsive Personality Disorder: A New Analysis. *CNS Spectr.* **2015**, *20*, 490–499. [CrossRef]
41. Paast, N.; Khosravi, Z.; Memari, A.H.; Shayestehfar, M.; Arbabi, M. Comparison of Cognitive Flexibility and Planning Ability in Patients with Obsessive Compulsive Disorder, Patients with Obsessive Compulsive Personality Disorder, and Healthy Controls. *Shanghai Arch. Psychiatry* **2016**, *28*, 28.
42. Buzzichelli, S.; Marzola, E.; Amianto, F.; Fassino, S.; Abbate-Daga, G. Perfectionism and Cognitive Rigidity in Anorexia Nervosa: Is There an Association? *Eur. Eat. Disord. Rev.* **2018**, *26*, 360–366. [CrossRef]
43. Hosseini, S.A.; Padhy, R.K. Body Image Distortion. In *StatPearls*; StatPearls Publishing: Treasure Island, FL, USA, 2022.
44. King, I.C.C. Body Image in Paediatric Burns: A Review. *Burns Trauma* **2018**, *6*. [CrossRef]
45. Damiano, S.R.; Hart, L.M.; Paxton, S.J. Development and Validation of Parenting Measures for Body Image and Eating Patterns in Childhood. *J. Eat. Disord.* **2015**, *3*, 5. [CrossRef]
46. Shoraka, H.; Amirkafi, A.; Garrusi, B. Review of Body Image and Some of Contributing Factors in Iranian Population. *Int. J. Prev. Med.* **2019**, *10*. [CrossRef]
47. Bearman, S.K.; Presnell, K.; Martinez, E.; Stice, E. The Skinny on Body Dissatisfaction: A Longitudinal Study of Adolescent Girls and Boys. *J. Youth Adolesc.* **2006**, *35*, 217–229. [CrossRef]

48. Ciwoniuk, N.; Wayda-Zalewska, M.; Kucharska, K. Distorted Body Image and Mental Pain in Anorexia Nervosa. *Int. J. Environ. Res. Public. Health* **2023**, *20*, 718. [CrossRef]
49. Díaz-Marsá, M.; Alberdi-Páramo, Í.; Niell-Galmés, L. Nutritional Supplements in Eating Disorders. *Actas Esp. Psiquiatr.* **2017**, *45*, 26–36.
50. Nitsch, K.; Prajs, E.; Kurpisz, J.; Tyburski, E. Obraz Ciała i Jego Zaburzenia. Aspekty Teoretyczne w Kontekście Wybranych Jednostek Psychopatologicznych. *Psychiatr. Psychol. Klin.* **2012**, *3*, 176–182.
51. Guillaume, S.; Gorwood, P.; Jollant, F.; Van den Eynde, F.; Courtet, P.; Richard-Devantoy, S. Impaired Decision-Making in Symptomatic Anorexia and Bulimia Nervosa Patients: A Meta-Analysis. *Psychol. Med.* **2015**, *45*, 3377–3391. [CrossRef]
52. Couturier, J.; Kimber, M.; Szatmari, P. Efficacy of Family-based Treatment for Adolescents with Eating Disorders: A Systematic Review and Meta-analysis. *Int. J. Eat. Disord.* **2013**, *46*, 3–11. [CrossRef]
53. le Grange, D.; Eisler, I. Family Interventions in Adolescent Anorexia Nervosa. *Child Adolesc. Psychiatr. Clin. N. Am.* **2009**, *18*, 159–173. [CrossRef]
54. Lock, J. Evaluation of Family Treatment Models for Eating Disorders. *Curr. Opin. Psychiatry* **2011**, *24*, 274–279. [CrossRef]
55. Eisler, I.; Simic, M.; Blessitt, E.; Dodge, L.; MCCAED Team. Maudsley Service Manual for Child and Adolescent Eating Disorders. 2016. Available online: https://mccaed.slam.nhs.uk/wp-content/uploads/2019/11/Maudsley-Service-Manual-for-Child-and-Adolescent-Eating-Disorders-July-2016.pdf (accessed on 9 May 2023).
56. Pilecki, M.W.; Sałapa, K.; Józefik, B. Socio-Cultural Context of Eating Disorders in Poland. *J. Eat. Disord.* **2016**, *4*, 11. [CrossRef] [PubMed]
57. Tchanturia, K.; Davies, H.; Harrison, A.; Fox, J.R.E.; Treasure, J.; Schmidt, U. Altered Social Hedonic Processing in Eating Disorders. *Int. J. Eat. Disord.* **2012**, *45*, 962–969. [CrossRef]
58. Momeñe, J.; Estévez, A.; Griffiths, M.D.; Macía, P.; Herrero, M.; Olave, L.; Iruarrizaga, I. Eating Disorders and Intimate Partner Violence: The Influence of Fear of Loneliness and Social Withdrawal. *Nutrients* **2022**, *14*, 2611. [CrossRef] [PubMed]
59. Harney, M.B.; Fitzsimmons-Craft, E.E.; Maldonado, C.R.; Bardone-Cone, A.M. Negative Affective Experiences in Relation to Stages of Eating Disorder Recovery. *Eat. Behav.* **2014**, *15*, 24–30. [CrossRef] [PubMed]
60. Elwyn, R. A Lived Experience Response to the Proposed Diagnosis of Terminal Anorexia Nervosa: Learning from Iatrogenic Harm, Ambivalence and Enduring Hope. *J. Eat. Disord.* **2023**, *11*, 2. [CrossRef]
61. Rymarczyk, K. The Role of Personality Traits, Sociocultural Factors, and Body Dissatisfaction in Anorexia Readiness Syndrome in Women. *J. Eat. Disord.* **2021**, *9*, 51. [CrossRef]
62. Levine, M.P.; Smolak, L.; Hayden, H. The Relation of Sociocultural Factors to Eating Attitudes and Behaviors among Middle School Girls. *J. Early Adolesc.* **1994**, *14*, 471–490. [CrossRef]
63. Jones, D.C.; Crawford, J.K. The Peer Appearance Culture During Adolescence: Gender and Body Mass Variations. *J. Youth Adolesc.* **2006**, *35*, 243–255. [CrossRef]
64. Bozzola, E.; Spina, G.; Agostiniani, R.; Barni, S.; Russo, R.; Scarpato, E.; Di Mauro, A.; Di Stefano, A.V.; Caruso, C.; Corsello, G.; et al. The Use of Social Media in Children and Adolescents: Scoping Review on the Potential Risks. *Int. J. Environ. Res. Public. Health* **2022**, *19*, 9960. [CrossRef]
65. Dufour, R.; Novack, K.; Picard, L.; Chadi, N.; Booij, L. The Use of Technology in the Treatment of Youth with Eating Disorders: A Scoping Review. *J. Eat. Disord.* **2022**, *10*, 182. [CrossRef]
66. Wiberg, A.-C.; Ghaderi, A.; Danielsson, H.B.; Safarzadeh, K.; Parling, T.; Carlbring, P.; Jansson, M.; Welch, E. Internet-Based Cognitive Behavior Therapy for Eating Disorders—Development and Feasibility Evaluation. *Internet Interv.* **2022**, *30*, 100570. [CrossRef]
67. Vaccaro, C.M.; Guarino, G.; Conte, D.; Ferrara, E.; Ragione, L.D.; Bracale, R. Social Networks and Eating Disorders during the COVID-19 Pandemic. *Open Med.* **2021**, *16*, 1170–1174. [CrossRef]
68. Frieiro, P.; González-Rodríguez, R.; Domínguez-Alonso, J. Self-esteem and Socialisation in Social Networks as Determinants in Adolescents' Eating Disorders. *Health Soc. Care Community* **2022**, *30*, e4416–e4424. [CrossRef]
69. Bonfanti, R.C.; Lo Coco, G.; Salerno, L.; Di Blasi, M. The Thin Ideal and Attitudes towards Appearance as Correlates of Exercise Addiction among Sporty People during the COVID-19 Pandemic. *Behav. Sci.* **2022**, *12*, 187. [CrossRef]
70. Clarke, J.; Peyre, H.; Alison, M.; Bargiacchi, A.; Stordeur, C.; Boizeau, P.; Mamou, G.; Crépon, S.G.; Alberti, C.; Léger, J. Abnormal Bone Mineral Density and Content in Girls with Early-Onset Anorexia Nervosa. *J. Eat. Disord.* **2021**, *9*, 9. [CrossRef]
71. Misra, M.; Klibanski, A. Bone Health in Anorexia Nervosa. *Curr. Opin. Endocrinol. Diabetes Obes.* **2011**, *18*, 376. [CrossRef]
72. Robinson, L.; Micali, N.; Misra, M. Eating Disorders and Bone Metabolism in Women. *Curr. Opin. Pediatr.* **2017**, *29*, 488. [CrossRef]
73. Miller, K.K.; Lee, E.E.; Lawson, E.A.; Misra, M.; Minihan, J.; Grinspoon, S.K.; Gleysteen, S.; Mickley, D.; Herzog, D.; Klibanski, A. Determinants of Skeletal Loss and Recovery in Anorexia Nervosa. *J. Clin. Endocrinol. Metab.* **2006**, *91*, 2931–2937. [CrossRef]
74. Vestergaard, P.; Emborg, C.; Støving, R.K.; Hagen, C.; Mosekilde, L.; Brixen, K. Fractures in Patients with Anorexia Nervosa, Bulimia Nervosa, and Other Eating Disorders—A Nationwide Register Study. *Int. J. Eat. Disord.* **2002**, *32*, 301–308. [CrossRef]
75. Jada, K.; Djossi, S.K.; Khedr, A.; Neupane, B.; Proskuriakova, E.; Mostafa, J.A. The Pathophysiology of Anorexia Nervosa in Hypothalamic Endocrine Function and Bone Metabolism. *Cureus* **2021**, *13*, e20548. [CrossRef]
76. Legroux-Gerot, I.; Vignau, J.; Biver, E.; Pigny, P.; Collier, F.; Marchandise, X.; Duquesnoy, B.; Cortet, B. Anorexia Nervosa, Osteoporosis and Circulating Leptin: The Missing Link. *Osteoporos. Int.* **2010**, *21*, 1715–1722. [CrossRef] [PubMed]

77. Legroux-Gérot, I.; Vignau, J.; D'Herbomez, M.; Collier, F.; Marchandise, X.; Duquesnoy, B.; Cortet, B. Evaluation of Bone Loss and Its Mechanisms in Anorexia Nervosa. *Calcif. Tissue Int.* **2007**, *81*, 174–182. [CrossRef] [PubMed]
78. Legroux, I.; Cortet, B. Factors Influencing Bone Loss in Anorexia Nervosa: Assessment and Therapeutic Options. *RMD Open* **2019**, *5*, e001009. [CrossRef] [PubMed]
79. Zipfel, S.; Löwe, B.; Reas, D.L.; Deter, H.-C.; Herzog, W. Long-Term Prognosis in Anorexia Nervosa: Lessons from a 21-Year Follow-up Study. *Lancet* **2000**, *355*, 721–722. [CrossRef]
80. Misra, M.; Miller, K.K.; Cord, J.; Prabhakaran, R.; Herzog, D.B.; Goldstein, M.; Katzman, D.K.; Klibanski, A. Relationships between Serum Adipokines, Insulin Levels, and Bone Density in Girls with Anorexia Nervosa. *J. Clin. Endocrinol. Metab.* **2007**, *92*, 2046–2052. [CrossRef]
81. Legroux-Gerot, I.; Vignau, J.; Collier, F.; Cortet, B. Factors Influencing Changes in Bone Mineral Density in Patients with Anorexia Nervosa-Related Osteoporosis: The Effect of Hormone Replacement Therapy. *Calcif. Tissue Int.* **2008**, *83*, 315–323. [CrossRef]
82. Rosa-Caldwell, M.E.; Eddy, K.T.; Rutkove, S.B.; Breithaupt, L. Anorexia Nervosa and Muscle Health: A Systematic Review of Our Current Understanding and Future Recommendations for Study. *Int. J. Eat. Disord.* **2023**, *56*, 483–500. [CrossRef]
83. Solenberger, S.E. Exercise and Eating Disorders: A 3-Year Inpatient Hospital Record Analysis. *Eat. Behav.* **2001**, *2*, 151–168. [CrossRef]
84. McLoughlin, D.M.; Spargo, E.; Wassif, W.S.; Newham, D.J.; Peters, T.J.; Lantos, P.L.; Russell, G.F.M. Structural and Functional Changes in Skeletal Muscle in Anorexia Nervosa. *Acta Neuropathol.* **1998**, *95*, 540–632. [CrossRef]
85. Fernández-del-Valle, M.; Larumbe-Zabala, E.; Morande-Lavin, G.; Perez Ruiz, M. Muscle Function and Body Composition Profile in Adolescents with Restrictive Anorexia Nervosa: Does Resistance Training Help? *Disabil. Rehabil.* **2016**, *38*, 346–353. [CrossRef]
86. Leitner, L.M.; Wilson, R.J.; Yan, Z.; Gödecke, A. Reactive Oxygen Species/Nitric Oxide Mediated Inter-Organ Communication in Skeletal Muscle Wasting Diseases. *Antioxid. Redox Signal.* **2017**, *26*, 700–717. [CrossRef]
87. Hodson, N.; Philp, A. The Importance of MTOR Trafficking for Human Skeletal Muscle Translational Control. *Exerc. Sport Sci. Rev.* **2019**, *47*, 46. [CrossRef]
88. Shimkus, K.L.; Jefferson, L.S.; Gordon, B.S.; Kimball, S.R. Repressors of MTORC1 Act to Blunt the Anabolic Response to Feeding in the Soleus Muscle of a Cast-immobilized Mouse Hindlimb. *Physiol. Rep.* **2018**, *6*, e13891. [CrossRef]
89. Alaaraj, N.; Soliman, A.; Hamed, N.; Alyafei, F.; De Sanctis, V. Understanding the Complex Role of Mtorc as an Intracellular Critical Mediator of Whole-Body Metabolism in Anorexia Nervosa: A Mini Review. *Acta Bio Medica Atenei Parm.* **2021**, *92*, e2021170.
90. Tripathi, D.; Kant, S.; Pandey, S.; Ehtesham, N.Z. Resistin in Metabolism, Inflammation, and Disease. *FEBS J.* **2020**, *287*, 3141–3149. [CrossRef]
91. Cui, H.; Moore, J.; Ashimi, S.S.; Mason, B.L.; Drawbridge, J.N.; Han, S.; Hing, B.; Matthews, A.; McAdams, C.J.; Darbro, B.W. Eating Disorder Predisposition Is Associated with ESRRA and HDAC4 Mutations. *J. Clin. Investig.* **2013**, *123*, 4706–4713. [CrossRef]
92. Horard, B.; Vanacker, J.M. Estrogen Receptor-Related Receptors: Orphan Receptors Desperately Seeking a Ligand. *J. Mol. Endocrinol.* **2003**, *31*, 349–357. [CrossRef]
93. Bookout, A.L.; Jeong, Y.; Downes, M.; Ruth, T.Y.; Evans, R.M.; Mangelsdorf, D.J. Anatomical Profiling of Nuclear Receptor Expression Reveals a Hierarchical Transcriptional Network. *Cell* **2006**, *126*, 789–799. [CrossRef]
94. Singh, B.K.; Sinha, R.A.; Ohba, K.; Yen, P.M. Role of Thyroid Hormone in Hepatic Gene Regulation, Chromatin Remodeling, and Autophagy. *Mol. Cell. Endocrinol.* **2017**, *458*, 160–168. [CrossRef]
95. Rangwala, S.M.; Li, X.; Lindsley, L.; Wang, X.; Shaughnessy, S.; Daniels, T.G.; Szustakowski, J.; Nirmala, N.R.; Wu, Z.; Stevenson, S.C. Estrogen-Related Receptor α Is Essential for the Expression of Antioxidant Protection Genes and Mitochondrial Function. *Biochem. Biophys. Res. Commun.* **2007**, *357*, 231–236. [CrossRef]
96. Wallace, M.A.; Hock, M.B.; Hazen, B.C.; Kralli, A.; Snow, R.J.; Russell, A.P. Striated Muscle Activator of Rho Signalling (STARS) Is a PGC-1α/Oestrogen-related Receptor-α Target Gene and Is Upregulated in Human Skeletal Muscle after Endurance Exercise. *J. Physiol.* **2011**, *589*, 2027–2039. [CrossRef] [PubMed]
97. Ranhotra, H.S. Up-Regulation of Orphan Nuclear Estrogen-Related Receptor Alpha Expression during Long-Term Caloric Restriction in Mice. *Mol. Cell. Biochem.* **2009**, *332*, 59–65. [CrossRef]
98. Giguére, V. To ERR in the Estrogen Pathway. *Trends Endocrinol. Metab.* **2002**, *13*, 220–225. [CrossRef] [PubMed]
99. Sender, R.; Fuchs, S.; Milo, R. Revised Estimates for the Number of Human and Bacteria Cells in the Body. *PLoS Biol.* **2016**, *14*, e1002533. [CrossRef] [PubMed]
100. Rosenbaum, M.; Knight, R.; Leibel, R.L. The Gut Microbiota in Human Energy Homeostasis and Obesity. *Trends Endocrinol. Metab.* **2015**, *26*, 493–501. [CrossRef]
101. Lozupone, C.A.; Stombaugh, J.I.; Gordon, J.I.; Jansson, J.K.; Knight, R. Diversity, Stability and Resilience of the Human Gut Microbiota. *Nature* **2012**, *489*, 220–230. [CrossRef]
102. Borgo, F.; Riva, A.; Benetti, A.; Casiraghi, M.C.; Bertelli, S.; Garbossa, S.; Anselmetti, S.; Scarone, S.; Pontiroli, A.E.; Morace, G. Microbiota in Anorexia Nervosa: The Triangle between Bacterial Species, Metabolites and Psychological Tests. *PLoS ONE* **2017**, *12*, e0179739. [CrossRef]
103. Mack, I.; Cuntz, U.; Grämer, C.; Niedermaier, S.; Pohl, C.; Schwiertz, A.; Zimmermann, K.; Zipfel, S.; Enck, P.; Penders, J. Weight Gain in Anorexia Nervosa Does Not Ameliorate the Faecal Microbiota, Branched Chain Fatty Acid Profiles and Gastrointestinal Complaints. *Sci. Rep.* **2016**, *6*, 26752. [CrossRef]

104. Mörkl, S.; Lackner, S.; Müller, W.; Gorkiewicz, G.; Kashofer, K.; Oberascher, A.; Painold, A.; Holl, A.; Holzer, P.; Meinitzer, A. Gut Microbiota and Body Composition in Anorexia Nervosa Inpatients in Comparison to Athletes, Overweight, Obese, and Normal Weight Controls. *Int. J. Eat. Disord.* **2017**, *50*, 1421–1431. [CrossRef]
105. Morita, C.; Tsuji, H.; Hata, T.; Gondo, M.; Takakura, S.; Kawai, K.; Yoshihara, K.; Ogata, K.; Nomoto, K.; Miyazaki, K. Gut Dysbiosis in Patients with Anorexia Nervosa. *PLoS ONE* **2015**, *10*, e0145274. [CrossRef]
106. Ruusunen, A.; Rocks, T.; Jacka, F.; Loughman, A. The Gut Microbiome in Anorexia Nervosa: Relevance for Nutritional Rehabilitation. *Psychopharmacology* **2019**, *236*, 1545–1558. [CrossRef]
107. Armougom, F.; Henry, M.; Vialettes, B.; Raccah, D.; Raoult, D. Monitoring Bacterial Community of Human Gut Microbiota Reveals an Increase in Lactobacillus in Obese Patients and Methanogens in Anorexic Patients. *PLoS ONE* **2009**, *4*, e7125. [CrossRef]
108. Ghoshal, U.; Shukla, R.; Srivastava, D.; Ghoshal, U.C. Irritable Bowel Syndrome, Particularly the Constipation-Predominant Form, Involves an Increase in Methanobrevibacter Smithii, Which Is Associated with Higher Methane Production. *Gut Liver* **2016**, *10*, 932. [CrossRef]
109. Breton, J.; Legrand, R.; Akkermann, K.; Järv, A.; Harro, J.; Déchelotte, P.; Fetissov, S.O. Elevated Plasma Concentrations of Bacterial ClpB Protein in Patients with Eating Disorders. *Int. J. Eat. Disord.* **2016**, *49*, 805–808. [CrossRef]
110. Kleiman, S.C.; Watson, H.J.; Bulik-Sullivan, E.C.; Huh, E.Y.; Tarantino, L.M.; Bulik, C.M.; Carroll, I.M. The Intestinal Microbiota in Acute Anorexia Nervosa and during Renourishment: Relationship to Depression, Anxiety, and Eating Disorder Psychopathology. *Psychosom. Med.* **2015**, *77*, 969. [CrossRef]
111. Prochazkova, P.; Roubalova, R.; Dvorak, J.; Tlaskalova-Hogenova, H.; Cermakova, M.; Tomasova, P.; Sediva, B.; Kuzma, M.; Bulant, J.; Bilej, M.; et al. Microbiota, Microbial Metabolites, and Barrier Function in A Patient with Anorexia Nervosa after Fecal Microbiota Transplantation. *Microorganisms* **2019**, *7*, 338. [CrossRef]
112. Gouba, N.; Raoult, D.; Drancourt, M. Gut Microeukaryotes during Anorexia Nervosa: A Case Report. *BMC Res. Notes* **2014**, *7*, 33. [CrossRef]
113. Mörkl, S.; Lackner, S.; Meinitzer, A.; Mangge, H.; Lehofer, M.; Halwachs, B.; Gorkiewicz, G.; Kashofer, K.; Painold, A.; Holl, A.K. Gut Microbiota, Dietary Intakes and Intestinal Permeability Reflected by Serum Zonulin in Women. *Eur. J. Nutr.* **2018**, *57*, 2985–2997. [CrossRef]
114. Speranza, E.; Cioffi, I.; Santarpia, L.; Del Piano, C.; De Caprio, C.; Naccarato, M.; Marra, M.; De Filippo, E.; Contaldo, F.; Pasanisi, F. Fecal Short Chain Fatty Acids and Dietary Intake in Italian Women with Restrictive Anorexia Nervosa: A Pilot Study. *Front. Nutr.* **2018**, *5*, 119. [CrossRef]
115. Tang, W.K.; Sun, F.C.S.; Ungvari, G.S.; O'donnell, D. Oral Health of Psychiatric In-Patients in Hong Kong. *Int. J. Soc. Psychiatry* **2004**, *50*, 186–191. [CrossRef]
116. Kisely, S.; Baghaie, H.; Lalloo, R.; Johnson, N.W. Association between Poor Oral Health and Eating Disorders: Systematic Review and Meta-Analysis. *Br. J. Psychiatry* **2015**, *207*, 299–305. [CrossRef] [PubMed]
117. Ramon, T.; Grinshpoon, A.; Zusman, S.P.; Weizman, A. Oral Health and Treatment Needs of Institutionalized Chronic Psychiatric Patients in Israel. *Eur. Psychiatry* **2003**, *18*, 101–105. [CrossRef] [PubMed]
118. Hasan, S.; Ahmed, S.; Panigrahi, R.; Chaudhary, P.; Vyas, V.; Saeed, S. Oral Cavity and Eating Disorders: An Insight to Holistic Health. *J. Fam. Med. Prim. Care* **2020**, *9*, 3890.
119. Scharner, S.; Stengel, A. Alterations of Brain Structure and Functions in Anorexia Nervosa. *Clin. Nutr. Exp.* **2019**, *28*, 22–32. [CrossRef]
120. DeBate, R.D.; Tedesco, L.A.; Kerschbaum, W.E. Knowledge of Oral and Physical Manifestations of Anorexia and Bulimia Nervosa among Dentists and Dental Hygienists. *J. Dent. Educ.* **2005**, *69*, 346–354. [CrossRef]
121. Lourenço, M.; Azevedo, Á.; Brandão, I.; Gomes, P.S. Orofacial Manifestations in Outpatients with Anorexia Nervosa and Bulimia Nervosa Focusing on the Vomiting Behavior. *Clin. Oral Investig.* **2018**, *22*, 1915–1922. [CrossRef]
122. Johansson, A.-K.; Mjanger Øvretvedt, T.; Reinholtsen, K.K.; Johansson, A. Eating Disorders: An Analysis of Self-Induced Vomiting, Binge Eating, and Oral Hygiene Behavior. *Int. J. Clin. Pract.* **2022**, *2022*, 6210372. [CrossRef]
123. Jagielska, G.; Kacperska, I. Outcome, Comorbidity and Prognosis in Anorexia Nervosa. *Psychiatr. Pol.* **2017**, *51*, 205–218. [CrossRef]
124. Roberts, M.W.; Li, S.-H. Oral Findings in Anorexia Nervosa and Bulimia Nervosa: A Study of 47 Cases. *J. Am. Dent. Assoc.* **1987**, *115*, 407–410. [CrossRef]
125. Favaro, A.; Ferrara, S.; Santonastaso, P. The Spectrum of Eating Disorders in Young Women: A Prevalence Study in a General Population Sample. *Psychosom. Med.* **2003**, *65*, 701–708. [CrossRef]
126. Paszynska, E.; Hernik, A.; Slopien, A.; Roszak, M.; Jowik, K.; Dmitrzak-Weglarz, M.; Tyszkiewicz-Nwafor, M. Risk of Dental Caries and Erosive Tooth Wear in 117 Children and Adolescents' Anorexia Nervosa Population—A Case-Control Study. *Front. Psychiatry* **2022**, *13*, 874263. [CrossRef]
127. Panico, R.; Piemonte, E.; Lazos, J.; Gilligan, G.; Zampini, A.; Lanfranchi, H. Oral Mucosal Lesions in Anorexia Nervosa, Bulimia Nervosa and EDNOS. *J. Psychiatr. Res.* **2018**, *96*, 178–182. [CrossRef]
128. Mascitti, M.; Coccia, E.; Vignini, A.; Aquilanti, L.; Santarelli, A.; Salvolini, E.; Sabbatinelli, J.; Mazzanti, L.; Procaccini, M.; Rappelli, G. Anorexia, Oral Health and Antioxidant Salivary System: A Clinical Study on Adult Female Subjects. *Dent. J.* **2019**, *7*, 60. [CrossRef]
129. Vergata, R.T. Oral Manifestations of Eating Disorders in Adolescent Patients. A Review. *Eur. J. Paediatr. Dent.* **2021**, *22*, 155.

130. Aranha, A.C.; Eduardo, C.d.P.; Cordás, T.A. Eating Disorders. Part I: Psychiatric Diagnosis and Dental Implications. *J. Contemp. Dent. Pract.* **2008**, *9*, 73–81. [CrossRef]
131. Hermont, A.P.; Oliveira, P.A.D.; Martins, C.C.; Paiva, S.M.; Pordeus, I.A.; Auad, S.M. Tooth Erosion and Eating Disorders: A Systematic Review and Meta-Analysis. *PLoS ONE* **2014**, *9*, e111123. [CrossRef]
132. Rangé, H.; Pallier, A.; Ali, A.; Huas, C.; Colon, P.; Godart, N. Risk Factors for Oral Health in Anorexia Nervosa: Comparison of a Self-Report Questionnaire and a Face-to-Face Interview. *Int. J. Environ. Res. Public. Health* **2021**, *18*, 4212. [CrossRef]
133. Nijakowski, K.; Walerczyk-Sas, A.; Surdacka, A. Regular Physical Activity as a Potential Risk Factor for Erosive Lesions in Adolescents. *Int. J. Environ. Res. Public. Health* **2020**, *17*, 3002. [CrossRef]
134. Pleple, A.; Lalanne, C.; Huas, C.; Mattar, L.; Hanachi, M.; Flament, M.F.; Carchon, I.; Jouen, F.; Berthoz, S.; Godart, N. Nutritional Status and Anxious and Depressive Symptoms in Anorexia Nervosa: A Prospective Study. *Sci. Rep.* **2021**, *11*, 771. [CrossRef]
135. Lobera, I.J.; Ríos, P.B. Choice of Diet in Patients with Anorexia Nervosa. *Nutr. Hosp.* **2009**, *24*, 682–687.
136. Garrido-Martínez, P.; Domínguez-Gordillo, A.; Cerero-Lapiedra, R.; Burgueño-García, M.; Martínez-Ramírez, M.-J.; Gómez-Candela, C.; Cebrián-Carretero, J.-L.; Esparza-Gómez, G. Oral and Dental Health Status in Patients with Eating Disorders in Madrid, Spain. *Med. Oral Patol. Oral Cirugia Bucal* **2019**, *24*, e595. [CrossRef] [PubMed]
137. Silverstein, L.S.; Haggerty, C.; Sams, L.; Phillips, C.; Roberts, M.W. Impact of an Oral Health Education Intervention among a Group of Patients with Eating Disorders (Anorexia Nervosa and Bulimia Nervosa). *J. Eat. Disord.* **2019**, *7*, 29. [CrossRef] [PubMed]
138. Marzola, E.; Nasser, J.A.; Hashim, S.A.; Shih, P.-A.B.; Kaye, W.H. Nutritional Rehabilitation in Anorexia Nervosa: Review of the Literature and Implications for Treatment. *BMC Psychiatry* **2013**, *13*, 290. [CrossRef] [PubMed]
139. Andrewes, D.G.; O'Connor, P.; Mulder, C.; McLennan, J.; Derham, H.; Weigall, S.; Say, S. Computerised Psychoeducation for Patients with Eating Disorders. *Aust. N. Z. J. Psychiatry* **1996**, *30*, 492–497. [CrossRef]
140. Ng, L.W.C.; Ng, D.P.; Wong, W.P. Is Supervised Exercise Training Safe in Patients with Anorexia Nervosa? A Meta-Analysis. *Physiotherapy* **2013**, *99*, 1–11. [CrossRef]
141. Rizk, M.; Kern, L.; Lalanne, C.; Hanachi, M.; Melchior, J.; Pichard, C.; Mattar, L.; Group, E.; Berthoz, S.; Godart, N. High-intensity Exercise Is Associated with a Better Nutritional Status in Anorexia Nervosa. *Eur. Eat. Disord. Rev.* **2019**, *27*, 391–400. [CrossRef]
142. Dalle Grave, R.; Calugi, S.; El Ghoch, M.; Conti, M.; Fairburn, C.G. Inpatient Cognitive Behavior Therapy for Adolescents with Anorexia Nervosa: Immediate and Longer-Term Effects. *Front. Psychiatry* **2014**, *5*, 14. [CrossRef]
143. Steinglass, J.; Albano, A.M.; Simpson, H.B.; Carpenter, K.; Schebendach, J.; Attia, E. Fear of Food as a Treatment Target: Exposure and Response Prevention for Anorexia Nervosa in an Open Series. *Int. J. Eat. Disord.* **2012**, *45*, 615–621. [CrossRef]
144. Fisher, C.A.; Skocic, S.; Rutherford, K.A.; Hetrick, S.E. Family Therapy Approaches for Anorexia Nervosa. *Cochrane Database Syst. Rev.* **2018**, *10*, CD004780. [CrossRef]
145. Barber, S.; Thornicroft, G. Reducing the Mortality Gap in People with Severe Mental Disorders: The Role of Lifestyle Psychosocial Interventions. *Front. Psychiatry* **2018**, *9*, 463. [CrossRef]
146. Hart, S.; Field, T.; Hernandez-Reif, M.; Nearing, G.; Shaw, S.; Schanberg, S.; Kuhn, C. Anorexia Nervosa Symptoms Are Reduced by Massage Therapy. *Eat. Disord.* **2001**, *9*, 289–299. [CrossRef]
147. Fogarty, S.; Smith, C.A.; Touyz, S.; Madden, S.; Buckett, G.; Hay, P. Patients with Anorexia Nervosa Receiving Acupuncture or Acupressure; Their View of the Therapeutic Encounter. *Complement. Ther. Med.* **2013**, *21*, 675–681. [CrossRef]
148. American Dietetic Association. Position of the American Dietetic Association: Nutrition Intervention in the Treatment of Anorexia Nervosa, Bulimia Nervosa, and Eating Disorders Not Otherwise Specified (EDNOS). *J. Am. Diet. Assoc.* **2001**, *101*, 810–819. [CrossRef]
149. Rock, C.L. Nutritional Rehabilitation for Anorexia Nervosa. In *The Treatment of Eating Disorders: A Clinical Handbook*; Guilford Press: New York, NY, USA, 2010; pp. 187–197.
150. Wilson, G.T.; Grilo, C.M.; Vitousek, K.M. Psychological Treatment of Eating Disorders. *Am. Psychol.* **2007**, *62*, 199. [CrossRef]
151. Gómez Candela, C.; Palma Milla, S.; Miján de la Torre, A.; Rodríguez Ortega, P.; Matía Martín, P.; Loria Kohen, V.; Campos del Portillo, R.; Virgili Casas, M.; Martínez Olmos, M.Á.; Mories Álvarez, M.T. Consenso Sobre La Evaluación y El Tratamiento Nutricional de Los Trastornos de La Conducta Alimentaria: Anorexia Nerviosa. *Nutr. Hosp.* **2018**, *35*, 11–48. [CrossRef]
152. De La Guía, G.D.T. *Guía de Práctica Clínica Sobre Trastornos de La Conducta Alimentaria*; Hospital Psiquiátrico "Román Alberca": Murcia, Spain, 2009.
153. Association, A.P. Practice Guideline for the Treatment of Patients with Eating Disorders. *Am. J. Psychiatry* **2006**, *163*, 4–54.
154. Reiter, C.S.; Graves, L. Nutrition Therapy for Eating Disorders. *Nutr. Clin. Pract.* **2010**, *25*, 122–136. [CrossRef]
155. Wakefield, A.; Williams, H. *Practice Recommendations for the Nutritional Management of Anorexia Nervosa in Adults*; Dietitians Association of Australia: Deakin, Australia, 2009.
156. National Collaborating Centre for Mental Health (UK). *Eating Disorders: Core Interventions in the Treatment and Management of Anorexia Nervosa, Bulimia Nervosa and Related Eating Disorders*; National Collaborating Centre for Mental Health: Leicester, UK, 2004.
157. McMaster, C.M.; Wade, T.; Franklin, J.; Hart, S. Development of Consensus-based Guidelines for Outpatient Dietetic Treatment of Eating Disorders: A Delphi Study. *Int. J. Eat. Disord.* **2020**, *53*, 1480–1495. [CrossRef]
158. Trocki, O.; Shepherd, R.W. Change in Body Mass Index Does Not Predict Change in Body Composition in Adolescent Girls with Anorexia Nervosa. *J. Acad. Nutr. Diet.* **2000**, *100*, 457. [CrossRef]

159. Bartrina, J.A.; Majem, L.S. Objetivos Nutricionales Para La Población Española: Consenso de La Sociedad Española de Nutrición Comunitaria 2011. *Rev. Esp. Nutr. Comunitaria Span. J. Community Nutr.* **2011**, *17*, 178–199.
160. Capo-chichi, C.D.; Guéant, J.-L.; Lefebvre, E.; Bennani, N.; Lorentz, E.; Vidailhet, C.; Vidailhet, M. Riboflavin and Riboflavin-Derived Cofactors in Adolescent Girls with Anorexia Nervosa. *Am. J. Clin. Nutr.* **1999**, *69*, 672–678. [CrossRef] [PubMed]
161. Grinspoon, S.; Herzog, D.; Klibanski, A. Mechanisms and Treatment Options for Bone Loss in Anorexia Nervosa. *Psychopharmacol. Bull.* **1997**, *33*, 399–404. [PubMed]
162. Ortega Anta, M.R.; Aparicio Vizuete, A.; Lopez Sobaler, A.M. *Educación Nutricional*; Gil Hernández, Á., Ed.; Tratado de Nutrición; Editorial Medica Panamericana: Madrid, Spain, 2005; pp. 514–554.
163. Dwyer, J.T. Role of the Nutritionist in Eating Disorders. *New Dir. Ment. Health Serv.* **1986**, *1986*, 21–28. [CrossRef]
164. Ozier, A.D.; Henry, B.W. Position of the American Dietetic Association: Nutrition Intervention in the Treatment of Eating Disorders. *J. Am. Diet. Assoc.* **2011**, *111*, 1236–1241. [CrossRef]
165. Mayer, L.E.S.; Schebendach, J.; Bodell, L.P.; Shingleton, R.M.; Walsh, B.T. Eating Behavior in Anorexia Nervosa: Before and after Treatment. *Int. J. Eat. Disord.* **2012**, *45*, 290–293. [CrossRef]
166. Nomdedeu, C.L. Educación Nutricional de La Población General y de Riesgo. In *Nutrición y Metabolismo en Trastornos de la Conducta Alimentaria*; Fundación Dialnet: La Rioja, Spain, 2004; pp. 459–476.
167. Quiles Marcos, Y.; Leon Zarceno, E.; Lopez Lopez, J.A. Effectiveness of Exercise-based Interventions in Patients with Anorexia Nervosa: A Systematic Review. *Eur. Eat. Disord. Rev.* **2021**, *29*, 3–19. [CrossRef]
168. Melissa, R.; Lama, M.; Laurence, K.; Sylvie, B.; Jeanne, D.; Odile, V.; Nathalie, G. Physical Activity in Eating Disorders: A Systematic Review. *Nutrients* **2020**, *12*, 183. [CrossRef]
169. Vancampfort, D.; Vanderlinden, J.; De Hert, M.; Soundy, A.; Adamkova, M.; Skjaerven, L.H.; Catalan-Matamoros, D.; Lundvik Gyllensten, A.; Gomez-Conesa, A.; Probst, M. A Systematic Review of Physical Therapy Interventions for Patients with Anorexia and Bulemia Nervosa. *Disabil. Rehabil.* **2014**, *36*, 628–634. [CrossRef]
170. Le Grange, D.; Eisler, I. The Link between Anorexia Nervosa and Excessive Exercise: A Review. *Eur. Eat. Disord. Rev.* **1993**, *1*, 100–119. [CrossRef]
171. Stice, E. Risk and Maintenance Factors for Eating Pathology: A Meta-Analytic Review. *Psychol. Bull.* **2002**, *128*, 825. [CrossRef]
172. Leuenberger, A. Endorphins, Exercise, and Addictions: A Review of Exercise Dependence. 2006. Available online: https://impulse.pubpub.org/pub/ff9iap4s (accessed on 9 May 2023).
173. Cook, B.; Engel, S.; Crosby, R.; Hausenblas, H.; Wonderlich, S.; Mitchell, J. Pathological Motivations for Exercise and Eating Disorder Specific Health-related Quality of Life. *Int. J. Eat. Disord.* **2014**, *47*, 268–272. [CrossRef]
174. Strober, M.; Freeman, R.; Morrell, W. The Long-term Course of Severe Anorexia Nervosa in Adolescents: Survival Analysis of Recovery, Relapse, and Outcome Predictors over 10–15 Years in a Prospective Study. *Int. J. Eat. Disord.* **1997**, *22*, 339–360. [CrossRef]
175. Ruiz, J.R.; Sui, X.; Lobelo, F.; Morrow, J.R.; Jackson, A.W.; Sjöström, M.; Blair, S.N. Association between Muscular Strength and Mortality in Men: Prospective Cohort Study. *BMJ* **2008**, *337*, a439. [CrossRef]
176. Chen, L.; Nelson, D.R.; Zhao, Y.; Cui, Z.; Johnston, J.A. Relationship between Muscle Mass and Muscle Strength, and the Impact of Comorbidities: A Population-Based, Cross-Sectional Study of Older Adults in the United States. *BMC Geriatr.* **2013**, *13*, 74. [CrossRef]
177. Trombetti, A.; Reid, K.F.; Hars, M.; Herrmann, F.R.; Pasha, E.; Phillips, E.M.; Fielding, R.A. Age-Associated Declines in Muscle Mass, Strength, Power, and Physical Performance: Impact on Fear of Falling and Quality of Life. *Osteoporos. Int.* **2016**, *27*, 463–471. [CrossRef]
178. Martin, S.P.K.; Bachrach, L.K.; Golden, N.H. Controlled Pilot Study of High-Impact Low-Frequency Exercise on Bone Loss and Vital-Sign Stabilization in Adolescents with Eating Disorders. *J. Adolesc. Health* **2017**, *60*, 33–37. [CrossRef]
179. Calogero, R.M.; Pedrotty, K.N. The Practice and Process of Healthy Exercise: An Investigation of the Treatment of Exercise Abuse in Women with Eating Disorders. *Eat. Disord.* **2004**, *12*, 273–291. [CrossRef]
180. Carei, T.R.; Fyfe-Johnson, A.L.; Breuner, C.C.; Brown, M.A. Randomized Controlled Clinical Trial of Yoga in the Treatment of Eating Disorders. *J. Adolesc. Health* **2010**, *46*, 346–351. [CrossRef]
181. Borden, A.; Cook-Cottone, C. Yoga and Eating Disorder Prevention and Treatment: A Comprehensive Review and Meta-Analysis. *Eat. Disord.* **2020**, *28*, 400–437. [CrossRef]
182. Rizzuto, L.; Hay, P.; Noetel, M.; Touyz, S. Yoga as Adjunctive Therapy in the Treatment of People with Anorexia Nervosa: A Delphi Study. *J. Eat. Disord.* **2021**, *9*, 111. [CrossRef] [PubMed]
183. del Valle, M.F.; Pérez, M.; Santana-Sosa, E.; Fiuza-Luces, C.; Bustamante-Ara, N.; Gallardo, C.; Villasenor, A.; Graell, M.; Morandé, G.; Romo, G.R. Does Resistance Training Improve the Functional Capacity and Well Being of Very Young Anorexic Patients? A Randomized Controlled Trial. *J. Adolesc. Health* **2010**, *46*, 352–358. [CrossRef] [PubMed]
184. Fernandez-del-Valle, M.; Larumbe-Zabala, E.; Villaseñor-Montarroso, A.; Cardona Gonzalez, C.; Diez-Vega, I.; Lopez Mojares, L.M.; Perez Ruiz, M. Resistance Training Enhances Muscular Performance in Patients with Anorexia Nervosa: A Randomized Controlled Trial. *Int. J. Eat. Disord.* **2014**, *47*, 601–609. [CrossRef] [PubMed]
185. Trott, M.; Jackson, S.E.; Firth, J.; Jacob, L.; Grabovac, I.; Mistry, A.; Stubbs, B.; Smith, L. A Comparative Meta-Analysis of the Prevalence of Exercise Addiction in Adults with and without Indicated Eating Disorders. *Eat. Weight Disord.-Stud. Anorex. Bulim. Obes.* **2021**, *26*, 37–46. [CrossRef]

186. Galmiche, M.; Déchelotte, P.; Lambert, G.; Tavolacci, M.P. Prevalence of Eating Disorders over the 2000–2018 Period: A Systematic Literature Review. *Am. J. Clin. Nutr.* **2019**, *109*, 1402–1413. [CrossRef]
187. Ward, Z.J.; Rodriguez, P.; Wright, D.R.; Austin, S.B.; Long, M.W. Estimation of Eating Disorders Prevalence by Age and Associations with Mortality in a Simulated Nationally Representative US Cohort. *JAMA Netw. Open* **2019**, *2*, e1912925. [CrossRef]
188. Keski-Rahkonen, A. Epidemiology of Binge Eating Disorder: Prevalence, Course, Comorbidity, and Risk Factors. *Curr. Opin. Psychiatry* **2021**, *34*, 525–531. [CrossRef]
189. Grilo, C.M.; Ivezaj, V.; Lydecker, J.A.; White, M.A. Toward an Understanding of the Distinctiveness of Body-Image Constructs in Persons Categorized with Overweight/Obesity, Bulimia Nervosa, and Binge-Eating Disorder. *J. Psychosom. Res.* **2019**, *126*, 109757. [CrossRef]
190. Perrotta, G. Neural Correlates in Eating Disorders: Definition, Contexts and Clinical Strategies. *J. Pub. Heath Cat.* **2019**, *2*, 137–148.
191. Hay, P. Current Approach to Eating Disorders: A Clinical Update. *Intern. Med. J.* **2020**, *50*, 24–29. [CrossRef]
192. Solmi, M.; Radua, J.; Stubbs, B.; Ricca, V.; Moretti, D.; Busatta, D.; Carvalho, A.F.; Dragioti, E.; Favaro, A.; Monteleone, A.M. Risk Factors for Eating Disorders: An Umbrella Review of Published Meta-Analyses. *Braz. J. Psychiatry* **2020**, *43*, 314–323. [CrossRef]
193. McCabe, M.; Tatangelo, G.; Watson, B.; Fuller-Tyszkiewicz, M.; Rodgers, R.F.; Aimé, A.; Mellor, D.; Granero-Gallegos, A.; Strodl, E.; Caltabiano, M. Development and Testing of a Model for Risk and Protective Factors for Eating Disorders and Higher Weight among Emerging Adults: A Study Protocol. *Body Image* **2019**, *31*, 139–149. [CrossRef]
194. Herle, M.; De Stavola, B.; Hübel, C.; Abdulkadir, M.; Ferreira, D.S.; Loos, R.J.F.; Bryant-Waugh, R.; Bulik, C.M.; Micali, N. A Longitudinal Study of Eating Behaviours in Childhood and Later Eating Disorder Behaviours and Diagnoses. *Br. J. Psychiatry* **2020**, *216*, 113–119. [CrossRef]
195. Graell, M.; Morón-Nozaleda, M.G.; Camarneiro, R.; Villaseñor, Á.; Yáñez, S.; Muñoz, R.; Martínez-Núñez, B.; Miguélez-Fernández, C.; Muñoz, M.; Faya, M. Children and Adolescents with Eating Disorders during COVID-19 Confinement: Difficulties and Future Challenges. *Eur. Eat. Disord. Rev.* **2020**, *28*, 864–870. [CrossRef]
196. Cooper, M.; Reilly, E.E.; Siegel, J.A.; Coniglio, K.; Sadeh-Sharvit, S.; Pisetsky, E.M.; Anderson, L.M. Eating Disorders during the COVID-19 Pandemic and Quarantine: An Overview of Risks and Recommendations for Treatment and Early Intervention. *Eat. Disord.* **2022**, *30*, 54–76. [CrossRef]
197. Fernández-Aranda, F.; Munguía, L.; Mestre-Bach, G.; Steward, T.; Etxandi, M.; Baenas, I.; Granero, R.; Sánchez, I.; Ortega, E.; Andreu, A. COVID Isolation Eating Scale (CIES): Analysis of the Impact of Confinement in Eating Disorders and Obesity—A Collaborative International Study. *Eur. Eat. Disord. Rev.* **2020**, *28*, 871–883. [CrossRef]
198. Murray, S.B. Updates in the Treatment of Eating Disorders in 2018: A Year in Review in Eating Disorders: The Journal of Treatment & Prevention. *Eat. Disord.* **2019**, *27*, 6–17.
199. Hale, M.D.; Logomarsino, J.V. The Use of Enteral Nutrition in the Treatment of Eating Disorders: A Systematic Review. *Eat. Weight Disord.-Stud. Anorex. Bulim. Obes.* **2019**, *24*, 179–198. [CrossRef]
200. Agras, W.S. Cognitive Behavior Therapy for the Eating Disorders. *Psychiatr. Clin.* **2019**, *42*, 169–179. [CrossRef]
201. Bohon, C. Binge Eating Disorder in Children and Adolescents. *Child Adolesc. Psychiatr. Clin.* **2019**, *28*, 549–555. [CrossRef]
202. Puls, H.; Schmidt, R.; Hilbert, A. Therapist Adherence and Therapeutic Alliance in Individual Cognitive-behavioural Therapy for Adolescent Binge-eating Disorder. *Eur. Eat. Disord. Rev.* **2019**, *27*, 182–194. [CrossRef] [PubMed]
203. Flynn, M.; Austin, A.; Lang, K.; Allen, K.; Bassi, R.; Brady, G.; Brown, A.; Connan, F.; Franklin-Smith, M.; Glennon, D. Assessing the Impact of First Episode Rapid Early Intervention for Eating Disorders on Duration of Untreated Eating Disorder: A Multi-centre Quasi-experimental Study. *Eur. Eat. Disord. Rev.* **2021**, *29*, 458–471. [CrossRef] [PubMed]
204. Le Grange, D.; Eckhardt, S.; Dalle Grave, R.; Crosby, R.D.; Peterson, C.B.; Keery, H.; Lesser, J.; Martell, C. Enhanced Cognitive-Behavior Therapy and Family-Based Treatment for Adolescents with an Eating Disorder: A Non-Randomized Effectiveness Trial. *Psychol. Med.* **2022**, *52*, 2520–2530. [CrossRef] [PubMed]
205. Davis, L.E.; Attia, E. Recent Advances in Therapies for Eating Disorders. *F1000Research* **2019**, *8*. [CrossRef]
206. Linardon, J.; Shatte, A.; Messer, M.; Firth, J.; Fuller-Tyszkiewicz, M. E-Mental Health Interventions for the Treatment and Prevention of Eating Disorders: An Updated Systematic Review and Meta-Analysis. *J. Consult. Clin. Psychol.* **2020**, *88*, 994. [CrossRef]
207. Kakhramonovich, T.P. Epidemiology of Pysichiatric Disorders. *Tex. J. Med. Sci.* **2022**, *12*, 102–105.
208. Todisco, P.; Meneguzzo, P.; Garolla, A.; Antoniades, A.; Vogazianos, P.; Tozzi, F. Impulsive Behaviors and Clinical Outcomes Following a Flexible Intensive Inpatient Treatment for Eating Disorders: Findings from an Observational Study. *Eat. Weight Disord.-Stud. Anorex. Bulim. Obes.* **2021**, *26*, 869–877. [CrossRef]
209. Dahlenburg, S.C.; Gleaves, D.H.; Hutchinson, A.D. Treatment Outcome Research of Enhanced Cognitive Behaviour Therapy for Eating Disorders: A Systematic Review with Narrative and Meta-Analytic Synthesis. *Eat. Disord.* **2019**, *27*, 482–502. [CrossRef]
210. Joshua, P.R.; Lewis, V.; Kelty, S.F.; Boer, D.P. Is Schema Therapy Effective for Adults with Eating Disorders? A Systematic Review into the Evidence. *Cogn. Behav. Ther.* **2023**, *52*, 1–19. [CrossRef]
211. Smith, E.; Simpson, S.; McIntosh, V.V.W. Introduction to Schema Therapy for Eating Disorders. In *Schema Therapy for Eating Disorders*; Routledge: Abingdon, UK, 2019; pp. 3–11. ISBN 0429295715.
212. Thomas, J.J.; Becker, K.R.; Kuhnle, M.C.; Jo, J.H.; Harshman, S.G.; Wons, O.B.; Keshishian, A.C.; Hauser, K.; Breithaupt, L.; Liebman, R.E. Cognitive-behavioral Therapy for Avoidant/Restrictive Food Intake Disorder: Feasibility, Acceptability, and Proof-of-concept for Children and Adolescents. *Int. J. Eat. Disord.* **2020**, *53*, 1636–1646. [CrossRef]

213. Linardon, J.; Kothe, E.J.; Fuller-Tyszkiewicz, M. Efficacy of Psychotherapy for Bulimia Nervosa and Binge-eating Disorder on Self-esteem Improvement: Meta-analysis. *Eur. Eat. Disord. Rev.* **2019**, *27*, 109–123. [CrossRef]
214. Butler, R.M.; Heimberg, R.G. Exposure Therapy for Eating Disorders: A Systematic Review. *Clin. Psychol. Rev.* **2020**, *78*, 101851. [CrossRef]
215. Citrome, L. Binge Eating Disorder Revisited: What's New, What's Different, What's next. *CNS Spectr.* **2019**, *24*, 4–13. [CrossRef]
216. Coelho, J.S.; Suen, J.; Clark, B.A.; Marshall, S.K.; Geller, J.; Lam, P.-Y. Eating Disorder Diagnoses and Symptom Presentation in Transgender Youth: A Scoping Review. *Curr. Psychiatry Rep.* **2019**, *21*, 107. [CrossRef]
217. Zickgraf, H.F.; Ellis, J.M.; Essayli, J.H. Disentangling Orthorexia Nervosa from Healthy Eating and Other Eating Disorder Symptoms: Relationships with Clinical Impairment, Comorbidity, and Self-Reported Food Choices. *Appetite* **2019**, *134*, 40–49. [CrossRef]
218. Van Zyl, N.; Andrews, L.; Williamson, H.; Meyrick, J. The Effectiveness of Psychosocial Interventions to Support Psychological Well-Being in Post-Operative Bariatric Patients: A Systematic Review of Evidence. *Obes. Res. Clin. Pract.* **2020**, *14*, 404–420. [CrossRef]
219. Fiorillo, A.; Luciano, M.; Pompili, M.; Sartorius, N. Reducing the Mortality Gap in People with Severe Mental Disorders: The Role of Lifestyle Psychosocial Interventions. *Front. Psychiatry* **2019**, *10*, 434. [CrossRef]
220. Barbui, C.; Purgato, M.; Abdulmalik, J.; Acarturk, C.; Eaton, J.; Gastaldon, C.; Gureje, O.; Hanlon, C.; Jordans, M.; Lund, C. Efficacy of Psychosocial Interventions for Mental Health Outcomes in Low-Income and Middle-Income Countries: An Umbrella Review. *Lancet Psychiatry* **2020**, *7*, 162–172. [CrossRef]
221. Unikel-Santoncini, C.; de León-Vázquez, C.D.; Rivera-Márquez, J.A.; Bojorquez-Chapela, I.; Méndez-Ríos, E. Dissonance-Based Program for Eating Disorders Prevention in Mexican University Students. *Psychosoc. Interv.* **2019**, *28*, 29–35. [CrossRef]
222. Pina, A.A.; Polo, A.J.; Huey, S.J. Evidence-Based Psychosocial Interventions for Ethnic Minority Youth: The 10-Year Update. *J. Clin. Child Adolesc. Psychol.* **2019**, *48*, 179–202. [CrossRef]
223. Derenne, J.L.; Beresin, E. V Body Image, Media, and Eating Disorders. *Acad. Psychiatry* **2006**, *30*, 257–261. [CrossRef] [PubMed]
224. Ríos, P.B. Nutritional Treatment in Eating Disorders. In *Relevant Topics in Eating Disorders*; IntechOpen: London, UK, 2012; ISBN 9535100017.
225. Napolitano, F.; Bencivenga, F.; Pompili, E.; Angelillo, I.F. Assessment of Knowledge, Attitudes, and Behaviors toward Eating Disorders among Adolescents in Italy. *Int. J. Environ. Res. Public. Health* **2019**, *16*, 1448. [CrossRef] [PubMed]
226. Nagata, J.M.; Ganson, K.T.; Austin, S.B. Emerging Trends in Eating Disorders among Sexual and Gender Minorities. *Curr. Opin. Psychiatry* **2020**, *33*, 562. [CrossRef]
227. Favaro, A.; Busetto, P.; Collantoni, E.; Santonastaso, P. The Age of Onset of Eating Disorders. In *Age of Onset of Mental Disorders*; Springer: Cham, Switzerland; Midtown Manhattan: New York, NY, USA, 2019; pp. 203–216.
228. Demmler, J.C.; Brophy, S.T.; Marchant, A.; John, A.; Tan, J.O.A. Shining the Light on Eating Disorders, Incidence, Prognosis and Profiling of Patients in Primary and Secondary Care: National Data Linkage Study. *Br. J. Psychiatry* **2020**, *216*, 105–112. [CrossRef] [PubMed]
229. Qian, J.; Wu, Y.; Liu, F.; Zhu, Y.; Jin, H.; Zhang, H.; Wan, Y.; Li, C.; Yu, D. An Update on the Prevalence of Eating Disorders in the General Population: A Systematic Review and Meta-Analysis. *Eat. Weight Disord.-Stud. Anorex. Bulim. Obes.* **2022**, *27*, 415–428. [CrossRef]
230. Linardon, J.; Gleeson, J.; Yap, K.; Murphy, K.; Brennan, L. Meta-Analysis of the Effects of Third-Wave Behavioural Interventions on Disordered Eating and Body Image Concerns: Implications for Eating Disorder Prevention. *Cogn. Behav. Ther.* **2019**, *48*, 15–38. [CrossRef]
231. Nightingale, B.A.; Cassin, S.E. Disordered Eating among Individuals with Excess Weight: A Review of Recent Research. *Curr. Obes. Rep.* **2019**, *8*, 112–127. [CrossRef]
232. Cameron, K.; Ogrodniczuk, J.; Hadjipavlou, G. Changes in Alexithymia Following Psychological Intervention: A Review. *Harv. Rev. Psychiatry* **2014**, *22*, 162–178. [CrossRef]
233. Vandereycken, W.; Depreitere, L.; Probst, M. Body-Oriented Therapy for Anorexia Nervosa Patients. *Am. J. Psychother.* **1987**, *41*, 252–259. [CrossRef]
234. Nozaki, T.; Takao, M.; Takakura, S.; Koreeda-Arimura, C.; Ishido, K.; Yamada, Y.; Kawai, K.; Takii, M.; Kubo, C. Psychopathological Features of Patients with Prolonged Anorexia Nervosa as Assessed by the Minnesota Multiphasic Personality Inventory. *Eat. Weight Disord.-Stud. Anorex. Bulim. Obes.* **2006**, *11*, 59–65. [CrossRef]
235. Probst, M.; Majeweski, M.L.; Albertsen, M.N.; Catalan-Matamoros, D.; Danielsen, M.; De Herdt, A.; Duskova Zakova, H.; Fabricius, S.; Joern, C.; Kjölstad, G. Physiotherapy for Patients with Anorexia Nervosa. *Adv. Eat. Disord. Theory Res. Pract.* **2013**, *1*, 224–238. [CrossRef]
236. Thien, V.; Thomas, A.; Markin, D.; Birmingham, C.L. Pilot Study of a Graded Exercise Program for the Treatment of Anorexia Nervosa. *Int. J. Eat. Disord.* **2000**, *28*, 101–106. [CrossRef]
237. Agne, A.; Quesnel, D.A.; Larumbe-Zabala, E.; Olmedillas, H.; Graell-Berna, M.; Perez-Ruiz, M.; Fernandez-del-Valle, M. Progressive Resistance Exercise as Complementary Therapy Improves Quality of Life and Body Composition in Anorexia Nervosa: A Randomized Controlled Trial. *Complement. Ther. Clin. Pract.* **2022**, *48*, 101576. [CrossRef]
238. Field, T.; Schanberg, S.; Kuhn, C.; Field, T.; Fierro, K.; Henteleff, T.; Mueller, C.; Yando, R.; Shaw, S.; Burman, I. Bulimic Adolescents Benefit from Massage Therapy. *Adolescence* **1998**, *33*, 555–564.

239. Field, T.M. Massage Therapy Effects. *Am. Psychol.* **1998**, *53*, 1270. [CrossRef] [PubMed]
240. Pawlow, L.A.; O'neil, P.M.; Malcolm, R.J. Night Eating Syndrome: Effects of Brief Relaxation Training on Stress, Mood, Hunger, and Eating Patterns. *Int. J. Obes.* **2003**, *27*, 970–978. [CrossRef] [PubMed]
241. Wanden-Berghe, R.G.; Sanz-Valero, J.; Wanden-Berghe, C. The Application of Mindfulness to Eating Disorders Treatment: A Systematic Review. *Eat. Disord.* **2010**, *19*, 34–48. [CrossRef] [PubMed]
242. Probst, M.; Van Coppenolle, H.; Vandereycken, W. Body Experience in Anorexia Nervosa Patients: An Overview of Therapeutic Approaches. *Eat. Disord.* **1995**, *3*, 145–157. [CrossRef]
243. Zipfel, S.; Giel, K.E.; Bulik, C.M.; Hay, P.; Schmidt, U. Anorexia Nervosa: Aetiology, Assessment, and Treatment. *Lancet Psychiatry* **2015**, *2*, 1099–1111. [CrossRef]

nutrients

Article

Introducing a Smart Toy in Eating Disorder Treatment: A Pilot Study

Dimitri Chubinidze [1,2], Zhuo Li [1], Petr Slovak [3], Julian Baudinet [1,4], Emmanuelle Dufour [2] and Kate Tchanturia [1,2,5,*]

1 Department of Psychological Medicine, Institute of Psychiatry, Psychology and Neuroscience (IoPPN), King's College London, London SE5 8AF, UK; dimitri.chubinidze@kcl.ac.uk (D.C.); zhuo.li@kcl.ac.uk (Z.L.); julian.baudinet@kcl.ac.uk (J.B.)
2 National Eating Disorders Service, South London and Maudsley NHS Foundation Trust, London SE5 8AZ, UK; emmanuelle.dufour@slam.nhs.uk
3 Department of Informatics, King's College London, London WC2B 4BG, UK; petr.slovak@kcl.ac.uk
4 Maudsley Centre for Child and Adolescent Eating Disorders (MCCAED), Maudsley Hospital, London SE5 8AZ, UK
5 Department of Psychology, Ilia State University, Tbilisi 0162, Georgia
* Correspondence: kate.tchanturia@kcl.ac.uk; Tel.: +44-077-0804-6640

Abstract: Individuals with eating disorders (EDs) often encounter challenges related to body image, emotional, and sensory difficulties during nutritional rehabilitation. To address these challenges, a novel technology-enabled smart toy, Purrble, designed for immediate assistance in emotion regulation, is being explored. A mixed-method approach involving workshops, diaries, and focus groups was employed to examine the feasibility of Purrble as a therapeutic tool and its impact on participants' daily routines, sensory experiences, and emotional states. The study results demonstrate the engagement and acceptability of this device. Qualitative analysis revealed that participants independently used and integrated Purrble into their emotional and sensory regulation practices. These pilot results support the potential for a shift in the delivery of adjunct therapeutic tools through technology, particularly for ED patients with complex presentations. Future research is necessary to further explore the psychological benefits of this intervention.

Keywords: eating disorders; anorexia nervosa; autism; emotion regulation; sensory sensitivity; technology-enabled interventions

Citation: Chubinidze, D.; Li, Z.; Slovak, P.; Baudinet, J.; Dufour, E.; Tchanturia, K. Introducing a Smart Toy in Eating Disorder Treatment: A Pilot Study. *Nutrients* **2024**, *16*, 467. https://doi.org/10.3390/nu16040467

Academic Editor: Andrea Vania

Received: 7 December 2023
Revised: 29 January 2024
Accepted: 3 February 2024
Published: 6 February 2024

1. Introduction

Eating disorders (EDs) are complex conditions that impact an individual's psychological and physiological well-being. Often compounded by a range of comorbid conditions, EDs significantly influence both clinical practice and the quality of life for those affected [1,2]. A noteworthy comorbidity is that between EDs and the autism spectrum condition (henceforth 'autism', as preferred by the autistic community) [3], presenting unique challenges in daily routines for those undergoing treatment. This is particularly evident in the realm of sensory disturbances and difficulties in emotion regulation (ER) [4–7].

Research indicates a relationship between sensory processing and eating behaviors. Individuals with autism as well as with ED may exhibit heightened sensitivities in sensory domains such as smell, taste, vision, and texture [8–15]. This heightened sensitivity can be exacerbated in comorbid cases with ED and autism and can result in difficulties adapting to a treatment setting and the active avoidance of specific foods [16].

Emotion regulation capacity is also considered a crucial aspect of ED management. Challenges in recognizing, identifying and expressing emotions, which are exaggerated in the context of comorbid autism, can exacerbate the complexities of EDs [17–22]. Sensory disturbances, such as difficulties in interpreting and tolerating sensations, further influence emotion regulation. This is particularly relevant in EDs, where misinterpretation of bodily signals can lead to inappropriate emotional responses [23].

Consequently, for people with EDs, it may be beneficial to provide adjunct therapeutic interventions that specifically target ER, as well as sensory sensitivities. Accordingly, there is an expanding scope of research focused on exploring creative and innovative methods to augment talking therapies with evidence-based adjunct interventions [24]. In the novel clinical pathway for EDs and autism, known as PEACE (Pathway for Eating Disorders and Autism developed from Clinical Experience), these avenues are being explored (see [25] for details of this pathway). Within PEACE, adjustments have been made to the clinical environment, communication, and clinical practices to support patients' sensory and emotional needs. This includes the development of psychoeducational materials and experiential activities aimed at enhancing sensory well-being [26,27]. Additionally, a sensory workshop has been developed to assist patients in creating a soothing sensory toolkit, which has been proven to be beneficial for patients [27,28]. It is worth mentioning that these adjunct interventions can benefit not only individuals with comorbid EDs or autism but every patient with an ED. The recognition of the need for additional activities and enhanced collaboration across clinical services has led to the acknowledgment that further development in these areas is essential. In response, our approach involved developing existing adjunct tools, such as the sensory well-being workshop, with the introduction of a technology-enabled tool—a socially assistive robot known as Purrble [29,30].

Purrble's design is theoretically grounded in Gross' extended process model of ER [31]. It specifically targets two distinct stages: (a) the attentional deployment stage, by redirecting the user's focus away from the emotion-eliciting situation towards interaction with the device; and (b) the response modulation stage, by promoting downregulation through pleasant tactile interaction [32].

The soothing effects of interacting with Purrble are further heightened due to the device being modeled on human–animal interactions. Research in various domains suggests that tactile stimulation is an adaptive strategy for modulating stress responses. This mechanism is thought to be key in the emotion-regulating effects of human–animal interactions [33]. It is also hypothesized to play a role in 'social touch' [34] and in the calming benefits of animal-like robots [35]. Moreover, the effectiveness of human–animal interactions in promoting soothing and emotion regulation is further evidenced by animal-assisted therapies. These therapies have been recognized for addressing a variety of psychological needs, as indicated in several studies [36–40].

There is a growing body of research suggesting that Purrble can offer significant benefits by improving ER in young people [29,32,41] and in highly anxious university students [42]. Although there is emerging evidence suggesting that Purrble could support ER across the lifespan, its use in mental health clinical settings, particularly in the context of EDs, has not yet been explored.

The primary aim of this pilot study is to investigate (a) the level of engagement and acceptability of Purrble within intensive treatment programs for EDs and (b) the exploration of the perceived impact of this device on participants' sensory and emotional well-being.

2. Materials and Methods

2.1. Participants

Participants in the study included adult and young patients with a confirmed DSM-5 [43] diagnosis of an ED (made by lead clinicians). All participants were recruited from ED services at the South London and Maudsley NHS Foundation Trust (SLaM) for young persons and adults. Anonymous feedback from participants was collected as part of routine clinical practice. The data generated for this study were approved by the SLaM Child and Adolescent Mental Health (CAMHS) Service Evaluation and Audit Committee (328) and by the SLaM Clinical Governance and Audit Committee (032019).

2.2. Intervention

Purrble is a compact, affordable sensory device shaped like a plush animal (Figure 1) designed to provide in-the-moment ER support in daily life [29,41]. The device is conceptualized as a creature that experiences anxiety, requiring care and comfort when agitated. Its emotional state is conveyed through a simulated heartbeat, created by inbuilt electronics that generate vibrations ranging from frantic and anxious to slow and calm. The toy's heartbeat starts off fast and can be calmed through stroking, denoting a state of relaxation. While the calming process generally takes less than a minute, the duration can vary depending on the user's interaction with the device.

Figure 1. Intervention tool—Purrble.

Due to its sensory characteristics, including its soft and cuddly design, neutral colors, and gentle sensory feedback, such as soft vibrations and purring sounds, as well as its compact and portable size, Purrble showed promise as an addition to patients' sensory toolkits. It can be used both in clinical settings and various other environments. Furthermore, this intervention is designed to require minimal clinical effort while offering additional therapeutic support and facilitating seamless integration across various clinical services.

2.3. Procedure

The study involved three stages: (1) a sensory wellbeing workshop (a single session) and Purrble distribution; (2) 10-day interaction period with the device; and (3) post-experience focus group.

Sensory well-being workshop and distribution of Purrble: Participants attended a one-hour session aimed to enhance their understanding of the sensory system and its role in self-regulation. It also provided strategies to improve sensory well-being and equipped participants with the tools and language to communicate their sensory needs. The workshop included a do-it-yourself (DIY) activity, where participants created a sensory item, such as a glitter jar or scented hand cream. The workshop was led by two or three facilitators and the first author. Details of the workshop's procedure and protocol are available in the pilot evaluation [27,28]. At the conclusion of the workshop, the smart toy Purrble was introduced to the participants as a sensory tool. It was provided without explicit training in emotion regulation and distributed to each participant for personal use.

Ten-Day Interaction with Purrble Documented Through Diary Recording: Post-workshop, participants engaged with Purrble for a period of 10 days. They documented their daily interactions in a specially designed diary (see Supplementary Material Diary Template S1). This diary gauged the frequency of device use, its perceived benefits on emotional state improvement, and sensory sensitivity. Participants rated their experiences daily using a 10-point visual analogue scale. The diary also included a section detailing specific contexts or situations in which Purrble was used. Additionally, it featured an area for open-ended feedback, allowing participants to share reflections on their interactions with the device.

Focus Group Session: Following the 10-day experience, participants were invited to a focus group session to share their experiences. The discussion was also aimed at exploring any concerns, suggesting modifications, and discussing participants' future intentions regarding the use of Purrble. The focus group, lasting 45 min, was facilitated by the first author and a member of the respective clinical team and audio recorded for analysis.

2.4. *Analysis*

Focus group recordings were transcribed verbatim by DC. Transcripts were read and re-read by authors DC and ZL to ensure thorough familiarization. In parallel, qualitative content from the participants' 10-day diaries was compiled. An initial set of codes was generated by DC, based on the content of the transcripts, and it was evaluated by KT. The codes were then generated from the data using NVivo14 software. DC conducted qualitative content analysis [44] to identify potential themes, which were subsequently reviewed by ZL and KT. This review process involved evaluating how well the themes encapsulated the coded data and their reflection of the entire dataset. Themes are reported together with supporting quotes in the Section 3. Quotes from participants are anonymized using participant numbers (e.g., P1, P2 . . .). Quotes from focus groups are marked with 'FG'.

In the Section 3, qualitative and quantitative findings are presented together, rather than separately, with quantitative data presented under relevant themes. This integrated approach will specifically highlight themes related to the usage patterns of the device and its perceived impact on ER and sensory sensitivity.

3. Results

In total, four workshops were conducted from June to September 2023, involving 26 patients across three ED services: inpatient and daycare programs for adults and the Intensive day Treatment Program (ITP) for young people (cf. [45] for programme details). The number of participants in each workshop ranged from 3 to 10. However, five participants (19.23%) did not return their diaries and were absent from the focus group. The reasons for non-participation among the five excluded participants varied. Three required admissions to a different ward due to worsening conditions, including self-harm and suicidal thoughts. One temporarily left London for vacation, and one chose to discontinue participation due to discomfort with Purrble, reminiscent of a negative experience with a similar toy. Clinical characteristics varied among these participants, with two having comorbid depression, one with comorbid anxiety, one with autism, and one participant having no comorbidities. Individuals who did not return diaries were excluded from the analysis, leaving a total of 21 cases.

Table 1 summarizes the participant demographics. The majority of the participants (n = 20) were female, with one individual identifying as non-binary. The mean age of participants was 21.9 years, with an age range of 13 to 37 years. The predominant ethnic background in the study was White British (n = 16), from other ethnic backgrounds (n = 5). Approximately 52.38% (n = 11) of the participants had a comorbid condition, and 38.1% (n = 8) had either a formal diagnosis of autism or high autistic traits, as assessed by the clinical team.

Table 1. Summary of Participants' Health and Demographic Characteristics.

Variables	*n* = 21
Age, mean (SD)	21.9 (7.4)
Gender, female *n* (%)	20 (95.2%)
Ethnicity, *n* (%)	
White British	16 (76.2%)
White Other	1 (4.8%)
Black British	1 (4.8%)
Asian	1 (4.8%)
Mixed	2 (9.5%)
Diagnosis, *n* (%)	
AN restrictive subtype	17 (80.9%)
AN binge-purge subtype	3 (14.3%)
AN atypical	1 (4.8%)
Duration of ED in years, mean (SD)	4.8 (5.2)
Missing, *n* (%)	4 (19%)
BMI on admission, mean (SD)	15.85 (2.95)
Comorbidity, *n* (%)	
EUPD	2 (9.5%)
Autism (including trait and diagnosis)	8 (38.1%)
MDD	1 (4.8%)

Abbreviations: AN—Anorexia Nervosa; ED—Eating Disorder; BMI—Body Mass Index; EUPD—Emotionally Unstable Personality Disorder; MDD—Major Depressive Disorder; SD—Standard Deviation.

On average, participants interacted with Purrble approximately 25 times over the 10-day period. While a rating scale was initially provided, the majority of participants chose to directly report the number of interactions with Purrble, and we estimated the daily usage based on the latter. The trend in their daily usage of Purrble and self-reported improvement in emotional state and sensory sensitivity over the 10-day period is reported in Figure 2. Four themes arose from participants' qualitative feedback in diaries and focus groups, including the following: (1) engagement and relationship dynamics; (2) acceptability and ubiquity in daily activities; (3) anxiety, distress, and discomfort management; and (4) sensory tuning.

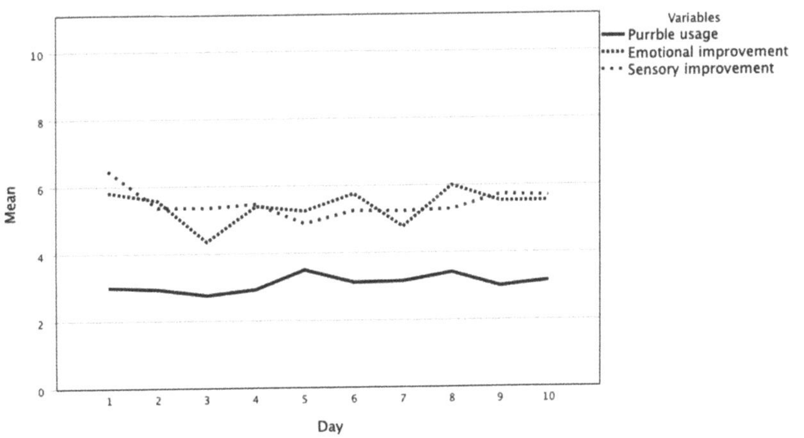

Figure 2. Trends in Purrble usage, emotional, and sensory improvements over 10-day experience.

3.1. Engagement and Relationship Dynamics

This theme focuses on the dynamics of participant engagement with the intervention device. It explores the evolving nature of engagement with Purrble and the developing dynamics of the relationship between participants and the device. Additionally, this section investigates themes such as empathetic engagement, bonding, and perceived responsibility.

Figure 2 shows that overall, participants' engagement with Purrble was consistent over the 10-day period. Qualitative analysis of diary entries and focus group discussions further identified that engagement evolved from curiosity into more focused and targeted interactions, particularly during moments requiring significant emotional support:

> *I was probably using Purrble more at the beginning, but I think that was more exploratory at the start. And then, as the days went on, it shifted to more acute situations where I felt, 'Okay, I really need to calm down'.* [FG]

> *Some days I used it more often than others, like when I encountered a stressful situation.* [FG]

The interactions between participants and Purrble illustrate an evolving, empathetic relationship characterized by sentiments of responsibility, caregiving, and mutual emotional regulation. This dynamic extended beyond mere functional use, as participants perceived Purrble as an emotionally responsive companion. For example, participant 20 expressed a sense of caregiving responsibility:

> *I sometimes feel like I have responsibility to look after it.* [P20]

Further, participant 8 described a shared emotional state with Purrble, finding intrinsic calmness in soothing the toy:

> *The only thing was that when Purrble's heart rate increased, it made me feel quite anxious as if my toy was anxious. However, when it was calm, I fell into a rhythm of soothing it when it purred, and that was calming.* [P8]

The depth of the bond between participants and Purrble is perhaps most poignantly illustrated in the creative expressions of the participants themselves. Participant 6 expressed their relationship with Purrble through a poem, reflecting the emotional connection and significance Purrble had during the experience. The poem reads as follows:

> *I love my Purrble. His name is Arlo.*
>
> *Now he comes everywhere I go.*
>
> *I hold him close before a meal,*
>
> *I love his softness I can feel.*
>
> *First his sound and feel gave me joy,*
>
> *Now just sitting next to me,*
>
> *He is more than just a toy.* [P6]

Despite participants gravitating towards this growing bond, it is important to acknowledge that not all perceived the empathetic relationship as beneficial. P18 voiced concerns about the potential emotional burden:

> *It felt like you kind of had to look after another person, and sometimes you don't feel like you can't even look after yourself.* [P18]

3.2. Acceptability and Ubiquity in Daily Activities

Through the analysis it emerged that Purrble was seamlessly incorporated into the participants' everyday practices, both at home and on the ward/in clinic. Its presence in frequently used areas, such as living rooms and bedrooms, emerged as a consistent theme. For instance, a focus group participant mentioned, '*I generally used Purrble when I was in the living room, either watching TV or engaged in other activities. It was usually there on the couch, close by me'*. This sentiment was echoed by another participant who remarked, *'Purrble has been a constant presence in my living room'* [FG], highlighting its pervasive role in their living spaces.

Additionally, participants highlighted Purrble's role as a portable companion, providing support in various settings. This included scenarios such as *'walking around the house'* [FG] and *'... using Purrble in the car while traveling'* [FG].

The acceptability of the device is further illustrated by its incorporation into various routine activities such as leisure, study, work, or as a mealtime companion. Participants mentioned diverse reasons for using the device, including comfort, relaxation and improved concentration.

Purrble served as a companion during leisure activities, such as watching TV, and played a role in evening calming routines. Its effectiveness in offering companionship and comfort, especially during evening times, was emphasized:

> *I normally use Purrble once or twice a day and mainly in the evening more often than throughout the day. So, especially before bed, it helped me calm down ... Helping me in relaxing before going to sleep.* [P16]

Focus group discussions further highlighted Purrble's effectiveness as a sleep aid:

> *I always find it difficult to sleep because my brain constantly races, going round in circles. Purrble gives me something to focus on, unlike when I focus on the TV, which often leads to me watching endless episodes of shows.* [FG]

In addition to its role in relaxation, participants mentioned Purrble's benefits in terms of heightening attentional focus and improving task management. A participant from the ITP noted its utility during academic activities:

> *I was using Purrble when I was doing homework and stuff because I usually get distracted. So, by having Purrble, it kind of helped me counter that stressful situation because there was a lot to do. Having Purrble there was like having something to focus on.* [FG]

In addition to predominantly home-based activities, other participants reported utilizing Purrble in clinical procedures and during psychological therapies, highlighting soothing support and facilitating verbalization during discussions of traumatic events:

> *It [Purrble] had been helpful during therapy. Yeah, kind of like a sensory soothing toolkit while talking about stuff which is difficult.* [FG]

> *I get really distressed during NG [nasogastric feeding], so I use the Purrble. I mean, I quite like the fact that the heartbeat slowed down. It didn't completely calm me down, but I think it did help. I think my heartbeat was really fast, and I think my heartbeat ended up slightly matching the heartbeat of the Purrble in a way.* [FG]

3.3. Anxiety, Distress and Discomfort Management

This theme, together with those that follow, provides an examination of Purrble's perceived impact in managing emotionally charged states and addressing sensory-related issues.

One of the most reported psychological states where participants utilized Purrble is anxiety. They emphasized its immediate use as a response mechanism to anxiety episodes:

> *A lot of times, it's still just hanging out when turned off, but then I'd actually seek it out for anxiety ...* [FG]

> *Once the anxiety started, I would take Purrble. So, it was usually after the anxiety had begun.* [FG]

> *For me, it was mainly when I felt anxious, which happens often, especially in the evenings. During the initial days, I'd just pick Purrble up whenever I felt that way and soothe myself by stroking him.* [FG]

One participant discussed the strategy of keeping Purrble close for immediate access in case of sudden anxiety:

There were moments when I felt anxious, and I'd carry it around with me. The location varied based on whether I was alone in the house or not. If my housemates were home, I tended to only use Purrble in my bedroom. But when I was on my own, I'd have Purrble close by, just in case I started to feel anxious and needed immediate access. [FG]

While participants shared experiences of Purrble's soothing effect in times of anxiety, some pointed out their need for more active forms of distraction during high-stress moments:

Well, sometimes when I'm more stressed, I think I need more active distractions than just sitting with Purrble. But, you know, maybe sitting with Purrble would've been good. [FG]

When I had a meeting, I was very angry and very distressed, and I just couldn't. I didn't want to soothe or comfort at all, so I was too angry, so I wasn't able to reach out for Purrble. [FG]

Therefore, some participants, particularly in times of high stress, preferred more proactive coping strategies and found it challenging to calm down with Purrble.

Purrble demonstrated effectiveness in a variety of situations, including feelings of loneliness:

When I woke up feeling anxious, Purrble helped me a bit; when I had an argument with my parents; when I felt sad or lonely; when I just needed comforting; needed something to distract me when I watched TV or read. When trying to get back to sleep. [P18]

Time spent in the ward can produce a sense of loneliness for inpatient service users. Participants reported the following:

When I felt sad or sometimes even empty, I'd turn to him. I remember when I had a dog, cuddling with him always uplifted my spirits. So, with Purrble, it felt similar, like he'd cheer me up when I was down. [P7]

I used Purrble when I was feeling sad, lonely, slightly anxious, when I wanted to feel comforted. [P3]

Participants also highlighted Purrble's role in managing somatic conditions such as pain and shaking:

After eating, I usually experience a lot of pain in my stomach. Purrble helped not only with my anxiety but also by providing a distraction. Just holding him kind of made me concentrate on him instead of the pain. My stomach gets really painful when I eat, so it helped to cuddle him against my stomach to take my mind off the pain. [FG]

When I get anxiety, I get really bad shakes in my hands, and I found that just holding Purrble and feeling its vibrations helped to calm my shakes. [P1]

Mealtime Companion

One common experience was that Purrble helped participants with managing meal-related stress. An emotionally charged scenario commonly faced by individuals with ED involves stress and discomfort during mealtimes. Participants reported Purrble's effectiveness in these specific contexts, including addressing pre- and post-mealtime challenges, as well as serving as a companion during meals.

A focus group participant shared their experience using Purrble to manage pre-mealtime anxiety:

I've mainly used him before mealtimes … I'm getting used to reaching for him before a meal, especially as I approach my second mealtime. That's when I tend to get most anxious, anticipating what's to come. [FG]

Participants also reported significant challenges associated with post-mealtime discomfort and diverse emotional states, areas where Purrble provided relief:

Yesterday was the first time that I actually went out to dinner and wasn't sick afterwards. By the time we got home, my anxiety was getting worse and worse, to the point where I was close to having a panic attack. So as soon as I got home, I reached out for Purrble. [FG]

Yes, I've used it after meals. In fact, that's probably the most common time I've gone to it. That's when I feel the highest level of stress. [FG]

In my experience, it especially helped when I was overwhelmed, stressed, or had a lot of anxiety after meals. In these situations, it helped to ground me and bring my heart rate down, hearing and feeling my Purrble's heart decrease and slow relaxed me and soothed me until I realized that my heart rate was slow, and I wasn't really having thoughts, or I forgot what I was overwhelmed about. [P15]

In some instances, participants incorporated Purrble into their actual mealtime routine:

I use it [Purrble] a lot before meals and sometimes during meals, so like have it next to me during the meal. [FG]

It was helpful in the dining room to have something with you. [FG]

Participants also reported using the device during intense distress and feeling overwhelmed by sensory cues during the mealtime. A participant from the Daycare service provided a vivid scenario of using Purrble in a case of distress and sensory sensitivity:

There was this one day where so much was happening. We were having a challenging meal . . .. Everything came to a head when a glass smashed. I was already cooking and feeling stressed, and then I knocked the glass over. I was so frustrated. I went upstairs, took a moment with Purrble, and then came back down. Only after that, I felt ready to clean up the glass and sort out the kitchen. [FG]

3.4. Sensory Tunning

The sensory characteristics of Purrble, such as its tactile and auditory feedback, played a significant role in ER. A participant from the focus group reflected on the calming effect of Purrble's heartbeat:

I found Purrble's heartbeat incredibly calming. I knew beforehand that rhythmical things help ground me, so I expected the heartbeat to have a decent impact on me. But the softness of it was also comforting. I often cover my face when I'm anxious, and with Purrble, I found myself doing the same, holding him close to my face because of his soft texture. It reminded me of the comfort I get from blankets. [FG]

Emotions, with their intensity and character, bring forth varying sensory demands. A recurrent attribute emphasized by participants concerning Purrble was its lack of squishiness, particularly significant during intense emotional experiences. For example, Participant 7, during an emotionally challenging clinical procedure (i.e., Naso Gastric Feeding), expressed a need for tactile engagement:

I realized I felt like I needed to squeeze something . . . But I felt like the problem was you couldn't squeeze [Purrble], because I felt like if I squeezed it, it would have broken or something.

To navigate this emotional distress, the participant intuitively combined Purrble with a dinosaur toy ('teddy') to fulfil this sensory need:

What I would do is I would put the Purrble next to my teddy so I could squeeze something and then feel at the same time. [P7]

Sensory tuning, or the capacity to align with Purrble's sensory output (i.e., heartbeat), emerged as a vital aspect for participants. The challenge of synchronizing Purrble's heartbeat with their emotional states was both a therapeutic tool and a challenge. P2 accentuated the importance of personalization in managing the device's heartbeat:

I found it challenging to lower Purrble's heartbeat. It would be ideal if there was an option to adjust settings, like deciding if or when the noises occur or controlling how long it takes for the heartbeat to slow down. I would prefer a shorter duration for the heartbeat to decelerate.

The sense of achievement felt by participants for successfully calming Purrble is indicative of its therapeutic benefits. This engagement is characterized by a determination to regulate the toy's heartbeat, as one participant expressed:

> *Sometimes it made me quite determined to stop its heartbeat. When it started purring, then I'd also feel calm.* [FG]

4. Discussion

The objective of this study was to investigate the engagement and acceptability of the socially assistive robot, Purrble, in people with EDs. A key novel aspect of this study was the introduction of the intervention in a mental health clinical setting for EDs. This is especially pertinent during the critical phase of nutritional rehabilitation, where the interplay of emotional regulation and sensory processing is vital in treatment efficacy and patient responsiveness. The incorporation of Purrble in such settings represents a progressive intersection of mental health technology and nutritional rehabilitation.

The pilot study's findings suggest that participant engagement with the device was consistent throughout the 10-day experiential period. This level of engagement and acceptance aligns with prior deployments of the device across diverse groups, including children [29,32] and highly anxious university students [42].

Qualitative analysis further suggests that participant interactions with Purrble developed into empathetic relationships, characterized by feelings of responsibility and caregiving. This dynamic is indicative of mechanisms that potentially underpin sustained engagement with the intervention. For instance, portraying the device as an anxious entity in need of care seemed to foster a sense of relationship and responsibility towards the Purrble's wellbeing, echoing findings from earlier studies involving Purrble [29,32,41]. However, it is crucial to acknowledge that while these relationships predominantly evoked positive feelings of comfort and care, they sometimes introduced an additional sense of responsibility. Such complex emotional responses hold particular importance in mental health clinical settings for patients with comorbid conditions, highlighting the necessity of a nuanced approach to the psychological effects of these interactions.

Additionally, our findings align with earlier research regarding the integration of the device in home environments, specifically within leisure and bedroom settings [46]. Uniquely in our study population, the Purrble device was seamlessly incorporated into the daily routines of participants, even in highly structured environments like ward settings and therapy sessions. This adaptability within both home and clinical contexts underscores the flexibility and potential of Purrble as a therapeutic tool in diverse settings and routines.

The patient-reported experiences in this study corroborate the immediate regulatory effects of the Purrble device evidenced in previous research [29,32,42]. However, our study's participant group uniquely articulated a wide range of psychological states and scenarios in which they employed the intervention. These ranged from somatic symptoms such as pain and shaking to emotional and cognitive states, including anxiety, stress, panic, distress, concentration difficulties, overwhelming emotions, sadness, loneliness, anger, and sensory overload.

A particularly noteworthy application emerged in relation to mealtime routines. Participants in our study reported a significant positive impact of Purrble's ability to provide sensory grounding and emotional regulation both before and after meals. The presence of Purrble was also noted to offer comfort and support during meals, a time often fraught with heightened anxiety for individuals with EDs. This finding underscores the potential of Purrble as a therapeutic tool not only in managing a broad spectrum of psychological states but also in specifically addressing the complex emotional and sensory challenges associated with mealtimes in this patient population.

Our findings support the effectiveness of the human–animal interaction design model in the context of ER, as hypothesized in a previous study [32]. They further confirm the calming effects of tactile engagement with animal-like robots [35], as well as the soothing and emotion-regulation benefits demonstrated by animal-assisted therapies [36–40].

Consistent with prior research [29,32], participants expressed appreciation for the sensory characteristics of the device, particularly the calming effects of tactile interaction. However, some participants highlighted difficulties in synchronizing their heartbeat with the device's heartbeat, a challenge that became more pronounced during periods of high stress. This difficulty in achieving sensory tuning, or attunement, highlights a need for more customizable sensory features in therapeutic tools or varied sensory properties for more proactive engagement. These findings resonate with those from previous sensory well-being workshops [27,28], emphasizing the importance of creating personalized sensory toolkits. Such tools are essential to effectively address the diverse and specific needs of patients with EDs, enhancing the efficacy of sensory-based therapeutic interventions.

4.1. Implications

The feedback from participants in this pilot study suggests significant implications for broader research and clinical applications, particularly in the context of ED treatment programs. It could be particularly beneficial as a complimentary resource for individuals on waitlists for ED treatment programs. Given the high demand for specialized care in this field, patients often face prolonged waiting periods, sometimes exceeding 18 months, before they can access treatment [47]. This extended wait period highlights the critical need for innovative interventions bridging the care gap during these crucial waiting periods, offering immediate support and potentially enhancing long-term treatment outcomes.

4.2. Limitations

While the study offers valuable initial insights, more rigorous, large-scale research is necessary to ascertain the full extent and strength of these psychological effects. Key questions remain, particularly regarding the potential for these effects to facilitate long-term behavioral changes. Additionally, it is yet to be determined whether the reported effects are substantial enough to effectuate clinically significant changes in emotion-coping mechanisms and strategies. This highlights the need for further research to explore the longevity and clinical relevance of implications reported, thereby contributing to a more comprehensive understanding of the intervention's impact in the context of mental health clinical settings and therapeutic practices.

Furthermore, in this pilot study, we have identified two methodological areas for improvement in future research. Firstly, our findings revealed a disparity between qualitative and quantitative data, particularly in terms of participant engagement with the device. To address this, future studies will employ more comprehensive measures to better capture the subtle changes in user experiences. The second aspect pertains to the diary design used for reporting interactions with Purrble. Participants opted to report their interactions directly rather than using the provided rating scale. This preference for direct reporting over the structured format indicates an opportunity to optimize the diary design for more user-friendly and accurate data collection in subsequent studies.

5. Conclusions

This study represents the first known investigation into the feasibility of a smart toy, Purrble, in mental health clinical settings. The exploratory data demonstrated how the intervention was integrated into the daily routines of participants in both home and clinical environments. The findings on engagement, acceptability, and qualitative effects are promising, demonstrating that participants were able to use the device and integrate it into their practices for regulating emotions and sensory sensitivity. Future research is needed to build upon these initial findings with larger studies that examine the psychological efficacy of this intervention. More broadly, pilot study results indicate the potential for a technology-enabled shift in the delivery of adjunct therapeutic tools, particularly for addressing the complex needs of patients with comorbid conditions.

44. Kleinheksel, A.J.; Rockich-Winston, N.; Tawfik, H.; Wyatt, T.R. Demystifying content analysis. *Am. J. Pharm. Educ.* **2020**, *84*, 7113. [CrossRef] [PubMed]
45. Baudinet, J.; Simic, M.; Griffiths, H.; Donnelly, C.; Stewart, C.; Goddard, E. Targeting maladaptive overcontrol with radically open dialectical behaviour therapy in a day programme for ado-lescents with restrictive eating disorders: An uncontrolled case series. *J. Eat. Disord.* **2020**, *8*, 68. [CrossRef] [PubMed]
46. Theofanopoulou, N.; Isbister, K.; Edbrooke-Childs, J.; Slovák, P. A Smart Toy Intervention to Promote Emotion Regulation in Middle Childhood: Feasibility Study. *JMIR Ment. Health* **2019**, *6*, e14029. [CrossRef] [PubMed]
47. NHS England. Children and Young People with an Eating Disorder Waiting Times. NHS England. 2020. Available online: https://www.england.nhs.uk/statistics/statistical-work-areas/cyped-waiting-times/ (accessed on 17 November 2023).

nutrients MDPI

Systematic Review

Randomized Controlled Trials to Treat Obesity in Military Populations: A Systematic Review and Meta-Analysis

Davide Gravina [1,2,*], Johanna Louise Keeler [1,3], Melahat Nur Akkese [1], Sevgi Bektas [1,4], Paula Fina [5], Charles Tweed [3,6], Gerd-Dieter Willmund [7], Janet Treasure [1,3] and Hubertus Himmerich [1,3]

1 Centre for Research in Eating and Weight Disorders, Institute of Psychiatry, Psychology and Neuroscience, King's College, London SE5 8AF, UK; johanna.keeler@kcl.ac.uk (J.L.K.); melahat_nur.akkese@kcl.ac.uk (M.N.A.); sevgi.bektas1@kcl.ac.uk (S.B.); janet.treasure@kcl.ac.uk (J.T.); hubertus.himmerich@kcl.ac.uk (H.H.)
2 Department of Clinical and Experimental Medicine, University of Pisa, 56127 Pisa, Italy
3 South London and Maudsley NHS Foundation Trust, Bethlem Royal Hospital, Monks Orchard Road, Beckenham BR3 3BX, UK; charles.tweed@slam.nhs.uk
4 Department of Psychology, Hacettepe University, Ankara 06800, Türkiye
5 Faculty of Psychology, Sigmund Freud University Vienna, Freudplatz 1, 1020 Vienna, Austria; paula@thefinas.eu
6 Royal Navy Reserve, London WC1N 1NP, UK
7 Bundeswehr Center for Military Mental Health, Military Hospital Berlin, 13, 10115 Berlin, Germany; gw@ptzbw.org
* Correspondence: davide.gravina@hotmail.it or davide.gravina@kcl.ac.uk

Abstract: In recent years, overweight and obesity have reached an alarmingly high incidence and prevalence worldwide; they have also been steadily increasing in military populations. Military personnel, as an occupational group, are often exposed to stressful and harmful environments that represent a risk factor for disordered eating, with major repercussions on both physical and mental health. This study aims to explore the effectiveness of weight loss interventions and assess the significance of current obesity treatments for these populations. Three online databases (PubMed, PsycInfo, and Web of Science) were screened to identify randomized controlled trials (RCTs) aiming to treat obesity in active-duty military personnel and veterans. Random-effects meta-analyses were conducted for body weight (BW) and body mass index (BMI) values, both longitudinally comparing treatment groups from pre-to-post intervention and cross-sectionally comparing the treatment group to controls at the end of the intervention. A total of 21 studies were included: 16 cross-sectional (BW: n = 15; BMI: n = 12) and 16 longitudinal (BW: n = 15; BMI: n = 12) studies were meta-analyzed, and 5 studies were narratively synthesized. A significant small overall BW and BMI reduction from baseline to post-intervention was observed (BW: $g = -0.10$; $p = 0.015$; BMI: $g = -0.32$; $p < 0.001$), together with a decreased BMI ($g = -0.16$; $p = 0.001$) and nominally lower BW ($g = -0.08$; $p = 0.178$) in the intervention group compared to controls at the post-intervention time-point. Despite limitations, such as the heterogeneity across the included interventions and the follow-up duration, our findings highlight how current weight loss interventions are effective in terms of BW and BMI reductions in military populations and how a comprehensive approach with multiple therapeutic goals should be taken during the intervention.

Keywords: obesity; obesity treatment; weight loss intervention; military population; active-duty military personnel; veterans; RCT; randomized controlled trial; meta-analysis

Citation: Gravina, D.; Keeler, J.L.; Akkese, M.N.; Bektas, S.; Fina, P.; Tweed, C.; Willmund, G.-D.; Treasure, J.; Himmerich, H. Randomized Controlled Trials to Treat Obesity in Military Populations: A Systematic Review and Meta-Analysis. *Nutrients* **2023**, *15*, 4778. https://doi.org/10.3390/nu15224778

Academic Editor: Arved Weimann

Received: 16 September 2023
Revised: 3 November 2023
Accepted: 5 November 2023
Published: 14 November 2023

1. Introduction

Obesity is a global epidemic. Overweight and obesity are currently two of the main public health concerns across the world, with more than 1 billion people worldwide being obese—650 million adults, 340 million adolescents, and 39 million children [1]. Worldwide obesity has nearly tripled since 1975, and this number is still dramatically increasing.

According to the World Obesity Atlas report from the World Obesity Federation, the majority of the global population (51%, or over 4 billion people) will be living with either overweight or obesity by 2035 based on current trends, and 1 in 4 people worldwide (nearly 2 billion) will meet or exceed a body mass index (BMI) of 30 kg/m^2 [2].

Moreover, obesity is a systemic disease impacting the well-being of the person as a whole, with consequences for both physical and mental health. Indeed, the World Health Organization (WHO) estimates that by 2025, approximately 167 million people—adults and children—will become less healthy because they are overweight or obese [1]. Overweight and obesity are major risk factors for many chronic diseases, including cardiovascular diseases such as hypertension, dyslipidemia, heart diseases, and stroke, which are the world's leading causes of death [3]. Being overweight can also lead to type 2 diabetes and musculoskeletal disorders such as osteoarthritis and represents a risk factor for the occurrence of some tumors, including endometrial, breast, ovarian, prostate, liver, gallbladder, kidney, and colon [1].

Obesity is one side of the double burden of nutritional problems, affecting all age groups and countries world-wide regardless of their developmental stage [4–7]. Today, more people are obese than underweight in every region of the world except sub-Saharan Africa and Asia [1]. As mentioned above, the issue has grown to epidemic proportions, with over 4 million people dying each year because of being overweight or obese and its consequences [1].

It is now widely accepted that occupational factors may play an important role in the occurrence of excessive body weight [8–10]. The main occupational risk factors are performing sedentary work and adverse lifestyle factors such as high levels of occupational stress [8,11–13]. Military personnel represent a population particularly exposed to a higher level of stress, with a higher risk of both obesity and mental health conditions such as major depression, post-traumatic stress disorder (PTSD), and disinhibited eating [14].

Indeed, the obesity epidemic has reached the military population [15,16]: possibly related to high levels of stress and harmful environmental factors, especially during military exercises, military missions, or during deployment and relocation [8,17,18]. In addition, there may be limitations on food selection or availability, especially among particular services such as the Navy and Marine Corps [19].

Another consideration is that military personnel are required to meet body weight and composition standards to remain in the military and to be eligible for their work. This might lead to an increased focus and concern on body weight, shape, and fitness. In fact, soldiers are required to maintain fitness standards that are not required for civilians to be ready for combat. Nonetheless, the armed forces population is experiencing similar patterns of increasing levels of overweight and obesity as observed in civilian society [20,21].

Between 2002 and 2015, the rate of obesity among U.S. active-duty military service members increased by 68%, such that now nearly two-thirds of U.S. active-duty across all branches meet criteria for overweight and obesity [15,22]; and, thereafter, the overall prevalence of obesity within the U.S. active component increased from 16.3% in 2015 to 17.9% in 2019 [23,24]. Obesity was significantly higher among Navy and Air Force military personnel [15], In particular, Navy personnel have the third highest rate of overweight and obesity among all service branches (64.6%), with 48.9% overweight and 15.7% obese [22]. Evidence from the UK shows slightly lower collective values, with 38% being overweight and 14% obese [20]. In the scientific literature, soldiers fulfilling the criteria of obesity constitute more than 15% in the US Army [23,25], 12% in the British Army [26,27], 13% in the Iranian Army [28,29], 6% in the Polish Air Force [30], and 44% in the Saudi Arabian Army [31].

Moreover, the spread of the COVID-19 pandemic in March 2020 led to restrictions, which may have contributed to a worsening problem of obesity within the active military. A recent study by Legg et al. [23] reported that the monthly prevalence of obesity in U.S. active component military members ranged from 15.0% in August 2020 to 19.3% in April 2021, confirming a further growth trend. Again, among the services, the Navy and Marine

Corps showed the largest absolute increase in mean monthly obesity prevalence from the pre-pandemic period to the pandemic period (0.78% and 0.77%, respectively) [23].

Not only does obesity within the military ranks negatively impact the professional perception of the military in terms of appearance, but it also compromises function. The psychological and physiological impacts on obese military personnel include problems with cardiorespiratory fitness and neuromuscular fitness [32], heat stress [33], sleep apnea [34], a higher risk of musculoskeletal injuries and load carriage [35,36], and also mental health problems such as depressive symptoms [37], PTSD [38], anxiety, and substance and alcohol abuse disorders [39,40].

Veterans also have higher levels of obesity: 78% of U.S. military veterans are overweight or obese [41], and 65% of female and 45% of male veterans report at least one symptom of Binge Eating Disorder (BED) [42]. Sedentary habits and suboptimal levels of physical activity [43] may combine with PTSD (seen in 11 and 30% of veterans) to increase disordered eating behaviors [44,45]. Thus, addressing eating behavior within this population is a high priority [46].

A variety of weight management programs have been developed to address the problem of obesity in military personnel, including prevention programs [47], lifestyle programs such as "LOOK AHEAD" [48,49], "LE3AN" [50], "MOVE", and "armyMOVE!" [51,52], nutrition-focusing programs [53,54], internet-based interventions [55–57], cognitive behavioral interventions [58,59], Navy weight management programs [60], pharmacological interventions [61], and surgical approaches [62,63].

Several systematic reviews of weight management programs have been undertaken. A systematic review in 2011 by Sanderson et al. [20], including 17 studies, found that interventions based on exercise, healthy eating information, behavioral modification, and structured follow-up were effective for weight reduction. Moreover, a systematic review conducted in 2017 summarized data from 38 studies and found weight loss for up to 12 months from dietary, physical activity, and weight management interventions among active-duty military personnel [64]. However, a systematic review in 2021 of seven studies that evaluated the effectiveness of weight loss interventions among U.S. active-duty military populations concluded that there is not a sufficient body of evidence to determine if interventions are effective [65].

At the time of writing this paper, there has been no meta-analysis published regarding weight management programs in military populations. Therefore, the present systematic review and meta-analysis aims to synthesize the results of studies investigating obesity treatments in military populations, in which the interventions have been delivered with a randomized controlled trial study design.

The main aim of the study is to address the effectiveness of the therapeutic intervention in terms of weight loss by comparing the pre-intervention body weight and BMI values of participants who received the treatment to the post-intervention values, as well as comparing the intervention group to controls.

A secondary aim of this study is to investigate the extent to which demographic and intervention characteristics, including age, duration of the intervention, body weight (BW) at baseline, and body mass index (BMI) at baseline, may be associated with changes in weight.

2. Materials and Methods

This systematic review and meta-analysis was registered in PROSPERO (CRD42023439107) and was conducted according to the Preferred Reporting Items for Systematic Reviews and Meta-Analyses (PRISMA) guidelines [66].

2.1. Search Strategy

Two reviewers (DG and PF) independently and systematically searched the following electronic databases: Web of Science, PubMed, and APA PsycInfo (Ovid) from inception until 7 September 2023. Searches included the following keywords: "obesity" or "adiposity" or "overweight" in combination with "military" or "military personnel" or "army" or "navy" or "military force" or "air force" or "soldiers" and "treatment" or "clinical trial" or "psychotherapy" or "psychological therapy" or "CBT" or "cognitive behavioral therapy" or "pharmacological" or "medication" or "drug". The full search strategy is listed in Supplementary Materials Text S1. Searches were supplemented by internet searches and manual hand searches through reference lists to identify potentially relevant additional studies.

2.2. Eligibility Criteria

To be included in the systematic review, studies needed to meet the following criteria:

- ➢ Human studies
- ➢ The studies included only individuals aged ≥ 18 years
- ➢ The studies involved military populations: Army, Navy, or Air Force personnel, Active duty-military personnel, veterans
- ➢ The topic of the studies focused on weight management interventions (any kind of treatment, e.g., pharmacological, psychotherapeutic, lifestyle, and nutritional) to treat obesity and overweight
- ➢ The studies assessed randomized controlled trials (RCTs) to test the treatment
- ➢ The studies are original articles
- ➢ The studies are published in the English, German, Italian, Spanish or French language
- ➢ Ongoing studies were eligible in order to maximize inclusion
- ➢ Articles from literature that met the following criteria were eligible for exclusion:
- ➢ The topic is not related to weight management interventions or obesity treatment
- ➢ The sample is different from the military population
- ➢ The study design is different from a randomized controlled trial
- ➢ Articles not published in English, German, Italian, Spanish, or French.
- ➢ Animal or pre-clinical studies
- ➢ Article type other than original articles (i.e., systematic reviews, narrative reviews, meta-analyses, cross-sectional studies, perspective papers, letters (without data), masters or doctoral theses, case reports).

2.3. Study Selection

The search process was conducted independently by two reviewers (DG and PF). Titles and abstracts of publications yielded by the searches were imported into EndNote, where duplicates were removed. Titles and abstracts of the remaining records were imported into Rayyan and assessed against the aforementioned eligibility criteria. Those deemed highly unlikely to be relevant were discarded. Full texts of the remaining articles were assessed against the eligibility criteria before inclusion (see Figure 1). A third reviewer (HH) supervised the process and resolved any disagreements.

2.4. Data Extraction

The first author (DG) extracted data from all included studies into an electronic summary table, which was checked by another reviewer (HH). The following data were extracted: publication identifiers (journal, year, first author); country of origin; study design; study objective; methodology; sample characteristics (mean age, sample size, body weight, BMI); clinical characteristics (type of intervention, duration of the intervention, main findings, outcome data). Authors were contacted for data not obtainable from the manuscript.

2.5. Risk of Bias Assessment

Two reviewers inspected each study to assess the risk of bias independently by using the Scottish Intercollegiate Guidelines Network (SIGN) Methodology Checklist for randomized controlled trials (Table S1), evaluating the internal validity of the RCTs and the overall assessment of the studies through several domains of bias. A third reviewer coordinated the process and resolved any disagreements. See Table S2 for the assessments of study quality.

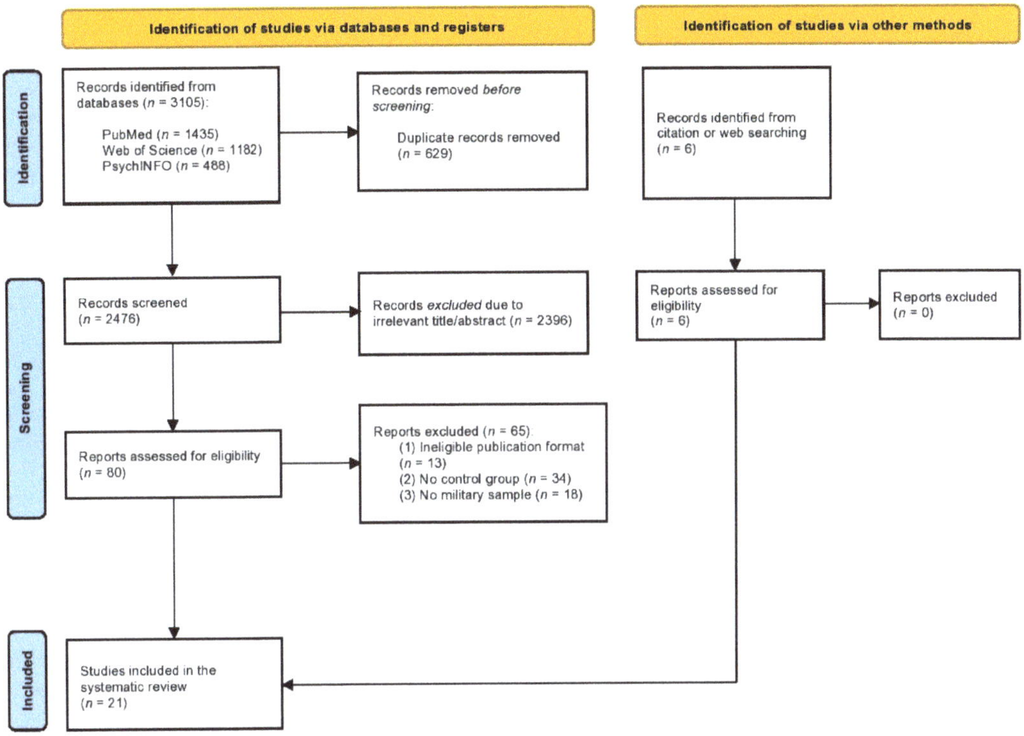

Figure 1. PRISMA flow diagram of the literature screening. Adapted from Page et al., 2021 [66].

2.6. Data Synthesis and Statistical Analysis

Individual meta-analyses were conducted in STATA (Release 17; StataCorp LP) using the 'meta-set' and 'meta-summarize' commands. The primary outcome measures were body weight (kg) and BMI (kg/m^2). Higgins I^2 was used to assess between-study heterogeneity, which was considered to be high when $I^2 > 75\%$. A random-effects model using the DerSimonian and Laird method [67] was used to pool the effect size (ES; Hedge's g) from the between-group difference for each study (Interventions vs. Controls, and Pre vs. Post) relative to the sample size. The ES expressed the difference between BW and BMI at baseline and at the last available follow-up timepoint for longitudinal meta-analyses (pre-to-post intervention). Regarding the cross-sectional meta-analyses, the ES expressed the difference between BW and BMI at post-intervention, where values were compared between treatment and control groups. Statistical significance of between-group differences was ascertained according to the $p < 0.05$ threshold.

We performed meta-regressions, using the 'metareg' command, to investigate the association between key demographic, study, and clinical variables (BMI at baseline, BW at baseline, mean age, duration of the intervention), and the ES of the meta-analysis comparing the pre-to-post intervention groups. The meta-regression analyses were not limited by the number of studies, as it is recommended that meta-regressions be conducted when there are 10 or more studies available for inclusion [68].

Alongside the evaluation of the primary outcome of the study, additional meta-analyses divided by sample and intervention type were conducted using the 'meta-set' and 'meta-summarize' commands, with the aim of evaluating the effectiveness of the different types of interventions included in the study.

Publication bias was assessed using Egger's test for small study effects, with funnel plots derived using the 'meta funnelplot' command. The Duval and Tweedie Trim and Fill method [69] was used successively if funnel plot asymmetry was detected to identify whether there were any missing studies.

3. Results

The literature search resulted in 3105 studies (Figure 1). After removing duplicates and reviewing titles and abstracts, 80 studies were retrieved for full-text review. A further 65 studies were excluded for the following reasons: ineligible publication format (n = 13), lack of a control group (n = 34), and not a military sample (n = 18). Finally, 21 studies [57,60,70–88] were included in this systematic review, 16 of which [57,71,72,74–78,80–85,87,88] were included in the meta-analysis.

3.1. *Characteristics of Included Studies and Participants*

A total of 4253 participants from 21 studies were included in this systematic review and meta-analysis. All studies had a randomized controlled clinical trial design. The duration of the interventions ranged from two weeks to twenty-four months, with a median of 26 weeks. The average age of participants ranged from 21.0 years to 61.8 years (mean ± SD: 45.6 ± 15.8), and most of the overall sample was represented by males (58%). The overall mean ± SD BW was 97.4 ± 21.2, and the overall mean ± SD BMI was 32.4 ± 5.6. The majority of the studies were conducted in the United States of America (n = 18); two studies were from Iran and one from Brazil. The military population studied included active-duty military soldiers (n = 7), Navy personnel (n = 3), Air Force personnel (n = 3), and veterans (n = 8).

Of these 21 included studies, sixteen studies [57,71,72,74–78,80–85,87,88] were included in the longitudinal meta-analyses investigating the effect sizes of BW [71,72,74–78,80,82–85,87,88] and BMI [57,71,72,77,78,80–85,87] comparing pre- to post-intervention values for the treatment group (11 studies were used in both the BW and BMI longitudinal meta-analyses). Sixteen studies [57,71,72,74–78,80–85,87,88] were also included in the cross-sectional meta-analyses comparing BW [57,71,72,74–78,80,82–85,87,88] and BMI [57, 71,72,77,78,80–85,87] data from the treatment group to controls at the post-intervention time-point (11 studies were used in both the BW and BMI cross-sectional meta-analyses).

Regarding the longitudinal meta-analyses, the pooled mean ± SD age for the intervention group was 42.5 ± 15.8 for the meta-analysis investigating BW (reported by 12 studies) and 35.8 ± 11.7 for the meta-analysis investigating BMI (reported by 9 studies). The pooled mean ± SD BW was 93.8 ± 20.7 for the pre-intervention group and 91.9 ± 20.0 for the post-intervention group, as reported by 15 studies. The pooled mean ± SD BMI reported by 12 studies was 30.6 ± 5.2 and 29.9 ± 5.2 for the pre-intervention and post-intervention groups, respectively.

Regarding the cross-sectional meta-analyses, the pooled mean ± SD age was 42.5 ± 15.8 for the treatment group and 44.0 ± 16.2 for the control group (reported by 12 studies) in the meta-analysis investigating BW; the pooled mean ± SD age for the cross-sectional meta-analyses investigating BMI was 35.8 ± 11.7 for the treatment group and 35.9 ± 11.6 for the control group (reported by 9 studies). The pooled mean ± SD BW was 91.9 ± 20.0 in the treatment group and 95.3 ± 19.9 in the control group, as reported in 15 studies. The pooled mean ± SD BMI was 29.9 ± 5.2 and 30.7 ± 5.2 for the treatment and control groups, respectively, reported in 12 studies.

Findings from five studies [60,70,73,79,86] were narratively synthesized and not included in the meta-analysis due to the required data being unavailable (n = 3) or because the studies were still in progress (n = 2). The characteristics of each study are summarized in Table 1.

Table 1. Study and sample characteristics for studies included in the systematic review and meta-analysis.

Study, Country	Year	Population	Sample (N)	Age (Years, M ± SD)	Study Design	Intervention	BW (M ± SD) T0 T1	BMI (M ± SD) T0 T1	Duration (Weeks)
Afari et al., [60] USA (Study ongoing)	2019	US Navy	178	29.7 ± 6.9	RCT	ACT + SS	94.8 ± 18.8 NR	33.1 ± 3.9 NR	8
Boutelle et al., [70] USA (Study ongoing)	2023	US Veterans	129	47.1 ± 11.3	RCT	CHARGE	NR NR	34.8 ± 4.7 NR	20
Damschroder et al., [71] USA	2014	US Veterans	481	55.0 ± 10.0	RCT	ASPIRE	113.2 ± 23.2 111.1 ± 25.1	36.6 ± 6.2 35.9 ± 7.3	52
Dennis et al., [72] USA	1999	US Navy	39	31.2 ± 6.5	RCT	Shipboard Weight Control Program	107.5 ± 11.0 100.3 ± 11.0	33.5 ± 2.8 31.2 ± 3.2	26
Erickson et al., [73] USA	2017	US Veterans	121	51.3 ± 9.2	RCT	LB Intervention	103.1 ± NR 101.8 ± NR	NR NR	52
Evans-Hudnall et al., [74] USA	2020	US Veterans	34	58.7 ± 9.1	RCT	HERO	112.8 ± 23.0 112.7 ± 17.5	36.7 ± 7.0 NR	16

Table 1. *Cont.*

Study, Country	Year	Population	Sample (N)	Age (Years, M ± SD)	Study Design	Intervention	BW (M ± SD) T0 T1	BMI (M ± SD) T0 T1	Duration (Weeks)
Goldberg et al., [75] USA	2013	US Veterans	109	52.0 ± 9.1	RCT	MOVE!	106.0 ± 21.7 105.9 ± 19.1	NR NR	26
Hoerster et al., [76] USA	2022	US Veterans	511	57.4 ± 13.9	RCT	D-ELITE	102.3 ± 14.5 100.4 ± 15.4	NR NR	52
Hosseini-Amiri et al., [77] Iran	2018	Active Soldiers	94	23.3 ± 1.6	RCT	EPPM	100.1 ±10.7 97.9 ±10.1	31.9 ± 2.7 30.9 ± 2.6	4
Hunter et al., [57] USA	2008	US Air Force active-duty personnel	446	34.0 ± 7.3	RCT	BIT	87.0 ± 15.2 86.4 ± 15.3	29.3 ± 3.0 29.1 ± 3.1	26
Krukowski et al., [78] USA	2018	Active duty-military personnel	248	34.6 ± 7.5	RCT	Look AHEAD ILI	89.0 ± 14.3 87.5 ± 14.9	30.4 ± 2.9 29.9 ± 3.2	52
Lutes et al., [79] USA	2017	US Veterans	332	55.9 ± 9.5	RCT	ASPIRE-SC	113.0 ± 22.4 111.4 ± NR	36.2 ± 6.0 NR	104

Table 1. *Cont.*

Study, Country	Year	Population	Sample (N)	Age (Years, M ± SD)	Study Design	Intervention	BW (M ± SD) T0 T1	BMI (M ± SD) T0 T1	Duration (Weeks)
McDoniel et al., [80] USA	2008	US Air Force active-duty personnel	54	28.0 ± 7.3	RCT	"Sensible Weight" Program	90.5 ± 14.4 89.1 ± 14.6	29.8 ± 2.4 29.1 ± 2.5	13
Parastouei et al., [81] Iran	2020	Military Personnel	60	41.5 ± 7.2	RCT	Synbiotic Supplementation	NR NR	32.1 ± 0.8 31.8 ± 0.9	8
Paravidino et al., [82] Brazil	2021	Military trainer of Naval Academy	72	21.0 ± 2.0	RCT	EFECT study	87.3 ± 9.6 86.4 ± 10.2	27.9 ± 2.1 27.6 ±2.1	2
Perez-Munoz et al., [83] USA	2023	Active-duty Military Women and TRICARE beneficiaries	430	30.6 ± 4.9	RCT	PPWL Intervention	74.2 ± 15.0 74.7 ± 15.0	27.6 ± 5.2 28.2 ± 5.6	26
Smith et al., [84] USA	2010	US Army Soldiers	113	28.4 ± 7.4	RCT	Meal-Replacement Program	97.2 ± 15.1 93.8 ± 15.5	32.8 ±3.0 32.0 ± 3.0	26
Smith et al., [85] USA	2012	Active-duty Soldiers	435	NR	RCT	Orlistat	99.6 ± 15.8 96.5 ± 16.5	33.3 ± 3.4 32.3 ± 3.7	26

Table 1. *Cont.*

Study, Country	Year	Population	Sample (N)	Age (Years, M ± SD)	Study Design	Intervention	BW (M ± SD) T0 T1	BMI (M ± SD) T0 T1	Duration (Weeks)
Staudter et al. [86] USA	2011	US Active-duty military	106	50.0 ± 9.3	RCT	Pedometer Intervention	87.6 ± 16.3 NR	32.5 ± 5.4 NR	12
Veverka et al., [87] USA	2003	US Air Force active-duty personnel	39	NR	RCT	Stages of Change Model	85.5 ± 12.9 85.9 ± 13.2	26.9 ± 3.3 26.7 ± 3.6	26
Voils et al., [88] USA	2017	US Veterans	222	61.8 ± 8.3	RCT	Maintenance Intervention	103.6 ± 20.4 105.2 ± 14.4	34.0 ± 6.1 NR	56

Abbreviations: BW: body weight; BMI: body mass index; T0: baseline time-point (pre-intervention); T1: post-intervention time-point; RCT: randomized controlled trial; ACT: Acceptance and Commitment Therapy; SS: ShipShape; NR: Not Reported; CHARGE: Controlling Hunger and ReGulating Eating for Veterans; ASPIRE-SC: The Aspiring for Lifelong Health program-Small Changes intervention; LB: Lifestyle Balance; HERO: Healthy Emotions and Improving Health Behavior Outcomes; D-ELITE: Evaluation of Lifestyle Interventions to Treat Elevated Cardiometabolic Risk in Primary Care; EPPM: Extended Parallel Process Model on knowledge, attitudes, and practices; BIT: Behavioral Internet Therapy; ILI: Intensive Lifestyle Intervention; EFECT: Physical Exercise and Compensatory Effects; TRICARE: The uniformed services health care program for active-duty service members and active provided by the United States Department of Defense (DoD); PPWL: Post-partum Weight Loss.

3.2. *Weight Loss Interventions of the Included Studies*

Across 21 studies, the majority of papers (n = 19) reported an intervention based on cognitive behavioral therapy or behavioral modification, except for two studies [81,84].

Diet education or nutritional modifications was offered by twelve of the interventions [57,72,73,75,78,80,81,83–85,87,88], as well as one study that provided a synbiotic supplementation [81] and two studies that provided a meal-replacement program [78,84]. In twelve studies [57,60,70–72,76,77,82,83,86–88], participants self-reported outcomes such as weight, whereas in twelve studies [60,70,71,73–75,78–81,84,85], specialists were available for monitoring, counselling, and outcome measurement.

Specific devices, such as a pedometer [86] and a calorimeter [80], were used in two studies, and in five studies, the interventions were delivered through an Internet-based program or Internet-assistance were provided during the intervention [57,76,78,86,87].

Additionally, two studies used a specific weight loss program for the Navy population, or Marines [60,72]. One study focused on a post-partum weight loss program [83], and another study explored a weight loss maintenance program to address the amount of weight regain and retention during follow-up [88].

Finally, one study investigated the efficacy of the pharmacologic treatment Orlistat versus placebo [85].

A summary of the interventions provided by each study is shown in Supplementary Materials Table S3.

3.3. *Results for the Meta-Analysis Comparing Pre-to-Post Intervention for the Treatment Group*

Data from a total of 15 studies, including a total sample of 2666 participants (pre-intervention group n = 1431; post-intervention group n = 1235) were used to compare the pre-treatment values of the BW of participants who received the intervention with the post-treatment values. Similarly, a total sample of 1942 participants from 12 studies was used to compare BMI values (pre-intervention group n = 1028; post-intervention group n = 914). Forest plots for the BW and BMI meta-analyses are presented in Figures 2 and 3, respectively. A summary of comparative outcomes and heterogeneity for BW and BMI meta-analyses for the intervention group are shown in Table 2.

Table 2. Summary of overall outcomes and heterogeneity for cross-sectional and longitudinal BW and BMI meta-analyses, with analyses according to sample type (active-duty personnel and veterans).

Group	*N*	*SMD*	*95% CI*	*Z*	*p*	*Heterogeneity*
Pre-to-post intervention	(Pre, Post)					
Overall						
BW (n = 15)	1431, 1235	−0.10	−0.18, −0.02	−2.43	0.015 *	I^2 = 3.45%; p = 0.413
BMI (n = 12)	1028, 914	−0.32	−0.48, −0.15	−3.74	<0.001 *	I^2 = 61.2%; p < 0.001 **
Active-duty personnel						
BW (n = 10)	838, 726	−0.12	−0.23, −0.00	−2.04	0.041 *	I^2 = 12.6%; p = 0.327
BMI (n = 11)	868, 792	−0.35	−0.54, −0.16	−3.62	<0.001 *	I^2 = 64.3%; p <0.002 **
Veterans						
BW (n = 5)	593, 476	−0.09	−0.21, −0.04	−1.36	0.174	I^2 = 4.71%; p = 0.380
Treatment vs. controls	(Treatment, Control)					
Overall						
BW (n = 15)	1235, 1094	−0.08	−0.19, 0.03	−1.35	0.178	I^2 = 32.7%; p = 0.107
BMI (n = 12)	914, 752	−0.16	−0.26, −0.06	−3.23	0.001 *	I^2 = 0.00%; p = 0.711
Active-duty personnel						
BW (n = 10)	762, 603	−0.06	−0.21, 0.09	−0.77	0.439	I^2 = 31.1%; p = 0.159
BMI (n = 11)	792; 633	−0.17	−0.27, −0.06	−3.14	0.001 *	I^2 = 0.00%; p = 0.643
Veterans						
BW (n = 5)	473, 491	−0.10	−0.30, −0.09	−1.05	0.294	I^2 = 47.6%; p = 0.106

Notes. * Significant findings at $p < 0.05$; ** Significant findings at $p < 0.01$; BW: body weight; BMI: body mass index. N: Number of participants; SMD: Standardized Mean Difference; CI: Confidence Intervals; Z: Z-value.

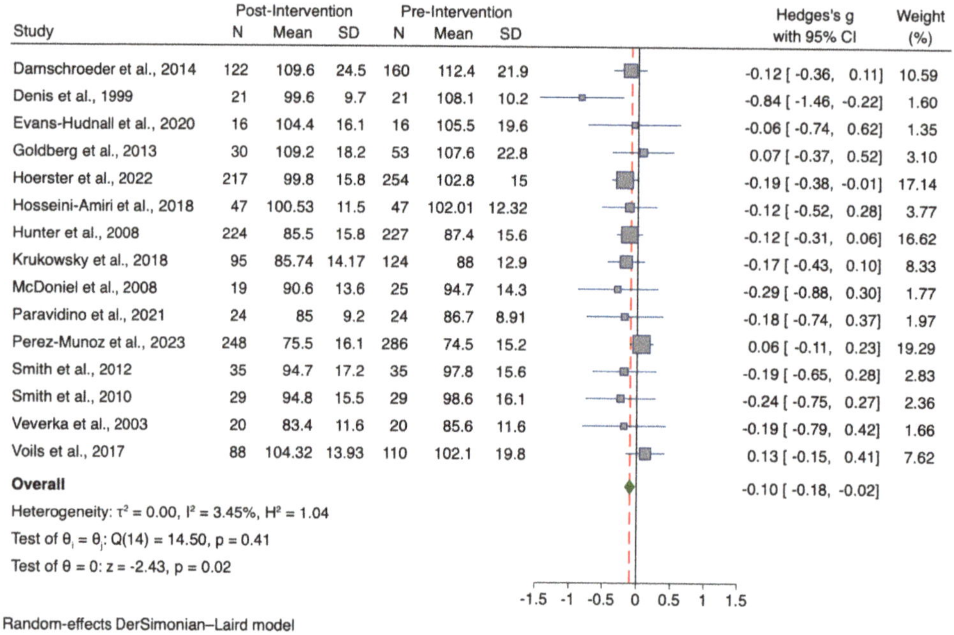

Figure 2. Forest plot of Hedge's *g* comparing pre-intervention and post-intervention BW values of the treatment group [57,71,72,74–78,80,82–85,87,88]. Zero indicates no effect, whereas values on the left of this line indicate a decrease in BW when comparing values at baseline and after treatment. The dashed line represents the overall effect size.

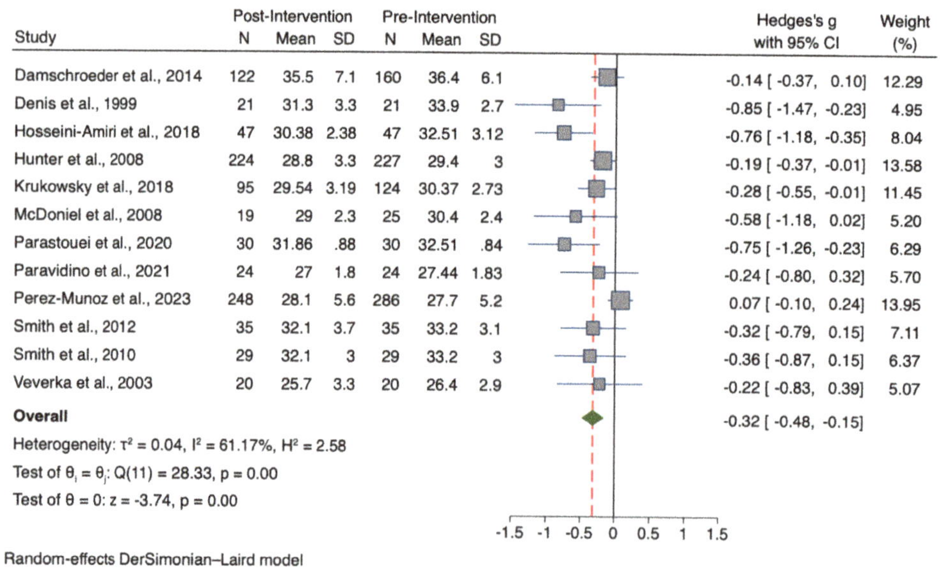

Figure 3. Forest plot of Hedge's *g* comparing pre-intervention and post-intervention BMI values of the treatment group [57,71,72,77,78,80–85,87]. Zero indicates no effect, whereas values on the left of this line indicate a decrease in BMI when comparing values at baseline and after treatment. The dashed line represents the overall effect size.

Participants who received the intervention showed overall lower post-treatment BW values than pre-treatment, with a small but significant effect (Hedges $g = -0.10$; 95% CI -0.02, -0.18; $p = 0.015$). Furthermore, results showed intervention effectiveness also in terms of BMI reduction, as a small-to-medium but significant overall reduction of post-treatment body mass index than baseline was observed ($g = -0.32$; 95% CI -0.15, -0.48; $p = 0.0002$).

Results for the Meta-Analyses Comparing Pre-to-Post Outcomes Per Sample Type

Additional meta-analyses were conducted, separating active military personnel from veterans, with the aim of evaluating the effectiveness of the weight loss interventions on these different samples analyzed individually (Table 2). Results showed a significant decrease in both BMI ($g = -0.35$; $p < 0.001$) and BW ($g = -0.12$; $p = 0.041$) in active-duty military personnel. In veterans, there was a small decrease in BW, although this was not significant ($g = -0.09$; $p = 0.173$). There was insufficient data to meta-analyze outcomes relating to BMI in veterans.

3.4. *Results for the Treatment Group vs. Controls Meta-Analyses*

A total of 15 studies were included in the meta-analysis comparing BW data for the interventional group to the comparison group, with a total sample of 2329 participants (intervention group, n = 1235; control group, n = 1094). Results showed an overall lower body weight in the treatment group compared to controls at the post-intervention time-point (T1), although statistical significance was not achieved ($g = -0.08$; 95% CI -0.19, 0.03; $p = 0.178$). The forest plots for BW and BMI are presented in Figures 4 and 5, respectively, and comparative outcomes and heterogeneity are summarized in Table 2.

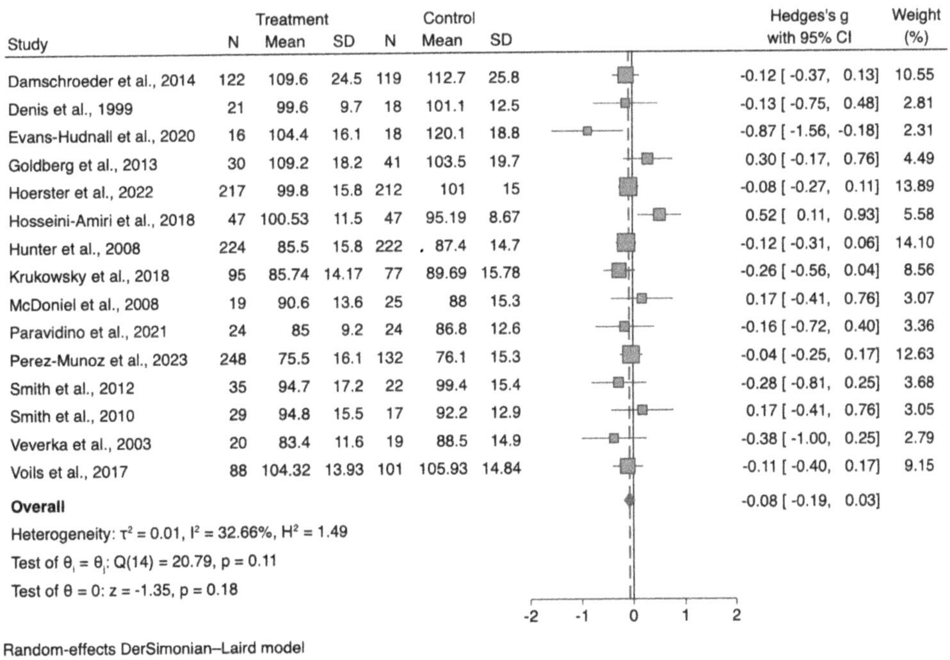

Figure 4. Forest plot of Hedge's *g* of BW values in the treatment group compared to controls [57,71, 72,74–78,80,82–85,87,88]. Zero indicates no effect, whereas values on the left of this line indicate a decrease in body weight when comparing values between the treatment and control groups at the post-intervention time-point. The dashed line represents the overall effect size.

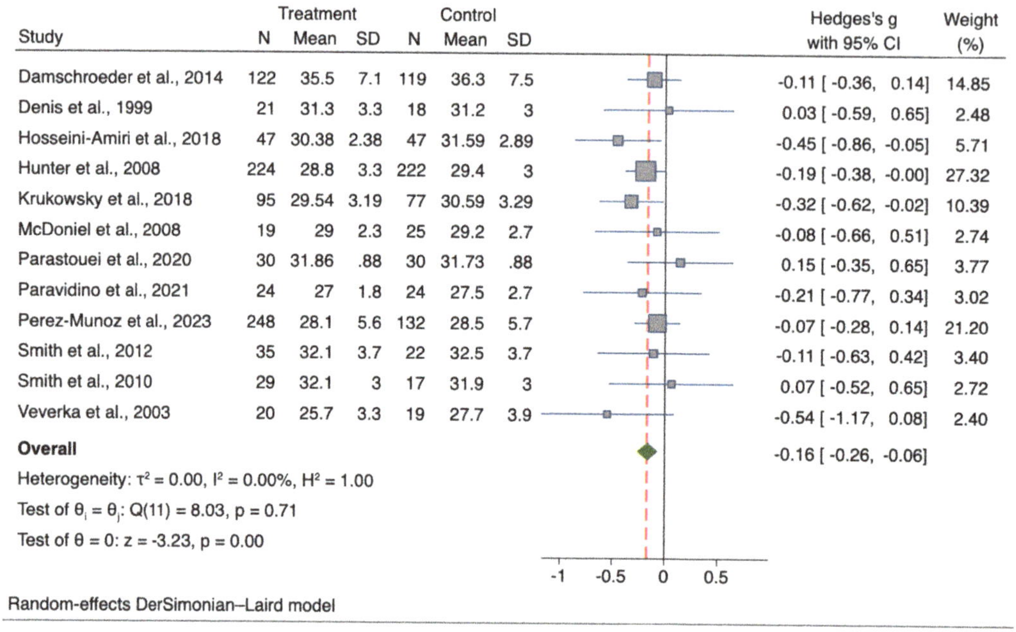

Figure 5. Forest plot of Hedge's g of BMI values in the treatment group compared to controls [57,71,72,77,78,80–85,87]. Zero indicates no effect, whereas values on the left of this line indicate a decrease in body mass index when comparing values between the treatment and control groups at the post-intervention time-point. The dashed line represents the overall effect size.

A total sample of 1666 participants from 12 studies was used to compare BMI values (intervention group n = 914; control group n = 752). At post-intervention, a significantly lower BMI in the treatment group was found in comparison with controls ($g = -0.16$; 95% CI -0.26, -0.06; $p = 0.001$).

Results for Treatment Group vs. Control Meta-Analyses Per Sample Type

Results showed that active-duty personnel in the treatment groups had a significantly lower BMI ($g = -0.17$; $p = 0.002$) and nominally lower BW compared to controls ($g = -0.06$; $p = 0.439$) at the post-intervention time-point. In veterans, BW was nominally lower in those given the intervention, although this failed to reach significance ($g = -0.10$; $p = 0.294$). Forest plots for each meta-analysis are provided in Supplemental Materials Figures S1–S6.

3.5. Meta-Regression Analyses

Meta-regression analyses were performed to test whether there was a significant relationship between various continuous variables and the effect sizes of BW and BMI in the group that received the weight loss intervention. Results are presented in Table 3. The continuous variables investigated in individual meta-regression analyses were the age of the sample, the duration of the intervention, the BW at baseline, and the BMI at baseline. None of these variables were significantly associated with the difference in BW or BMI between pre-intervention and post-intervention values (Table 3). An overview of the studies included in the meta-regression analyses is also provided in Table S4.

Table 3. Results of meta-regression analyses.

Group	*Variable*	*N Studies Included*	*β (SD)*	*95% CIs*	*p*
Intervention Group					
BW					
	Age	13	−0.002 (0.004)	−0.009, 0.005	0.609
	BW at baseline	15	0.005 (0.003)	−0.001, 0.011	0.135
	Duration of the intervention	15	−0.000 (0.003)	−0.005, 0.005	0.944
BMI					
	Age	10	0.002 (0.021)	−0.028, 0.015	0.545
	BMI at baseline	12	0.034 (0.028)	−0.021, 0.089	0.224
	Duration of the intervention	12	−0.008 (0.005)	−0.018, 0.003	0.142

Notes. BW: body weight; BMI: body mass index; N: Number; β (SD): Beta Coefficient (Standard Deviation); CIs: Confidence Intervals; *p*: *p*-value (significant findings at $p < 0.05$).

3.6. Sensitivity Analyses

All main meta-analyses showed low (BW longitudinal, $I^2 = 3.45\%$; BW cross-sectional, $I^2 = 32.7\%$; BMI cross-sectional, $I^2 = 0.0\%$) or moderate (BMI longitudinal, $I^2 = 61.2\%$) heterogeneity. The Egger's test for small study effects found that one meta-analysis had likely publication bias (BW cross-sectional: t = −0.18, $p = 0.859$; BMI cross-sectional: t = 0.04, $p = 0.968$; BW longitudinal: t = 1.41, $p = 0.160$; BMI longitudinal: t = 3.41, $p < 0.001$). When adjusting for publication bias using the trim and fill method, the BMI longitudinal meta-analysis remained statistically significant ($g = -0.20$; 95% CI −0.29, −0.11). See Figures S7–S10 for funnel plots.

Regarding the meta-analyses per sample type, low to moderate heterogeneity was also observed for both active-duty military personnel (BW longitudinal, $I^2 = 12.6\%$; BW cross-sectional, $I^2 = 31.1\%$; BMI longitudinal, $I^2 = 64.3\%$; BMI cross-sectional, $I^2 = 0.00\%$) and veterans (BW longitudinal, $I^2 = 4.71\%$; BW cross-sectional, $I^2 = 47.6\%$). The Egger's test for small study effects revealed that two meta-analyses had potential publication bias for active-duty military personnel (BW longitudinal: t = −2.27 $p = 0.023$; BMI longitudinal: t = −2.90, $p = 0.004$; BW cross-sectional: t = 0.24, $p = 0.813$; BMI cross-sectional: t = 0.10, $p = 0.917$) and veterans (BW longitudinal: t = 1.13 $p = 0.257$; BW cross-sectional: t = −0.94, $p = 0.345$). The Duval and Tweedie trim and fill method was used to identify funnel plot asymmetry and adjust to publication bias for both active-duty military (BW longitudinal: $g = -0.07$; CI −0.17, 0.02; BMI longitudinal: $g = -0.16$; CI −0.35, 0.03; BW cross-sectional: $g = -0.07$; CI −0.18, 0.04; BMI cross-sectional: $g = -0.18$; CI −0.29, −0.08) and veterans meta-analyses (BW longitudinal: $g = -0.17$; CI −0.32, −0.01; BW cross-sectional: $g = -0.05$; CI −0.16, 0.06). Funnel plots for each meta-analysis per sample type are provided in the Supplemental Material Figures S11–S16.

3.7. Results for Separate Meta-Analyses Divided by Intervention Type

Alongside the evaluation of the primary outcome of the study, additional meta-analyses were conducted with the aim of evaluating the effectiveness of the different types of interventions included in the study. The main intervention sub-groups were summarized as follows: behavioral and lifestyle intervention, diet and nutritional intervention, self-monitoring intervention, counseling-provided intervention, and Internet-based intervention. Pre-to-post BW and BMI values for the intervention group were analyzed both longitudinally and cross-sectionally compared to control at the post-intervention timepoint for each kind of intervention. The main findings showed that all the intervention subtypes showed a statistically significant decrease in BMI values for both longitudinal and cross-sectional meta-analyses, except for the counseling cross-sectional meta-analysis (Table 4). A comprehensive overview of all the results (number of studies included, Hedge's g, Confidence Intervals, *p*-values) for each meta-analysis is provided in Table S5.

Table 4. Results for longitudinal and cross-sectional BMI meta-analyses divided by intervention.

Type of Intervention	BMI Longitudinal Meta-Analyses				BMI Cross-Sectional Meta-Analyses			
	N	*Hedge's g*	*95% CI*	*p*	*N*	*Hedge's g*	*95% CI*	*p*
Behavioral and Lifestyle	10	−0.28	−0.45, −0.11	0.001 **	10	−0.18	−0.28, −0.08	< 0.001 **
Diet and Nutritional	9	−0.30	−0.50, −0.11	0.002 **	9	−0.15	−0.26, −0.03	0.010 *
Self-Monitoring	7	−0.26	−0.48, −0.04	0.021 *	7	−0.17	−0.28, −0.06	0.003 **
Counseling Provided	6	−0.30	−0.47, −0.14	0.001 **	6	−0.13	−0.29, 0.03	0.113
Internet-Based	3	−0.22	−0.37, −0.07	0.004 **	3	−0.25	−0.40, −0.09	0.002 **

Notes. * Significant findings at $p < 0.05$; ** Significant findings at $p < 0.01$; N = number of studies; g = Hedge's g Effect Size; 95% CI: Confidence Interval; *p* = *p*-value.

3.8. Narrative Synthesis of Additional findings

The results of studies excluded from the meta-analysis (n = 5) were narratively synthesized (see Supplementary Materials Text S2). The main findings were that behavioral interventions consisting of classes and individual nutritional counseling with a dietitian, as well as lifestyle change promotion using a pedometer, respectively, resulted in a greater and statistically significant decrease in average waist circumference, percent of body fat, and BMI in the treatment group compared to controls [73], and a significant change from baseline to post-intervention in BW, BMI, and percent of body fat [86].

However, an examination of the effectiveness of a second-year weight loss intervention based on behavioral modifications with non-clinician life-style coach sessions compared to usual care showed a significant weight regain in the participants, highlighting how the weight loss intervention was not sufficient to sustain the initial weight loss achieved during the follow-up [79].

4. Discussion

The present study is the first meta-analysis and systematic review of RCTs investigating the effectiveness of weight loss interventions to treat overweight and obesity in military populations, comparing the intervention group both longitudinally and cross-sectionally with controls. Our findings highlight how military personnel, such as active-duty military soldiers as well as veterans enrolled in weight loss intervention programs, achieve an overall body weight and BMI reduction from baseline to post-intervention. Moreover, at the post-intervention time-point, BMI was lower than controls with a small-to-moderate effect size. The effectiveness of the weight loss interventions was also found to be significant largely in active-duty military personnel but not in veterans. These findings indicate that weight management interventions in the military population are effective, updating previous research [20,65,89]. Most of the included interventions were based on behavior change strategies adopting a comprehensive approach with different therapeutic goals, combining lifestyle modification with diet and nutritional changes and physical activity therapy, in line with previous evidence [20].

The limited role of pharmacotherapy in this population reflects current practice [90]. Indeed, only one RCT study comparing Orlistat versus Placebo has been identified through screening and was included in the meta-analyses. However, it may be that the introduction of new medications such as Glucagon-Like Peptide-1 (GLP-1) receptor agonists may change practice: the American Association of Clinical Endocrinologist guidelines emphasize this point with consideration of those patients who have not responded to intensive lifestyle therapy, or have experienced weight regain after responding to lifestyle therapy, and those with more severe complications of obesity [91]. Adding a pharmacotherapeutic adjunct to behavioral and lifestyle therapy might turn out to be an effective strategy to stem the rising rate of obesity in the military population, especially for veterans [92], although further studies are needed in this regard.

Another relevant consideration of our research concerns the possible application of new therapeutic approaches based on the use of technology. Several forms of remote, internet-based interventions have been identified by our study, enhancing the increasing evidence that an internet-based weight management program could have relevant clinical implications, especially in the military population [55,93–95]. Indeed, among soldiers enrolled in the Army weight control program (AWCP), 29% (28% of males and 36% of females) reported that they would like access to an internet-based program to facilitate their weight loss efforts [96]. From this perspective, internet-based interventions may present a platform for a widespread approach to weight management [97] and offer a possible treatment option for the Navy or Marine Corps while aboard [98], in other cases when face-to-face interventions are not possible (such as during the restrictions due to the COVID-19 pandemic [23]), or to facilitate follow-up over time.

Another relevant consideration of these findings concerns the duration of the intervention: only two studies examined the effectiveness of the intervention after 12 months had passed [79,88]. In recent years, several meta-analyses of weight loss interventions conducted on the general population based on pharmacological (BMI_{ES}: −2.37, [99]), cognitive behavioral therapy (BMI_{ES}: −0.63; [100]), and lifestyle and dietary interventions (BMI_{ES}: −1.5; [101]) have been conducted, which also report successful weight loss induction but poor maintenance of weight loss [102,103]. Strategies used to maintain the weight loss include regular contact with a lifestyle counselor [95] or innovative strategies targeting participants at key moments of disengagement risk [104]. More research is needed to assess the long-term effectiveness of weight management interventions in military populations.

Even though we found that behavioral and lifestyle interventions, diet and nutritional interventions, self-monitoring interventions, counseling-provided interventions, and internet-based interventions are all effective in military populations, these approaches might not work for individual military personnel. For example, long-term face-to-face interventions are not feasible for military personnel with frequent different deployments or for sailors. Research should therefore address the specific needs of people with obesity serving in specific military branches and units.

Additionally, most studies have used body weight or BMI as outcome parameters to measure the success of weight loss programs. However, soldiers often endeavor to gain muscle mass in order to cope with the physical demands of their job or to appear physically fitter and more muscular, which can even become a mental health problem [105]. In fact, the comparatively smaller effect size of weight loss interventions in this population compared to general populations may be because of the different distribution of muscle mass and fat mass when considering the whole BMI composition in these populations. In this case, body weight and BMI metrics might be misleading indicators of health. Thus, BMI and body weight should not be the only outcomes of weight loss interventions; other parameters such as body fat percentage, waist-to-hip ratio, or abdominal circumference should also be considered [106].

5. Clinical Implications and Practical Recommendations

Recommendations for clinical practice can be outlined from the findings of this study (Table 5). First, it is recommended to use as comprehensive an approach as possible, as combining multiple therapeutic targets has been found to be the strategy with the greatest effectiveness. This outcome is in line with the latest update of the U.S. Veterans Health Administration (VHA) and Department of Defense (DoD) Clinical Practice Guidelines for the management of adult overweight and obesity, confirming how comprehensive lifestyle interventions (CLIs) have been, and continue to be, the foundation of weight loss management. CLIs combine three critical elements: nutritional, behavioral, and physical activity modification, with the goal of promoting a negative energy balance [107]. Successful interventions can be delivered with both individual and in-group sessions, but the strongest evidence was found to be for tailored approaches [108]. Indeed, both dietary strategies and physical activities are essential components of achieving weight loss. However, this goal can be achieved through various types of meal replacement and food regimens or different types of training sessions (such as aerobic or resistance/physical activities), and there is no one standardized program that universally represents the most effective approach. Thus, it is important to personalize the treatment strategy according to the patient and his or her medical comorbidities, if any, and to establish a collaborative plan that can also be useful in improving therapeutic adherence. Therefore, a successful approach might be to include the patient in an interdisciplinary approach, comprising a dietitian or nutritionist to deal with diet and food intake, a fitness coach for exercise, and a psychologist to provide psychoeducation to the patient by working on the emotional, cognitive, and social factors that influence their relationship with food and the environmental factors that may have a negative influence on lifestyle. This may have a positive influence on long-term weight management and may increase compliance with the dietary and exercise regimen [109].

It must also be considered that the specific military environment can play an important role in dietary and exercise restrictions. While both the Army and the Navy are focused on maintaining physical fitness and weight standards, the unique environments, operational demands, and challenges they face result in tailored approaches to weight loss interventions. The Navy might need to consider dietary and exercise adjustments that account for the specific challenges of being on a ship or naval base for a long period of time. This could involve more focus on balanced nutrition that remains accessible during long journeys and considering the limited space for storage. On the other hand, access to fresh versus ultra-processed foods (UPF) can significantly impact weight loss interventions in those populations. Access to fresh, whole foods is often preferred for weight loss and overall health due to their higher nutritional value and benefits. However, in certain military settings, such as deployments or naval environments, access to fresh food might be limited or impractical. Military interventions focusing on weight loss might need to strike a balance between providing nutrient-dense fresh foods whenever possible and incorporating healthier options within the constraints of the military environment. The next stage might be to develop qualitative studies to address the implications of access to fresh versus UPF in weight loss interventions, emphasizing the importance of adapting dietary guidelines to suit the practical challenges and unique circumstances faced by military personnel in different settings.

Interesting and promising evidence on the use of technology to deliver lifestyle modifications during a weight loss intervention can also be outlined from the present study. Although this represents a still-emerging topic, the use of web-based intervention and computer remote sessions is certainly a rapidly growing area. Although in-person dietary, physical activity, and behavioral interventions have the strongest evidence of effectiveness, the results of our study showed that in clinical practice, the use of internet-based interventions could be an effective strategy to enable the implementation of weight loss interventions when in-person sessions are not possible (such as due to geographical constraints or military service abroad).

Concerning the role of pharmacotherapy, our research highlighted how the use of medications is poorly employed for short-term weight loss interventions in active military populations. Only one RCT investigating the use of orlistat in addition to lifestyle modification in these populations was found. This reflects current clinical practice, as the use of pharmacological therapy is not the first-line approach, although it can certainly be an important tool for weight loss management in combination with CLIs, especially in the long term. Indeed, a systematic review and meta-analysis including 28 RCTs (n = 29,018) investigating the role of pharmacotherapy for the treatment of obesity in the general population concluded that the prescription of medications (such as liraglutide, naltrexone/bupropion, orlistat, or phentermine/topiramate) is recommended in addition to CLIs, especially for patients with a BMI of $\geq$30 kg/m^2, and as a long-term therapeutic strategy [110]. There is little support in the literature for the short-term use of medications for weight loss, and current international guidelines recommend the introduction of medications when CLIs have produced insufficient results [107]. The same applies to the use of bariatric surgery:

current U.S. VA/DoD Clinical Practical Guidelines point to bariatric surgery in general populations as a possible long-term treatment strategy in conjunction with CLIs for patients who are less than 65 years old and who have a BMI $\geq$ 40 kg/m^2, or patients who have an obesity-associated medical condition(s) with a BMI $\geq$ 35 kg/m^2 [107]. There is still a dearth of evidence for the long-term management of weight loss and maintenance interventions in military populations. Current guidelines recommend that if a patient has lost weight in a short-term treatment program, they should be placed in a program for weight maintenance over time [111]. Moreover, adjunctive pharmacotherapy could be combined with CLIs, as has been observed in the general population. Overall, further studies are currently needed to confirm the long-term effects of weight loss interventions in military populations before definitive recommendations can be made.

Table 5. A summary of the main clinical implications and practical recommendations for military populations based on the literature review.

Topic	Clinical Recommendations	Practical Implications	Level of Evidence	RCTs (n)
Short-term weight loss intervention for obesity (up to 6–12 months).	Individual or group-based comprehensive lifestyle intervention	Physical activity (aerobics, resistance, or high intensity); no sufficient evidence from RCTs regarding the superior effectiveness of one type, frequency, or intensity of physical activity.	High	18
		Dietary and nutritional interventions such as meal replacements promoting low caloric balance intake and healthy meal plans provided by a registered dietitian (when available) and individualized to each patient.	High	12
		Cognitive behavioral therapy, psychoeducational strategies, and motivational techniques for cognitive, emotional, and social factors that influence weight management.	High	12
		Structured outcome monitoring over time (clinical or self-monitoring): body weight, BMI, fat percentage, waist-to-hip ratio, abdominal circumference.	High	12
		Internet-based intervention when in-person programs are not available.	Good	5
		Behavioral therapy plus the use of technology (e.g., pedometer).	Weak	2
		Pharmacological intervention (e.g., Orlistat).	Weak	1
Long-term weight loss intervention for obesity	Military personnel who have lost weight should be enrolled in a comprehensive weight loss maintenance program.	Lack of evidence for weight maintenance programs in military populations.	Weak	2

Abbreviations: RCTs: randomized controlled trials; n: number; BMI: Body mass index. Algorithm: The level of evidence was evaluated based on the number of available RCTs, rated as follows: "weak" for RCTs n $\leq$ 2; "good" for RCTs n = 3–5; "high" for RCTs n > 5.

6. Strengths and Limitations

To the best of our knowledge, this systematic review and meta-analysis is the first of its kind to comprehensively review the literature on RCTs to address the effectiveness of weight loss interventions in the military population. Selection bias was mitigated by involving two independent reviewers during the screening, data extraction, and quality assessment procedures. However, several limitations should be highlighted. First, there is wide heterogeneity in the type of intervention, and interventions may have included several active elements. In addition, a separate meta-analysis could not be conducted to investigate the effectiveness of the weight loss interventions in terms of BMI values in the sample of only veterans due to a lack of sufficient data, as well as for gender-specific sub-group analyses. Finally, although inclusion criteria allowed the selection of ongoing studies to maximize inclusion, as well as the selection of articles in different languages (English, German, Italian, Spanish, and French) to limit possible bias due to missing data and ensure a high level of generalizability, most of the articles after the screening process were from the United States.

7. Conclusions

This systematic review and meta-analysis found a small to moderate short-term effectiveness of weight loss interventions among the military population and a promising effectiveness of Internet-based programs. However, there is still a need for further research to evaluate the long-term effects and weight loss maintenance following the intervention.

Supplementary Materials: The following supporting information can be downloaded at: https://www.mdpi.com/article/10.3390/nu15224778/s1, Supplementary Materials: Text S1. Full search strategy. Text S2. Narrative synthesis of studies not included in meta-analyses. Table S1. Scottish Intercollegiate Guidelines Network (SIGN) Methodology Checklist for Randomized Controlled Trials1. Table S2. Quality assessment of the studies included in the meta-analysis and systematic review using the SIGN Methodology Checklist for Randomized Controlled Trials. Table S3. An overview of the interventions provided by each study. Table S4. An overview of studies included in the meta-regression analyses. Table S5. Results for separate meta-analyses divided by intervention type. Figure S1. Forest Plot for the BW longitudinal meta-analysis in active-duty military personnel. Figure S2. Forest Plot for the BMI longitudinal meta-analysis in active-duty military personnel. Figure S3. Forest Plot for the BW cross-sectional meta-analysis in active-duty military personnel. Figure S4. Forest Plot for the BMI cross-sectional meta-analysis in active-duty military personnel. Figure S5. Forest Plot for the BW longitudinal meta-analysis in veterans. Figure S6. Forest Plot for the BW cross-sectional meta-analysis in veterans. Figure S7. Funnel plot for the BW longitudinal meta-analysis. Figure S8. Funnel plot for the BMI longitudinal meta-analysis. Figure S9. Funnel plot for the BW cross-sectional meta-analysis. Figure S10. Funnel plot for the BMI cross-sectional meta-analysis. Figure S11. Funnel plot adjusted for publication bias for the BW longitudinal meta-analysis in active-duty military personnel, with imputed studies (n = 2). Figure S12. Funnel plot adjusted for publication bias for the BMI longitudinal meta-analysis in active-duty military personnel, with imputed studies (n = 5). Figure S13. Funnel plot for the BW cross-sectional meta-analysis in active-duty military personnel. Figure S14. Funnel plot adjusted for publication bias for the BMI cross-sectional meta-analysis in active-duty military personnel, with imputed study (n = 1). Figure S15. Funnel plot adjusted for publication bias for the BW longitudinal meta-analysis in veterans, with imputed study (n = 3). Figure S16. Funnel plot adjusted for publication bias for the BW cross-sectional meta-analysis in veterans, with imputed study (n = 2).

Author Contributions: D.G.: conceptualization, methodology, investigation, data curation, formal analysis, writing—original draft preparation, writing—review and editing; J.L.K.: methodology, investigation, formal analysis, writing—review and editing, supervision; M.N.A.: methodology; S.B.: methodology; P.F.: methodology; C.T.: visualization, validation; G.-D.W.: visualization, validation; J.T.: conceptualization, methodology, validation, visualization, writing—review and editing, supervision; H.H.: conceptualization, methodology, validation, visualization, writing—review and editing, supervision. All authors have read and agreed to the published version of the manuscript.

Funding: J.L.K. acknowledges financial support from a Medical Research Council (MRC)-funded Doctoral Training Partnership stipend (ref: MR/N013700/1). M.N.A. and S.B. acknowledge financial support from the Turkish Ministry of National Education for granting an official scholarship for her PhD education at King's College London University. The funder had no role in this project. J.T. has received salary support from the NIHR BRC at the South London and Maudsley NHS Foundation Trust (SLaM) and KCL. H.H. has received salary support from the NIHR BRC at the South London and Maudsley NHS Foundation Trust (SLaM) and KCL.

Institutional Review Board Statement: Not applicable.

Informed Consent Statement: Not applicable.

Data Availability Statement: Not applicable.

Acknowledgments: The authors would like to thank Liliana Dell'Osso for her support for this study.

Conflicts of Interest: The authors declare no conflict of interest.

References

1. WHO. *WHO European Regional Obesity Report 2022*; WHO Regional Office for Europe: Copenhagen, Denmark, 2022; Available online: https://www.who.int/europe/publications/i/item/9789289057738 (accessed on 18 April 2023).
2. World Obesity Federation, World Obesity Atlas. 2023. Available online: https://data.worldobesity.org/publications/?cat=19 (accessed on 18 April 2023).
3. Must, A.; Spadano, J.; Coakley, E.H.; Field, A.E.; Colditz, G.; Dietz, W.H. The disease burden associated with overweight and obesity. *JAMA* **1999**, *282*, 1523–1529. [CrossRef] [PubMed]
4. Flegal, K.M.; Carroll, M.D.; Ogden, C.L.; Johnson, C.L. Prevalence and trends in obesity among US adults, 1999–2000. *JAMA* **2002**, *288*, 1723–1727. [CrossRef] [PubMed]
5. Onubi, O.J.; Marais, D.; Aucott, L.; Okonofua, F.; Poobalan, A.S. Maternal obesity in Africa: A systematic review and meta-analysis. *J. Public Health* **2016**, *38*, e218–e231. [CrossRef]
6. Rivera, J.Á.; de Cossío, T.G.; Pedraza, L.S.; Aburto, T.C.; Sánchez, T.G.; Martorell, R. Childhood and adolescent overweight and obesity in Latin America: A systematic review. *Lancet Diabetes Endocrinol.* **2014**, *2*, 321–332. [CrossRef] [PubMed]
7. Tanofsky-Kraff, M.; Sbrocco, T.; Theim, K.R.; Cohen, L.A.; Mackey, E.R.; Stice, E.; Henderson, J.L.; McCreight, S.J.; Bryant, E.J.; Stephens, M.B. Obesity and the US military family. *Obesity* **2013**, *21*, 2205–2220. [CrossRef]
8. Gaździńska, A.; Jagielski, P.; Turczyńska, M.; Dziuda, Ł.; Gaździński, S. Assessment of Risk Factors for Development of Overweight and Obesity among Soldiers of Polish Armed Forces Participating in the National Health Programme 2016–2020. *Int. J. Environ. Res. Public Health* **2022**, *19*, 3069. [CrossRef]
9. Allman-Farinelli, M.A.; Chey, T.; Merom, D.; E Bauman, A. Occupational risk of overweight and obesity: An analysis of the Australian Health Survey. *J. Occup. Med. Toxicol.* **2010**, *5*, 14. [CrossRef]
10. Schulte, P.A.; Wagner, G.R.; Downes, A.; Miller, D.B. A framework for the concurrent consideration of occupational hazards and obesity. *Ann. Occup. Hyg.* **2008**, *52*, 555–566.
11. Chau, J.Y.; van der Ploeg, H.P.; Merom, D.; Chey, T.; Bauman, A.E. Cross-sectional associations between occupational and leisure-time sitting, physical activity and obesity in working adults. *Prev. Med.* **2012**, *54*, 195–200. [CrossRef]
12. Moore, C.J.; Cunningham, S.A. Social position, psychological stress, and obesity: A systematic review. *J. Acad. Nutr. Diet.* **2012**, *112*, 518–526. [CrossRef]
13. Torres, S.J.; Nowson, C.A. Relationship between stress, eating behavior, and obesity. *Nutrition* **2007**, *23*, 887–894. [CrossRef] [PubMed]
14. Rice, A.; Lavender, J.M.; Shank, L.M.; Neyland, M.K.H.; Markos, B.; Repke, H.; Haynes, H.; Gallagher-Teske, J.; Schvey, N.A.; Sbrocco, T.; et al. Associations among alexithymia, disordered eating, and depressive symptoms in treatment-seeking adolescent military dependents at risk for adult binge-eating disorder and obesity. *Eat Weight Disord.* **2022**, *27*, 3083–3093. [CrossRef] [PubMed]
15. Smith, T.J.; Marriott, B.P.; Dotson, L.; Bathalon, G.P.; Funderburk, L.; White, A.; Hadden, L.; Young, A.J. Overweight and obesity in military personnel: Sociodemographic predictors. *Obesity* **2012**, *20*, 1534–1538. [CrossRef]
16. Shiozawa, B.; Madsen, C.; Banaag, A.; Patel, A.; Koehlmoos, T. Body mass index effect on health service utilization among active duty male United States Army soldiers. *Mil. Med.* **2019**, *184*, 447–453. [CrossRef] [PubMed]
17. Bray, R.M.; Camlin, C.S.; Fairbank, J.A.; Dunteman, G.H.; Wheeless, S.C. The effects of stress on job functioning of military men and women. *Armed Forces Soc.* **2001**, *27*, 397–417. [CrossRef] [PubMed]
18. Chukwura, C.L.; Santo, T.J.; Waters, C.N.; Andrews, A. 'Nutrition is out of our control': Soldiers' perceptions of their local food environment. *Public Health Nutr.* **2019**, *22*, 2766–2776. [CrossRef]
19. Carlton, J.R.; Manos, G.H.; Van Slyke, J.A. Anxiety and abnormal eating behaviors associated with cyclical readiness testing in a naval hospital active duty population. *Mil. Med.* **2005**, *170*, 663–667. [CrossRef]
20. Sanderson, P.W.; Clemes, S.A.; Biddle, S.J. The correlates and treatment of obesity in military populations: A systematic review. *Obes. Facts* **2011**, *4*, 229–237. [CrossRef]
21. McLaughlin, R.; Wittert, G. The obesity epidemic: Implications for recruitment and retention of defence force personnel. *Obes. Rev.* **2009**, *10*, 693–699. [CrossRef]
22. Miggantz, E.L.; Materna, K.; Herbert, M.S.; Golshan, S.; Hernandez, J.; Peters, J.; Delaney, E.; Webb-Murphy, J.; Wisbach, G.; Afari, N. Characteristics of active duty service members referred to the navy's weight-management program. *Mil. Med.* **2023**, *188*, e174–e181. [CrossRef]
23. Legg, M.; Stahlman, S.; Chauhan, A.; Patel, D.; Hu, Z.; Wells, N. Obesity prevalence among active component service members prior to and during the COVID-19 pandemic, January 2018–July 2021. *MSMR* **2022**, *29*, 8–16. [PubMed]

24. Armed Forces Health Surveillance Division. Department of Defense Health of the Force 2019. 2020. Available online: https://www.health.mil/Military-Health-Topics/Combat-Support/Armed-Forces-Health-Surveillance-Division/Re-ports-and-Publications (accessed on 20 April 2023).
25. Hruby, A.; Hill, O.T.; Bulathsinhala, L.; McKinnon, C.J.; Montain, S.J.; Young, A.J.; Smith, T.J. Trends in overweight and obesity in soldiers entering the US Army, 1989–2012. *Obesity* **2015**, *23*, 662–670. [CrossRef] [PubMed]
26. Sanderson, P.W.; Clemes, S.A.; Biddle, S.J. Prevalence and socio-demographic correlates of obesity in the British Army. *Ann. Hum. Biol.* **2014**, *41*, 193–200. [CrossRef] [PubMed]
27. Sundin, J.; Fear, N.T.; Wessely, S.; Rona, R.J. Obesity in the UK Armed Forces: Risk factors. *Mil. Med.* **2011**, *176*, 507–512. [CrossRef] [PubMed]
28. Quertier, D.; Goudard, Y.; Goin, G.; Régis-Marigny, L.; Sockeel, P.; Dutour, A.; Pauleau, G.; De La Villéon, B. Overweight and obesity in the French army. *Mil. Med.* **2022**, *187*, e99–e105. [CrossRef]
29. Salimi, Y.; Taghdir, M.; Sepandi, M.; Zarchi, A.-A.K. The prevalence of overweight and obesity among Iranian military personnel: A systematic review and meta-analysis. *BMC Public Health* **2019**, *19*, 162. [CrossRef]
30. Gaździńska; Jagielski, P.; Baran, P. Evaluation of nutritional status and the level of physical fitness of military flying personnel staying at the training camp. *Pol. J. Aviat. Med. Psychol.* **2018**, *24*, 12–18. [CrossRef]
31. Al-Qahtani, D.A.; Imtiaz, M.L.; Shareef, M.M. Obesity and cardiovascular risk factors in Saudi adult soldiers. *Saudi Med. J.* **2005**, *26*, 1260–1268.
32. Fogelholm, M.; Malmberg, J.; Suni, J.; Santtila, M.; Kyröläinen, H.; Mäntysaari, M. Waist circumference and BMI are independently associated with the variation of cardio-respiratory and neuromuscular fitness in young adult men. *Int. J. Obes.* **2006**, *30*, 962–969. [CrossRef]
33. Yokota, M.; Bathalon, G.P.; Berglund, L.G. Assessment of male anthropometric trends and the effects on simulated heat stress responses. *Eur. J. Appl. Physiol.* **2008**, *104*, 297–302. [CrossRef]
34. Loube, D.I.; Loube, A.A.; Mitler, M.M. Weight loss for obstructive sleep apnea: The optimal therapy for obese patients. *J. Am. Diet. Assoc.* **1994**, *94*, 1291–1295. [CrossRef] [PubMed]
35. Charteris, J. Load stress; carrier strain: Implications for military and recreational back packing. *Ergonomics* **2000**, *1*, 25–47.
36. Lyons, J.; Allsopp, A.; Bilzon, J. Influences of body composition upon the relative metabolic and cardiovascular demands of load-carriage. *Occup. Med.* **2005**, *55*, 380–384. [CrossRef] [PubMed]
37. Kress, A.M.; Peterson, M.R.; Hartzell, M.C. Association between obesity and depressive symptoms among US Military active duty service personnel 2002. *J. Psychosom. Res.* **2006**, *60*, 263–271. [CrossRef]
38. Vieweg, W.V.R.; Julius, D.A.; Benesek, J.; Satterwhite, L.; Fernandez, A.; Feuer, S.J.; Pandurangi, A.K. Posttraumatic stress disorder and body mass index in military veterans: Preliminary findings. *Prog. Neuro-Psychopharmacol. Biol. Psychiatry* **2006**, *30*, 1150–1154. [CrossRef]
39. Cypel, Y.S.; DePhilippis, D.; Davey, V. Substance Use in US Vietnam War Era Veterans and Nonveterans: Results from the Vietnam Era Health Retrospective Observational Study. *Subst. Use Misuse* **2023**, *58*, 858–870. [CrossRef]
40. Lesnewich, L.M.; Lu, S.-E.; Weinreb, K.S.; Sparks, S.O.; Litke, D.R.; Helmer, D.A.; Pigeon, W.R.; McAndrew, L.M. Associations between risky alcohol use, disability, and problem-solving impairment among Veterans with Gulf War Illness: Secondary data analysis of a randomized clinical trial. *J. Psychosom. Res.* **2023**, *170*, 111336. [CrossRef]
41. Development USDoVAOoRa. VA Research on Obesity, The U.S. Department of Veterans Affairs 2014. Available online: https://www.research.va.gov/topics/obesity (accessed on 23 April 2023).
42. Litwack, S.D.; Mitchell, K.S.; Sloan, D.M.; Reardon, A.F.; Miller, M.W. Eating disorder symptoms and comorbid psychopathology among male and female veterans. *Gen. Hosp. Psychiatry* **2014**, *36*, 406–410. [CrossRef]
43. Littman, A.J.; Damschroder, L.J.; Verchinina, L.; Lai, Z.; Kim, H.M.; Hoerster, K.D.; Klingaman, E.A.; Goldberg, R.W.; Owen, R.R.; Goodrich, D.E. National evaluation of obesity screening and treatment among veterans with and without mental health disorders. *Gen. Hosp. Psychiatry* **2015**, *37*, 7–13. [CrossRef]
44. Breland, J.Y.; Donalson, R.; Dinh, J.V.; Maguen, S. Trauma exposure and disordered eating: A qualitative study. *Women Health* **2018**, *58*, 160–174. [CrossRef]
45. Tanielian, T.; Jaycox, L.; Schell, T.; Marshall, G.; Burnam, M.; Eibner, C.; Karney, B.; Meredith, L.; Ringel, J.; Vaiana, M. *Invisible Wounds of War: Summary and Recommendations for Addressing Psychological and Cognitive Injuries*; RAND: Santa Monica, CA, USA, 2008.
46. Herrmann, T.; Preib, E.; French, M.; Beckstrom, J.; Nazarenko, E.; Lackner, R.; Marchand, W.R.; Yabko, B. Veterans' experiences with mindfulness-based eating: A mixed methods study on MB-SAVOR. *Complementary Ther. Clin. Pract.* **2022**, *47*, 101548. [CrossRef] [PubMed]
47. Spieker, E.A.; Sbrocco, T.; Theim, K.R.; Maurer, D.; Johnson, D.; Bryant, E.; Bakalar, J.L.; Schvey, N.A.; Ress, R.; Seehusen, D.; et al. Preventing Obesity in the Military Community (POMC): The development of a clinical trials research network. *Int. J. Environ. Res. Public Health* **2015**, *12*, 1174–1195. [CrossRef] [PubMed]
48. Krukowski, R.A.; Hare, M.E.; Talcott, G.W.; Gladney, L.A.; Johnson, K.C.; Richey, P.A.; Kocak, M.; Keller, P.L.; Hryshko-Mullen, A.; Klesges, R.C. Dissemination of the Look AHEAD intensive lifestyle intervention in the United States Air Force: Study rationale, design and methods. *Contemp. Clin. Trials* **2015**, *40*, 232–239. [CrossRef] [PubMed]

49. Group, L.A.R. The Look AHEAD study: A description of the lifestyle intervention and the evidence supporting it. *Obesity* **2006**, *14*, 737–752.
50. Earles, J.E.; Kerr, B.; James, L.C.; Folen, R.A. Clinical effectiveness of the LE 3 AN program: A military healthy lifestyle program. *J. Clin. Psychol. Med. Settings* **2007**, *14*, 51–57. [CrossRef]
51. Kahwati, L.C.; Lance, T.X.; Jones, K.R.; Kinsinger, L.S. RE-AIM evaluation of the Veterans Health Administration's MOVE! weight management program. *Transl. Behav. Med.* **2011**, *1*, 551–560. [CrossRef]
52. Piche, B.M.; Stankorb, S.M.; Salgueiro, M. Attitudes, beliefs, and behaviors of active duty soldiers attending the Army MOVE! weight management program. *Mil. Med.* **2014**, *179*, 906–912. [CrossRef]
53. Shahnazari, M.; Ceresa, C.; Foley, S.; Fong, A.; Zidaru, E.; Moody, S. Nutrition-focused wellness coaching promotes a reduction in body weight in overweight US veterans. *J. Acad. Nutr. Diet.* **2013**, *113*, 928–935. [CrossRef]
54. Arsenault, J.E.; Singleton, M.C.; Funderburk, L.K. Use of the Go-for-Green nutrition labeling system in military dining facilities is associated with lower fat intake. *J. Acad. Nutr. Diet.* **2014**, *114*, 1067–1071. [CrossRef]
55. Stewart, T.; Han, H.; Allen, H.R.; Bathalon, G.; Ryan, D.H.; Newton, R.L.; Williamson, D.A. HEALTH: Efficacy of an internet/population-based behavioral weight management program for the US Army. *J. Diabetes Sci. Technol.* **2011**, *5*, 178–187. [CrossRef]
56. Farage, G.; Simmons, C.; Kocak, M.; Klesges, R.C.; Talcott, G.W.; Richey, P.; Hare, M.; Johnson, K.C.; Sen, S.; Krukowski, R. Assessing the contribution of self-monitoring through a commercial weight loss app: MEdiation and predictive modeling study. *JMIR mHealth uHealth* **2021**, *9*, e18741. [CrossRef] [PubMed]
57. Hunter, C.M.; Peterson, A.L.; Alvarez, L.M.; Poston, W.C.; Brundige, A.R.; Haddock, C.K.; Van Brunt, D.L.; Foreyt, J.P. Weight management using the internet: A randomized controlled trial. *Am. J. Prev. Med.* **2008**, *34*, 119–126. [CrossRef] [PubMed]
58. Bowles, S.V.; Picano, J.; Epperly, T.; Myer, S. The LIFE program: A wellness approach to weight loss. *Mil. Med.* **2006**, *171*, 1089–1094. [CrossRef]
59. Davis, M.K. A comprehensive weight-loss program for soldiers. *Mil. Med.* **1996**, *161*, 84–88. [CrossRef] [PubMed]
60. Afari, N.; Cuneo, J.G.; Herbert, M.; Miller, I.; Webb-Murphy, J.; Delaney, E.; Peters, J.; Materna, K.; Miggantz, E.; Godino, J.; et al. Design for a cohort-randomized trial of an acceptance and commitment therapy-enhanced weight management and fitness program for Navy personnel. *Contemp. Clin. Trials Commun.* **2019**, *15*, 100408. [CrossRef]
61. Nadolsky, K.Z. Rationale for Utilization of Obesity Pharmacotherapy in the Active Duty Population. *Mil. Med.* **2018**, *183*, 45–50. [CrossRef]
62. Mailloux, O.; Tassé, N.; Tchernof, A.; Nadeau, M.; Dawe, P.; Beckett, A.; Biertho, L. Bariatric Surgery Should Be Offered to Active-Duty Military Personnel: A Retrospective Study of the Canadian Armed Forces' Experience. *Obes. Surg.* **2023**, *33*, 1092–1098. [CrossRef]
63. Walter, F.A.; Hoyt, T.; Martinez, H.; Dziura, J. Preoperative psychological assessment and weight loss outcomes in bariatric surgery patients at a military treatment facility: A retrospective profile analysis. *Mil. Med.* **2022**, *187*, e1169–e1175. [CrossRef]
64. Malkawi, A.M.; Meertens, R.M.; Kremers, S.P.J.; Sleddens, E.F.C. Dietary, physical activity, and weight management interventions among active-duty military personnel: A systematic review. *Mil. Med. Res.* **2018**, *5*, 43. [CrossRef]
65. Csizmar, G.T.; Irwin, M. Efficacy of weight loss interventions in United States active duty military populations: A systematic review. *Mil. Med.* **2021**, *186*, 1093–1099. [CrossRef]
66. Page, M.J.; McKenzie, J.E.; Bossuyt, P.M.; Boutron, I.; Hoffmann, T.C.; Mulrow, C.D.; Shamseer, L.; Tetzlaff, J.M.; Akl, E.A.; Brennan, S.E.; et al. The PRISMA 2020 statement: An updated guideline for reporting systematic reviews. *Int. J. Surg.* **2021**, *88*, 105906. [CrossRef] [PubMed]
67. DerSimonian; Laird, N. Meta-analysis in clinical trials. *Control. Clin. Trials* **1986**, *7*, 177–188. [CrossRef]
68. Higgins, J.P.T.; Chandler, J.; Cumpston, M.; Li, T.; Page, M.J.; Welch, V.A. *Cochrane Handbook for Systematic Reviews of Interventions*, Version 6.4; (updated August 2023). Cochrane, 2023. Available online: www.training.cochrane.org/handbook. (accessed on 3 November 2023).
69. Duval, S.; Tweedie, R. Trim and fill: A simple funnel-plot–based method of testing and adjusting for publication bias in meta-analysis. *Biometrics* **2000**, *56*, 455–463. [CrossRef] [PubMed]
70. Boutelle, K.N.; Afari, N.; Obayashi, S.; Eichen, D.M.; Strong, D.R.; Peterson, C.B. Design of the CHARGE study: A randomized control trial evaluating a novel treatment for Veterans with binge eating disorder and overweight and obesity. *Contemp. Clin. Trials* **2023**, *130*, 107234. [CrossRef] [PubMed]
71. Damschroder, L.J.; Lutes, L.D.; Kirsh, S.; Kim, H.M.; Gillon, L.; Holleman, R.G.; Goodrich, D.E.; Lowery, J.C.; Richardson, C.R. Small-Changes Obesity Treatment among Veterans 12-Month Outcomes. *Am. J. Prev. Med.* **2014**, *47*, 541–553. [CrossRef]
72. Dennis, K.E.; Pane, K.W.; Adams, B.K.; Qi, B.B. The impact of a shipboard weight control program. *Obes. Res.* **1999**, *7*, 60–67. [CrossRef]
73. Erickson, Z.D.; Kwan, C.L.; Gelberg, H.A.; Arnold, I.Y.; Chamberlin, V.; Rosen, J.A.; Shah, C.; Nguyen, C.T.; Hellemann, G.; Aragaki, D.R.; et al. A randomized, controlled multisite study of behavioral interventions for veterans with mental illness and antipsychotic medication-associated obesity. *J. Gen. Intern. Med.* **2017**, *32*, 32–39. [CrossRef]
74. Evans-Hudnall, G.; O Odafe, M.; Johnson, A.; Armenti, N.; O'neil, J.; Lawson, E.; Trahan, L.H.; Rassu, F.S. Using an Adjunctive Treatment to Address Psychological Distress in a National Weight Management Program: Results of an Integrated Pilot Study. *Mil. Med.* **2020**, *185*, E1662–E1670. [CrossRef]

75. Goldberg, R.W.; Reeves, G.; Tapscott, S.; Medoff, D.; Dickerson, F.; Goldberg, A.P.; Ryan, A.S.; Fang, L.J.; Dixon, L.B.; Muralidharan, A.; et al. "MOVE!": Outcomes of a weight loss program modified for veterans with serious mental illness. *Psychiatr. Serv.* **2013**, *64*, 737–744. [CrossRef]
76. Hoerster, K.D.; Hunter-Merrill, R.; Nguyen, T.; Rise, P.; Barón, A.E.; McDowell, J.; Donovan, L.M.; Gleason, E.; Lane, A.; Plumley, R.; et al. Effect of a Remotely Delivered Self-directed Behavioral Intervention on Body Weight and Physical Health Status among Adults with Obesity: The D-ELITE Randomized Clinical Trial. *JAMA* **2022**, *328*, 2230–2241. [CrossRef]
77. Hosseini-Amiri, M.; Aliyari, S.; Zareiyan, A.; Dabbagh-Moghadam, A. The Effects of Extended Parallel Process Model on Obese Soldiers' Knowledge, Attitudes, and Practices about Obesity Management: A Randomized Controlled Clinical Trial. *Iran. J. Nurs. Midwifery Res.* **2018**, *23*, 458–464. [PubMed]
78. Krukowski, R.A.; Hare, M.E.; Talcott, G.W.; Johnson, K.C.; Richey, P.A.; Kocak, M.; Balderas, J.; Colvin, L.; Keller, P.L.; Waters, T.M.; et al. Dissemination of the Look AHEAD Intensive Lifestyle Intervention in the United States Military: A Randomized Controlled Trial. *Obesity* **2018**, *26*, 1558–1565. [CrossRef] [PubMed]
79. Lutes, L.D.; Damschroder, L.J.; Masheb, R.; Kim, H.M.; Gillon, L.; Holleman, R.G.; Goodrich, D.E.; Lowery, J.C.; Janney, C.; Kirsh, S.; et al. Behavioral treatment for veterans with obesity: 24-month weight outcomes from the ASPIRE-VA small changes randomized trial. *J. Gen. Intern. Med.* **2017**, *32*, 40–47. [CrossRef]
80. McDoniel, S.O.; Nelson, H.A.; Thomson, C.A. Employing RMR technology in a 90-day weight control program. *Obes. Facts.* **2008**, *1*, 298–304. [CrossRef] [PubMed]
81. Parastouei, K.; Saeidipoor, S.; Sepandi, M.; Abbaszadeh, S.; Taghdir, M. Effects of synbiotic supplementation on the components of metabolic syndrome in military personnel: A double-blind randomised controlled trial. *BMJ Mil. Health* **2022**, *168*, 362–367. [CrossRef]
82. Paravidino, V.B.; Mediano, M.F.F.; Silva, I.C.M.; Wendt, A.; Del Vecchio, F.B.; Neves, F.A.; Terra, B.d.S.; Gomes, E.A.C.; Moura, A.S.; Sichieri, R. Effect of physical exercise on spontaneous physical activity energy expenditure and energy intake in overweight adults (the EFECT study): A study protocol for a randomized controlled trial. *Trials* **2018**, *19*, 167. [CrossRef] [PubMed]
83. Pérez-Muñoz, A.; Hare, M.E.; Andres, A.; Klesges, R.C.; Wayne Talcott, G.; Little, M.A.; Waters, T.M.; Harvey, J.R.; Bursac, Z.; Krukowski, R.A. A Postpartum Weight Loss-focused Stepped-care Intervention in a Military Population: A Randomized Controlled Trial. *Ann. Behav. Med.* **2023**, *2023*, kaad014. [CrossRef]
84. Smith, T.J.; Sigrist, L.D.; Bathalon, G.P.; McGraw, S.; Karl, J.P.; Young, A.J. Efficacy of a meal-replacement program for promoting blood lipid changes and weight and body fat loss in US Army soldiers. *J. Am. Diet. Assoc.* **2010**, *110*, 268–273. [CrossRef]
85. Smith, T.J.; Crombie, A.; Sanders, L.F.; Sigrist, L.D.; Bathalon, G.P.; McGraw, S.; Young, A.J. Efficacy of orlistat 60 mg on weight loss and body fat mass in US Army soldiers. *J. Acad. Nutr. Diet.* **2012**, *112*, 533–540. [CrossRef]
86. Staudter, M.; Dramiga, S.; Webb, L.; Hernandez, D.; Cole, R. Effectiveness of pedometer use in motivating active duty and other military healthcare beneficiaries to walk more. *US Army Med. Dep. J.* **2011**, 108–119.
87. Veverka, D.V.; Anderson, J.; Auld, G.W.; Coulter, G.R.; Kennedy, C.; Chapman, P.L. Use of the stages of change model in improving nutrition and exercise habits in enlisted Air Force men. *Mil. Med.* **2003**, *168*, 373–379. [CrossRef] [PubMed]
88. Voils, C.I.; Olsen, M.K.; Gierisch, J.M.; McVay, M.A.; Grubber, J.M.; Gaillard, L.; Bolton, J.; Maciejewski, M.L.; Strawbridge, E.; Yancy, W.S., Jr. Maintenance of weight loss after initiation of nutrition training: A randomized trial. *Ann. Intern. Med.* **2017**, *166*, 463–471. [CrossRef] [PubMed]
89. Murray, J.; Aboul-Enein, B.H.; Bernstein, J.; Kruk, J. Selected weight management interventions for military populations in the United States: A narrative report. *Nutr. Health* **2017**, *23*, 67–74. [CrossRef] [PubMed]
90. Department of Veterans Affairs: Va/DoD Clinical Practice Guideline for Screening and Management of Overweight and Obesity Submitted 04/18/14. Available online: https://www.healthquality.va.gov/guidelines (accessed on 3 November 2023).
91. Garvey, W.T.; Mechanick, J.I.; Brett, E.M.; Garber, A.J.; Hurley, D.L.; Jastreboff, A.M.; Nadolsky, K.; Pessah-Pollack, R.; Plodkowski, R.; Reviewers of the AACE/ACE Obesity Clinical Practice Guidelines. American Association of Clinical Endocrinologists and American College of Endocrinology Comprehensive Clinical Practice Guidelines for medical care of patients with obesity. *Endocr. Pract.* **2016**, *22* (Suppl. S3), 1–203. [CrossRef]
92. Del Re, A.; Frayne, S.M.; Harris, A.H. Antiobesity medication use across the veterans health administration: Patient-level predictors of receipt. *Obesity* **2014**, *22*, 1968–1972. [CrossRef] [PubMed]
93. Rasu, R.S.; Hunter, C.M.; Peterson, A.L.; Maruska, H.M.; Foreyt, J.P. Economic evaluation of an Internet-based weight management program. *Am. J. Manag. Care* **2010**, *16*, e98–e104. [CrossRef]
94. Robbins, A.S.; Chao, S.Y.; Baumgartner, N.; Runyan, C.N.; Oordt, M.S.; Fonseca, V.P. A low-intensity intervention to prevent annual weight gain in active duty Air Force members. *Mil. Med.* **2006**, *171*, 556–561. [CrossRef]
95. Stewart, T.; May, S.; Allen, H.R.; Bathalon, G.P.; Lavergne, G.; Sigrist, L.; Ryan, D.; Williamson, D.A. Development of an internet/population-based weight management program for the US Army. *J. Diabetes Sci. Technol.* **2008**, *2*, 116–126. [CrossRef]
96. McGraw, S.M.; Bathalon, G.P.; Ellison, B.K.; Graff, J.D.; Burrell, L.M.; Carr, R.E.; Ross, C.M.; Williamson, D.A.; Young, A.J. Dieting practices of soldiers in the U.S. Army weight control program. *Med. Sci. Sports Exerc.* **2005**, *37*, S142.
97. Winett, R.A.; Tate, D.F.; Anderson, E.S.; Wojcik, J.R.; Winett, S.G. Long-term weight gain prevention: A theoretically based Internet approach. *Prev. Med.* **2005**, *41*, 629–641. [CrossRef]
98. McNulty, A.F. Prevalence and contributing factors of eating disorder behaviors in active duty service women in the Army, Navy, Air Force, and Marines. *Mil. Med.* **2001**, *166*, 53–58. [CrossRef] [PubMed]

99. Liu, Y.; Ruan, B.; Jiang, H.; Le, S.; Liu, Y.; Ao, X.; Huang, Y.; Shi, X.; Xue, R.; Fu, X.; et al. The Weight-loss Effect of GLP-1RAs Glucagon-Like Peptide-1 Receptor Agonists in Non-diabetic Individuals with Overweight or Obesity: A Systematic Review with Meta-Analysis and Trial Sequential Analysis of Randomized Controlled Trials. *Am. J. Clin. Nutr.* **2023**, *118*, 614–626. [CrossRef] [PubMed]
100. Mesarič, K.K.; Pajek, J.; Zakrajšek, B.L.; Bogataj, Š.; Kodrič, J. Cognitive behavioral therapy for lifestyle changes in patients with obesity and type 2 diabetes: A systematic review and meta-analysis. *Sci. Rep.* **2023**, *13*, 12793. [CrossRef] [PubMed]
101. Morgan-Bathke, M.; Baxter, S.D.; Halliday, T.M.; Lynch, A.; Malik, N.; Raynor, H.A.; Garay, J.L.; Rozga, M. Weight Management Interventions Provided by a Dietitian for Adults with Overweight or Obesity: An Evidence Analysis Center Systematic Review and Meta-Analysis. *J. Acad. Nutr. Diet.* **2023**, *123*, 520–545.e10. [CrossRef] [PubMed]
102. Hartmann-Boyce, J.; Theodoulou, A.; Oke, J.L.; Butler, A.R.; Bastounis, A.; Dunnigan, A.; Byadya, R.; Cobiac, L.J.; Scarborough, P.; Hobbs, F.R.; et al. Long-term effect of weight regain following behavioral weight management programs on cardiometabolic disease incidence and risk: Systematic review and meta-analysis. *Circ. Cardiovasc. Qual. Outcomes* **2023**, *16*, e009348. [CrossRef]
103. Theodoulou, A.; Hartmann-Boyce, J.; Gorenberg, J.; Oke, J.L.; Butler, A.R.; Bastounis, A.; Jebb, S.A.; Aveyard, P. Weight regain and mental health outcomes following behavioural weight management programmes: A systematic review with meta-analyses. *Clin. Obes.* **2023**, *13*, e12575. [CrossRef]
104. Crane, N.; Hagerman, C.; Horgan, O.; Butryn, M. Patterns and Predictors of Engagement with Digital Self-Monitoring during the Maintenance Phase of a Behavioral Weight Loss Program: Quantitative Study. *JMIR mHealth uHealth* **2023**, *18*, e45057. [CrossRef]
105. Applewhite, B.; Olivola, M.; Tweed, C.; Wesemann, U.; Himmerich, H. Body dysmorphic disorder, muscle dysmorphia, weight and shape dissatisfaction and the use of appearance-enhancing drugs in the military: A systematic review. *BMJ Mil. Health* **2022**, *8*, e002135. [CrossRef]
106. Clerc, P.G.; Mayer, S.B.; Graybill, S. Overweight BMI (25–29) in active duty military: Excess fat or more lean mass? A look at the evidence. *Mil. Med.* **2022**, *187*, 201–203. [CrossRef]
107. Mayer, S.B.; Graybill, S.; Raffa, S.D.; Tracy, C.; Gaar, E.; Wisbach, G.; Goldstein, M.G.; Sall, J. Synopsis of the 2020 US VA/DoD clinical practice guideline for the management of adult overweight and obesity. *Mil. Med.* **2021**, *186*, 884–896. [CrossRef]
108. National Institute for Health and Care Excellence. *Obesity: Identification, Assessment and Management*; NICE Guideline (CG189); National Institute for Health and Care Excellence: London, UK, 2014.
109. Yumuk, V.; Tsigos, C.; Fried, M.; Schindler, K.; Busetto, L.; Micic, D.; Toplak, H. European guidelines for obesity management in adults. *Obes. Facts* **2015**, *8*, 402–424. [CrossRef] [PubMed]
110. Khera, R.; Murad, M.H.; Chandar, A.K.; Dulai, P.S.; Wang, Z.; Prokop, L.J.; Loomba, R.; Camilleri, M.; Singh, S. Association of pharmacological treatments for obesity with weight loss and adverse events: A systematic review and meta-analysis. *JAMA* **2016**, *315*, 2424–2434. [CrossRef] [PubMed]
111. Hall, K.D.; Kahan, S. Maintenance of lost weight and long-term management of obesity. *Med. Clin.* **2018**, *102*, 183–197. [CrossRef] [PubMed]

nutrients

MDPI

Review

Diet Traps during Eating Disorders among Dentate Patients at an Oral Health Glance

Elzbieta Paszynska [1,*], Amadeusz Hernik [1], Hélène Rangé [2,3], Bennett T. Amaechi [4], Georgiana S. Gross [4] and Malgorzata Pawinska [5]

1 Department of Integrated Dentistry, Poznan University of Medical Sciences (PUMS), 60-812 Poznan, Poland
2 Department of Periodontology, Faculty of Odontology, University of Rennes, CHU de Rennes, F-35000 Rennes, France
3 Institut NUMECAN (Nutrition Metabolism and Cancer), INSERM, INRAE, University of Rennes, F-35000 Rennes, France
4 Department of Comprehensive Dentistry, School of Dentistry, University of Texas Health San Antonio, San Antonio, TX 78229, USA
5 Department of Integrated Dentistry, Medical University in Bialystok, 15-276 Bialystok, Poland
* Correspondence: paszynska@ump.edu.pl; Tel.: +48-61-8547250

Abstract: Persons suffering from eating disorders (ED) may often experience a recurrence/persistence symptoms despite the completion of psychiatric therapy. In most cases, their general health status is linked to current nutritional behaviors. Medical professionals, general practitioners (GPs), dieticians, and dentists may see those patients in their practices. At the same time, due to low sense of illness, some patients may delay or never seek professional medical care. The aim of this article is to analyze the main ED types according to dietary behaviors causing oral health problems and discuss oral health complications in affected dentate patients. The second objective is to update oral preventive measures and technological innovations together with active agents for oral hygiene care that might effectively support oral health maintenance during the presence of long-term symptoms. The research method involved a review of clinical reports as a synthesis of the electronic research in the Pubmed, Web of Science, and Google Scholar databases. Based on the research, ED patients were found to present related incidences of oral complications. Studies have reported that the possible course of an ED and comorbidities may be an imbalance in the oral environment. The results showed an association between biological (malnutrition, etc.), behavioral (binge eating episodes, vomiting, acidic diet, poor oral hygiene), and pharmacotherapeutic (addiction, hyposalivation) factors that may threaten oral health. Early diagnosis of the past and present symptoms is essential to eliminate and take control of destructive behaviors. Oral changes need to be tackled with medical insight, and additionally, the perception of dietary interactions is recommended.

Keywords: eating disorders; nutrition; oral health; oral hygiene; risk factors

Citation: Paszynska, E.; Hernik, A.; Rangé, H.; Amaechi, B.T.; Gross, G.S.; Pawinska, M. Diet Traps during Eating Disorders among Dentate Patients at an Oral Health Glance. *Nutrients* **2023**, *15*, 4414. https://doi.org/10.3390/nu15204414

Academic Editors: Hubertus Himmerich and Khadijeh Mirzaei

Received: 8 September 2023
Revised: 4 October 2023
Accepted: 7 October 2023
Published: 18 October 2023

1. Introduction

Medical professionals, including general practitioners (GPs), dieticians, and dentists, may see eating disorder (ED) patients in their practices. Patients attending medical or dental appointments usually show up for a routine check-up, and EDs can be diagnosed by accident. Those affected by eating disorders are young and often unaware of the seriousness of the complications that can occur as a result of a lack of diagnosis, untreated disease, and dietary errors. Romanos (2012) points out that oral symptoms can appear asymptomatically [1]. Implementing treatment standards, and providing professional medical and psychological help, can happen faster if there is knowledge on and awareness of dietary traps. This increases the chance of not only preventing dangerous health complications manifested in the oral cavity but also improving the prognosis of disease treatment.

The aim of this article is to analyze the main ED types according to the dietary choices that cause oral health symptoms and to discuss the side effects associated with EDs in affected dentate patients. The second goal is to update available oral preventive measures and technological innovations together with active oral hygiene measures and care that might effectively maintain oral health during the presence of long-term ED symptoms. The research method involved a review of the clinical reports on EDs as a synthesis of the electronic research in the Pubmed, Web of Science, Google Scholar databases, addressing patients residing in industrialized countries.

2. Eating Disorder Characteristics

2.1. Types of Eating Disorders

The most important classifications that describe diagnostic criteria are the American Psychiatric Association (APA) guidelines, provided in the Diagnostic and Statistical Manual of Mental Disorders (DSM 5th edition, updated in 2013), and the International Classification of Diseases (ICD-11), approved by the World Health Organization (updated 2019) [2,3]. According to ICD-11, this group of diseases falls under behavioral syndromes associated with physiological disorders and physical factors. According to the APA, eating disorders are characterized by ongoing abnormal behaviors related to feeding and eating, followed by changes in food intake or impaired food absorption, and this in turn causes disorders of physical health or social functioning. The symptoms of the disorder allow a clear diagnosis since the diagnosis of a particular disease entity excludes another. Division of EDs according to DSM-5 and ICD-11 subtypes is presented in Figure 1, [2,4].

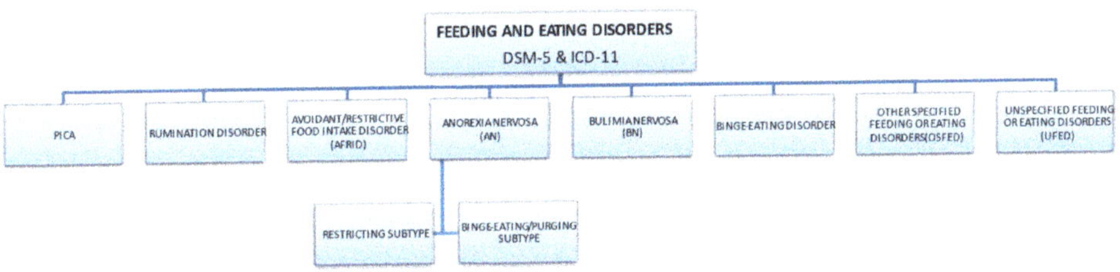

Figure 1. Division of EDs according to DSM-5 and ICD-11 subtypes.

2.2. ED Risk Factors

Eating disorders are characterized by multiple risk factors, which can be biological, psychological, social, and cultural determinants. Individual theoretical models were considered, referring to socio-cultural determinants (ideal body model, gender role), aspects of personal vulnerability (character and behavioral features, including genetic), family and interpersonal background (disorders occurring in the family, e.g., lack of hierarchy, unclear roles, existence of strong ties, turbulent way of solving problems, perfectionism or overprotection of the family, influences of the environment from the same age group), and influence of traumatic life events (e.g., physical, sexual violence) [5,6]. Risk factors predisposing to the disorder in childhood trigger the first onset of the disease in adolescence [7].

Eating disorders often begin with the selective eating of certain foods or following of specific diets, such as vegetarian diets, that seem easy to accept socially [8]. The image of idealized figures created by social media can also lead to behaviors that promote eating disorders [9]. A contributing factor is perfectionism, an ideal that is an important reference point for the individual [10–12]. The following risk groups can be recognized: athletes—including more often female athletes, such as those in cycling, judo, gymnastics, and athletics—and people working in modeling, dance training, the military, catering, and show business [13,14]. Biological backgrounds, e.g., disorders of the serotonergic system, are considered in genetic analysis or as correlations with metabolic and immunological features (including glycemic, lipid) and analysis of the fetal period [7]. Meta-analyses indicate the potential importance of single-nucleotide polymorphisms (SNPs) including serotonin receptor gene and serotonin transporter gene (5-HTR2A, 5-HTT), catechol-O-methyltransferase (COMT) polymorphism (Val158Met), and brain-derived neurotrophic factor (BDNF) polymorphism (Val66Met) [15,16]. The heritability in families with mono-dizygotic twins is estimated to be between 48–74% [17]. Genome-wide association studies (GWASs) have identified a chromosome 12 (12q13.2) locus that is relevant to the inheritance of the disease [18].

Researchers' attention is drawn to changes in both the hunger and satiety centers as well as to the associated regulation of appetite and metabolism. This includes anorexigenic (appetite-reducing) and orexigenic (appetite-stimulating) substances secreted centrally in the central nervous system and peripherally (by adiponectin and enterohormones). Substances that regulate appetite in the short term (e.g., cholecystokinin and other proteins) and long term (hormones leptin, insulin) and enterohormones that interact with the brain–gut axis (e.g., ghrelin, obestatin, neuropeptide B, vaspin phoenixin, spexin, kisspeptin) are involved in the processes of regulating appetite and metabolism [19–22].

2.3. Epidemiology

Current data show that about 2.9 million people worldwide suffer from anorexia nervosa (AN), with the prevalence in European countries being at the levels of 1–4% (women) and 0.3–0.7% (men) [23,24]. Regarding bulimia nervosa (BN), up to 3% of females and more than 1% of males suffer from this disorder during their lifetime [25]. According to current epidemiological data, EDs are characterized by a long-term course and a high mortality rate [6]. Thus, EDs are an important issue for modern medicine, especially since the incidence rate in Western-civilization countries has been almost declining since the 1970s and the full-blown course is being observed in patients at an increasingly younger age [6,25–27].

3. Common Dietary Intakes and Behavioral Habits Related to Oral Health in ED Patients

Persons with current or ongoing symptoms of EDs rarely seek out a dietitian for help with their erratic eating habits. These people often visit a medical doctor or dentist to help alleviate a health complication brought on by their dysfunctional eating habits. This section will discuss the food types and practices of patients with ongoing eating disorders. Several types of eating disorders have been described and diagnosed.

Anorexia nervosa: Persons with anorexia nervosa exhibit dietary behaviors that are prompted by their fear of weight gain and distorted body image. Their dietary intakes are consumed in very small portions throughout the day. The food intake is followed by extreme exercise to "burn off" the calories consumed. This form of AN is called the "Restricted Type". A subset of these patients will binge but purge after a small meal intake to further reduce caloric intake. Most of them purge by inducing vomiting, while others use laxatives. Foods commonly used by these patients are lower-calorie foods generally lower in sugar than the food seen in other types of eating disorders [28–30].

Bulimia Nervosa: This form of an ED is presented by eating not only very low-calorie foods. Often, high-calorie food is consumed and followed by purging behavior, particularly vomiting. These foods are consumed rapidly, usually in less than 15–20 min before abdominal discomfort sets in. In other cases, laxatives, fasting, and intense exercise are used as well. The vomiting group had a significantly higher intake of diet cola drinks (low-calorie carbonated drinks) in studies [28–30].

Binge Eating Disorder: This eating behavior is most destructive to the oral cavity and upper gastrointestinal tract. Individuals involved select high-calorie foods and follow with purging behavior. If purging is not activated, these people may use laxatives or diuretics and endure extreme exercise activity (Table 1).

Table 1. List of binge foods reported by patients: composition and NOVA categories [31].

	Carbohydrate %	Fat%	Protein%	NOVA Category
Ice cream	47	46	7	4
Doughnut	40–45	45–50	2–3	4
Pita bread and hummus	50	39	11	4
Raisin bagels	80	6	8	4
Cookies	60	38	2	4
Nuts	12	75	13	1 or 4
Diet soda	Artificial sweetener	0	0	4
Potato chips	35	60	5	4
Chocolate cake	57	40	3	4
Cherry yoghurt	75	10	15	4
Pizza	48	37	15	4

NOVA: a classification of food products that assigns the level of processing by the food industry. Examples are described as categories such as minimally processed and ultra-processed. Numerical categories defined: Group 1 comprises foods that are unprocessed or minimally processed. Examples: salt, sugar, oil, and fats. Group 2 is defined as comprising foods from Group 1 with minimal processing with culinary ingredients. Examples are oils, butter, lard, sugar, and salt. Group 3 comprises processed foods, i.e., products manufactured by industry with added substances to increase palatability or shelf stability. Examples include bottled vegetables or legumes preserved in brine and vinegar, fruits in syrup, meat products and canned fish, smoked fish, freshly baked bread, and simple cheeses to which salt is added. Group 4: Ultra-processed foods formulated with ingredients exclusive to industry. Examples are soft drinks, sweets, salty or fatty foods, packaged snacks, candy, pasties, breakfast cereals, and energy drinks.

Further proof of deficiencies in dietary intakes by persons with ongoing eating disorders of any type was described in the following studies. A case-controlled study by Johansson et al. [28] listed the types of foods consumed during the active disease. Patients filled out a 'food recall' of their dietary intakes. The findings concluded that their diet was significantly higher in caffeine-containing cola soft drinks (carbonated drinks). Also, they noted that sugar-containing foods were reduced, along with a reduction in the lunch meal quantity. The NHANES study reviewed the nutritional adequacy of dietary intake in women with anorexia nervosa. In women from 19 to 30 years of age, a 24 h dietary recall was analyzed and compared to the Dietary Reference Intakes (DRIs). It was concluded that all women were deficient in energy, macronutrients, and micronutrients [29,30].

In conclusion, patients with EDs are at risk for malnutrition, which could have significant general and oral health consequences. It is recommended that these patients be assessed both psychologically and physically by health professionals. Counseling by Registered Dietitians is highly recommended to support improved dietary intakes.

4. Oral Complications of ED

4.1. Oral Hygiene

Until recently, studies have reported conflicting results regarding oral hygiene and periodontal health conditions in ED patients. On the one hand, the ED sufferer exhibits personality traits supposed to lead to overzealous toothbrushing. On the other hand, they suffer from depressive comorbidity with low interest in oral hygiene practices and a higher risk of periodontal diseases. Indeed, periodontal diseases include reversible and irreversible clinical forms of periodontal tissue destruction, known as plaque-induced gingivitis and periodontitis, respectively. The evidence-based pathogenesis of periodontal diseases emphasizes the role of malnutrition [32,33], substance abuse in particular tobacco smoking and alcohol consumption [34], and anxiety [35]. Unbalanced diets with a high consumption of carbohydrates [36,37] and deficiency in vitamins and minerals are frequent in EDs [38]. In addition, mood and anxiety disorders are commonly associated with EDs [39,40]. Hence, ED patients exhibit dietary habits and comorbidities at high risk of periodontal disease.

According to the joint classification of the American Academy of Periodontology and the European Federation of Periodontology, plaque-induced gingivitis is an inflammatory response of the gingival tissues resulting from bacterial plaque accumulation located at and below the gingival margin [41]. Managing gingivitis is a primary preventive strategy for periodontitis. Periodontitis is a chronic multifactorial inflammatory disease associated with dysbiotic plaque biofilms and characterized by the progressive destruction of the tooth-supporting apparatus [42]. Its primary features include the loss of periodontal tissue support, manifested through clinical attachment loss (CAL) and radiographically assessed alveolar bone loss; the presence of periodontal pocketing; and gingival bleeding. Without treatment, periodontitis leads to tooth loss.

Case-control studies comparing oral health conditions in EDs with controls have evaluated plaque control with conflicting results. Two studies in the 1990s using the Plaque Index (PI) from Silness and Loe [43], which is a partial recording system prone to bias, found no difference in plaque accumulation between ED and non-ED subjects [44,45]. The oral hygiene level was better in ED outpatients than in non-ED controls in two studies again using non-full-mouth recorded data [28,46]. Poorer plaque control in ED inpatients compared to non-ED controls was shown in two studies [47,48]. In a study with a subgroup analysis, according to the ED diagnosis, plaque control was worse in patients with AN compared to patients with BN and controls with a Plaque Control Record (PCR), measured at 79%, 64%, and 53%, $p < 0.01$, respectively [47]. Hyposalivation induced by the disease and the psychotropic medications [49] may make plaque control difficult and the lack of salivary mechanical and biological antimicrobial actions can favor plaque accumulation [50,51]. However, in studies, the toothbrushing frequency in people with EDs was not different [52] or even higher than in non-ED controls [47,53].

4.2. Periodontal Health

Several observational studies have assessed the periodontal status of ED patients, mostly those suffering from AN and BN. Gingivitis case definitions used were often heterogeneous among studies based on partial-mouth examination approaches with the Gingival Index (GI) [54], leading to no difference [45] and less gingival inflammation [28,46] in ED patients. However, the two studies with full-mouth bleeding scores showed a higher occurrence of plaque-induced gingivitis in ED patients [47,48]. In addition, two well-designed case-control studies found that ED patients present significantly fewer healthy (no gingival bleeding after probing) sextants, measured by the Community Periodontal Index and Treatment Needs (CPITN) [55], compared with non-ED controls, but found no difference in the number of sextants with periodontitis (probing pocket depth > 3mm) [48,56]. These results are in accordance with those of Pallier and collaborators, who observed a higher CAL, with significantly more sites exhibiting gingival recessions in the ED group in comparison with the control group, but no difference in the number of sites with probing pocket depth [47]. To sum up, ED sufferers are not at risk of periodontitis but are at high risk of plaque-induced gingivitis and gingival recessions.

Gingival recession is defined as the apical shift of the gingival margin with respect to the cementoenamel junction; it is associated with attachment loss and with the exposure of the root surface to the oral environment [57]. People with AN and BN have an increased prevalence of gingival recession compared to subjects without EDs of the same age [47,48]. Gingival recessions have multiple maxillary and mandibular localizations. A whitish appearance of the free gingival margin of the recession is frequent as a sign of chemically induced tissue damage by intrinsic and extrinsic acidity. Atypical localizations such as the palatine surfaces of the upper molars are characteristic of ED patients [58]. Toothbrushing frequency is one of the main risk factors for gingival recession along with improper toothbrushing duration and force [57]. This could explain the more frequent generalized gingival recessions observed in ED patients.

Induced by both tooth wear and gingival recessions, dentinal hypersensitivity is also more frequently reported by patients suffering from EDs than controls [44,53].

4.3. Oral Mucosal Health

Oral mucosal lesions are often observed in EDs. Factors such as malnutrition and associated deficiencies of vitamins or micro- and macronutrients, dehydration, or pathological behaviors such as provoking vomiting, overbiting, and other parafunctions predispose to their occurrence [59]. Stress levels and addictions like smoking are also important [52].

When analyzing the published results, it is important to pay attention to the profile of the study group. Some differences will exist due to the fact that previous authors analyzed subjects with varying proportions of AN, BN, and EDNOS cases; disease durations; and ages. Panico et al. (2018) [60] and Lesar et al. (2022) [61] observed oral lesions in up to 94% of patients. Their locations can be on the lips, which are dry in as many as 76.2% [62], 93.2% [52], or 69% [28] of subjects. In addition, exfoliative cheilitis, angular cheilitis, and labial erythema are very frequently diagnosed [1,52,53,60,61,63]. The mucosa is anemic, thin, and prone to injury. As a result, it is pale, and more often, ulcerations and hemorrhagic lesions appear on it, the occurrence of which are significantly influenced by the provocation of vomiting and the highly acidic content of vomit [52,60–63].

Patients are more likely to report mouth irritation, diagnosed as Burning Mouth Syndrome (BMS) [1,53,63]. This disease can be determined by the dentist on the basis of subjective symptoms reported by the patient as it is usually not accompanied by any objective local changes in the oral mucosa. Johansson et al. (2012) [28] showed that the chance of experiencing episodes of burning mouth is 14.2 times higher in patients with eating disorders than in healthy individuals. This is suggested to be associated with mucosal atrophy, xerostomia, and the general disruption of oral homeostasis. Individuals with eating disorders have also been observed to have more common symptoms of cheek and lip biting, as well as impressions on the tongue and linea alba, associated with increased stress levels [52,60,61,63]. A weakened immune system and dysbiosis resulting from a breach in tissue continuity will promote a variety of bacterial, viral, and fungal infections, with the typical clinical presentation of these pathogens [64–67].

4.4. Dental Health

4.4.1. Dental Caries

Eating patterns in EDs that may affect the onset of dental caries include, on the one hand, avoiding entire food groups and certain macronutrients without a medical reason, participating in fad diets to lose weight, and intentionally skipping meals, which may result in nutritional deficiency [68,69]. The low intake of nutrients such as proteins; vitamins A, C, and D; and minerals may enhance their susceptibility to demineralization in carious processes [70]. On the other hand, alternative eating behaviors, such as slowing down the pace of eating, the consumption of large amounts of high-caloric and high-carbohydrate food during binge eating, and engaging in making yourself vomit to control body weight, may contribute to the retention of food debris, the formation of dental plaque on the tooth surface, the lowering of the pH of the oral cavity environment, and the promotion of demineralization, and thus favor dental caries [68].

Additional factors that may promote dental caries in ED patients are alterations in the composition of saliva [71]; a decrease in saliva's buffer capacity and secretion rate [52,53,72,73], which may be associated with the structural changes in the salivary glands [28,74]; self-induced vomiting or starvation; and the side effects of psychotropic drugs (i.e., antidepressants, appetite suppressants), diuretics or laxatives [49]. Undoubtedly, poor oral hygiene in patients with eating disorders, presumably also resulting from psychological reasons, may be a contributing factor to dental caries [47,75].

Numerous studies have examined the association between dental caries and EDs, applying commonly used caries assessment measures such as caries prevalence or caries severity based on the clinical or (less often) radiographic recordings and expressed by the sum of decayed (D), missing (M), and filled (F) teeth (DMFT index) or tooth surfaces (DMFS index) [28,47,52,53,74–78] or the International Caries Detection and Assessment System (ICDAS-II) system) [79]. However, the results obtained are contradictory and vary between

37% and 80% depending on the type of ED and the qualification criteria adopted for diagnosis [74,75,78–81]. Some authors demonstrated significantly higher values in terms of the DMFT index and its D and M components in comparison with control groups [47,53,75].

Relationships between the caries index and vomiting frequency are also inconsistent although it is clear that gastric acid along with high-carbohydrate food or beverages may initiate the carious process [82].

It is important to emphasize that there is no single ED-associated factor, but rather, all the above-discussed contributing risk factors may have inputs in ED. The following factors should be considered: cariogenic diet, oral hygiene level, remineralization exposure, and taking certain types of medication [81].

4.4.2. Erosive Tooth Wear

Erosive tooth wear (dental erosion), the irreversible loss of tooth structure due to chemical dissolution by acids not of bacterial origin, is one of the major oral complications of eating disorders [83]. Erosive tooth wear (ETW) is caused by the frequent contact of teeth with acid from either the stomach (gastric acid) or drinking and eating acidic drinks and foods, particularly outside meal times [84]. EDs, particularly BN, can potentially increase the risk for ETW since they may be associated with bingeing on acidic foods and/or drinks followed by vomiting after every meal [28–31,85]. Some of these patients use these beverages both in the place of normal meals as well as during intense exercise to lose calories [31]. These characteristics are associated with the frequent contact of the teeth with gastric or dietary acids over an extended period of time, with the consequent wearing away of the dental hard tissue through acid demineralization, initially affecting the enamel, and with the progression to an advanced stage, dentin is exposed (Figures 2 and 3). The exposure of dentin may result in dentin hypersensitivity in response to external stimuli of a cold, hot, tactile, or osmotic nature. The acid of gastric juice brought up due to vomiting, the pH of which can be as low as 1, causes the wear of the palatal surfaces of upper incisors (Figure 2), and with lesion progression, the lingual surfaces of premolars and molars become affected, and in more advanced stages, the process extends to the occlusal surfaces of molars and to the facial surfaces of all teeth [85,86]. Erosion due to dietary acid, with pH ranging from 2.7 to 3.8, has no specific distribution pattern, but depends on factors such as the method of application. Thus, EDs in combination with vomiting are associated with an increased occurrence, severity, and risk of dental erosion [85]. The reported prevalence of ETW among eating disorder patients, particularly those with BN, varies among countries and ranges from 42 to 98% [85].

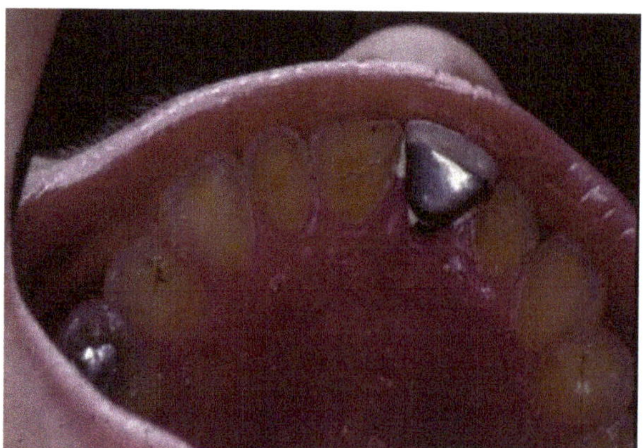

Figure 2. Palatal surfaces of maxillary teeth affected by erosive tooth wear (Courtsey Amaechi BT, UTHSA, San Antonio, TX, USA).

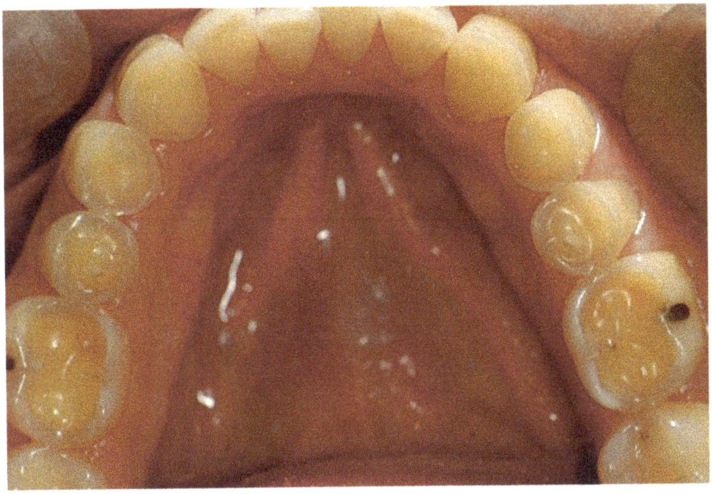

Figure 3. Occlusal surfaces of mandibular teeth affected by erosive teeth wear (Courtsey Amaechi BT, UTHSA, San Antonio, TX, USA).

Main ED's symptoms and oral effects are presented in Figure 4.

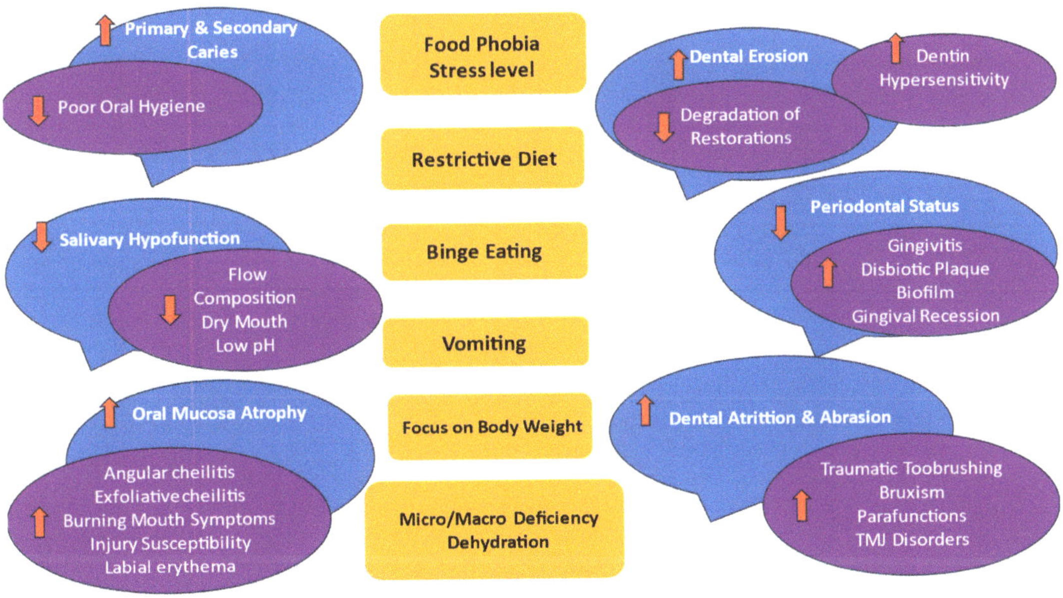

Figure 4. EDs' main symptoms (in yellow) and oral effects (in blue and purple). The following factors may influence any oral complications: subtype of AN, BN, and EDNOS according to nutritional behaviors; vomiting frequency; disease duration; age and sex of the patient; general health status; pharmacotherapy and their side effects to salivation; and psychosocial profile; as well as cariogenic/acidic diet, individual oral hygiene, and remineralization exposure. Abbreviations: ↑↓ means the ED symptoms increase the oral effect/decrease the oral effect.

5. Current Approaches to Managing ED

Due to the fact that the effects of EDs on both soft and hard oral tissues are of a multifactorial nature, the dentist may be the first to observe the existence of EDs that may not have been revealed to the medical personnel during history taking. Considering the etiology and symptoms of EDs, management requires a multidisciplinary team involving clinical psychologists, dietitians, and dental professionals. For this reason, the following management should be followed.

5.1. Counselling Relating to Dietary Habits

As noted in this manuscript, there exist a wide variety of manifestations of eating disorders. Multiple therapeutic approaches, including psychotherapy, hospitalization, and nutritional counseling, have been employed. A focus on the dietary intake and nutritional education of individuals suffering from eating disorders is of high significance.

The most severe and aggressive form of an eating disorder is anorexia nervosa. A study of young women who had current disease were assessed as to their state of malnutrition. The study investigated and analyzed their dietary intake with a 4-day food diary recorded by each participant. The nutritional analysis found that their diets were deficient in both macro- and micronutrients. Their body weight and basic mass index (BMI) were lower than in healthy reference groups. The study concluded that all anorexia nervosa patients must undergo nutritional counseling. In severe cases, refeeding may include hospitalization with Nasogastric-tube feedings, nutrition education, and the supplementation of vitamins and minerals to support the refeeding plan [28–30].

A meta-analysis of nutritional counseling as part of the clinical approach to treating eating disorders demonstrated the benefit to vary among patients, with some benefitting while others did not, and this was attributed to the varied manifestations of the disease [28,29]. However, a consistent finding indicated that BMI and weight increased with dietetic input. The quality of the diet was positively impacted by nutritional advice. Further, practices and standards conclude that refeeding intervention is necessary and a dietitian/nutritionist is well able to intervene as part of the interprofessional team. Valuable to the outcome was the dietitian's expertise in selecting a feeding plan to include menus tailored to each patient's taste and cultural background [28,29].

In general, the first rule of dietetic therapy is to personalize the refeeding plan of the patient presenting with an eating disorder. Conducting an interview with the patient to determine how they "feel about food" may assist in the education process. Such a conversation may unveil certain aversions to food types, smell, or consistency. The dietitian/nutritionist can prescribe a menu plan that does not include these food items. Furthermore, the omitted food items may be a topic of discussion to dispel any misinformation that may exist [31,87,88]. To initiate the discussion, the dietitian/nutritionist must have the patient's agreement on the composition of the menu plan. Once agreed upon, the calorie level should be advanced one to two times per week. Weight goals of 2–3 lbs per week are recommended for underweight patients (BMI < 18) [31,88]. It should be noted that eating disorder patients can have normal or above-normal BMI. Nutritional therapy should include support for consistency in the patients' eating patterns. These menus should include three meals with 1–3 snacks per day. Due to the use of laxatives, purging episodes, and excess exercise, the monitoring of electrolytes, weights, vital signs, and recovery meal plan compliance is advised [31,88]. The recovery meal plan may be a vehicle to give the eating disorder patient "permission" to eat. The goal is to reintroduce food as a partner and not an adversary in the healing process. The MyPlate approach (Figure 5, Table 2) is the simplest place to start with the education process [89,90], and foods to include are:

- Proteins: Total 5 oz/day
- Grains: 6 servings/day
- Dairy or alternative to dairy: 3 servings/day
- Vegetables and fruits: 5 servings/day
- Fats and oils: 4 servings/day

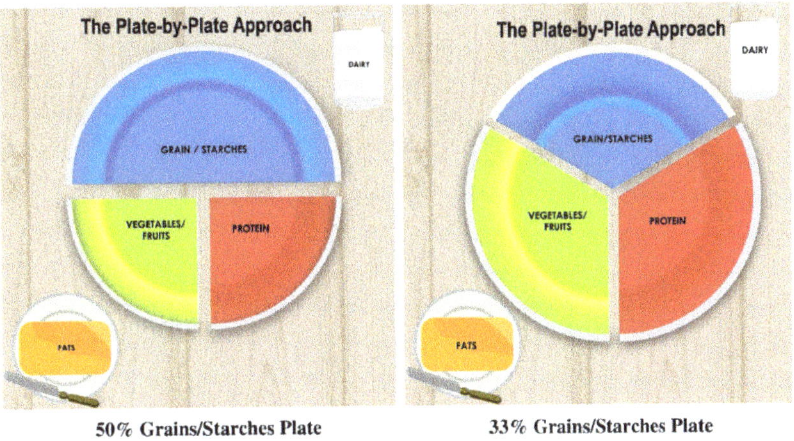

Figure 5. A typical example of the Plate-by-Plate Approach.

Table 2. Seven essential steps and strategies for implementing the Plate-by-Plate Strategy [89,90].

STEP 1	CHOOSE A 10-inch PLATE
STEP 2	PLATE ALL FOOD GROUPS
STEP 3	FILL THE PLATE UP
STEP 4	DECIDE HOW MANY MEALS AND SNACKS
STEP 5	INCLUDE VARIETY
STEP 6	DOES THE MEAL MAKE SENSE?
STEP 7	THE FINAL REVIEW: HOW DOES THE PLATE LOOK? • Are all the food groups present? Grains/Starches? Proteins? Fruit/Vegetables? Dairy? Fat? • Are the plate 50% grains/starches, 25% protein, 25% fruit/vegetables? (or 33% grains/starches, 33% protein, 33% fruit/vegetable, depending on which plate is recommended?) • Is the whole plate full? • Have you challenged your child?

The patient must keep an accurate food intake of dairy. The dietitian/nutritionist should review for balance, frequency of meal consumption, variety, and the appropriateness of portion sizes [89,90].

Standard practice for a dietitian's role in the supervision of a patient with an eating disorder is centered around the core dietetic skills of screening, professional responsibility, assessment, nutrition diagnosis, intervention, monitoring, and evaluation [91]. In summary, regular and careful assessments of nutritional intake along with nutritional counseling can reduce malnutrition in patients with eating disorders. To optimize the chance of full recovery from an eating disorder, a multidisciplinary approach is highly recommended. Currently, dietitians are a part of most eating disorder treatment teams.

5.2. Counseling Relating to the Prevention of Effects on Oral Soft Tissues

Faced with this increased risk of plaque-related gingivitis and gingival recession, prevention and early treatment require increased motivation for plaque control and the appropriate teaching of oral hygiene methods. An electric toothbrush is useful in patients with EDs to potentiate plaque removal while controlling brushing time and pressure on the periodontium due to the integrated timer and sensor [89].

Regular supportive periodontal care, two times a year, including professional mechanical supra-gingival plaque removal and lifestyle counselling is of the utmost importance for an efficacious preventive approach towards periodontal complications in ED patients.

5.3. Counseling Relating to the Prevention of Effects on Oral Hard Tissues (Dental Caries)

Individuals suffering from EDs are included in the high-caries-risk group [92]. Therefore, effective strategies for preventing dental caries should be applied (Table 3) [93].

While fluoride compounds, applied to remineralize dental hard tissues, need the salivary calcium and phosphate ions, in the case of ED hyposalivation, they may diminish the remineralization process. Apart from fluoride compounds, the cariostatic effect of fluoride-free formulations such as calcium phosphates, either in amorphous or crystalline form (hydroxyapatite), has also been used with advantageous results in several long-term randomized clinical trials [94–99]. The mode of action of hydroxyapatite involves providing mineral ions required for remineralization. These deposits form a protective layer on the enamel surface, eliminate the harmful effects of cariogenic acids, and prevent the adhesion of plaque-forming bacteria to the tooth surface [100].

Table 3. The main strategies for preventing dental caries.

Control the bacterial biofilm	via physical methods (toothbrush, interdental and tongue brushes, dental floss, irrigators) along with antibacterial preparations if temporarily needed (toothpaste or mouthwashes containing active agents, e.g., chlorhexidine, stannous salts, or zinc salts)
Increase in the resistance of hard dental tissues to the demineralizing effect of bacterial acids	via the topical use of remineralizing agents (at home and in-office application)
Modification of the diet	limiting the consumption of fermentable carbohydrates; avoiding sticky/acid products and replacing them with caries-protective foods such as raw vegetables, nuts, and cheese
Stimulation of salivary flow	with consistent food and sugar-free chewing gums containing sucrose substitutes i.e., xylitol
Altering medication-induced hyposalivation if needed [101–103]	
Regular dental check-ups every 3 months to monitor the oral condition and motivate the patient	

5.4. Counseling Relating to the Prevention of Effects on Oral Hard Tissues (Erosive Tooth Wear)

Once dental erosion is detected, there is a need for a full case history, which should include dietary history, medical history, dental hygiene habits, and lifestyle history. This would establish the risk factors and help in the development of individualized counseling. Erosive tooth wear management in ED patients must involve education (counselling) on risk factors and the prevention of occurrence and protection of affected surfaces from further damage, and in cases of severe structural damage, restoration may be required.

The counseling of ED patients should start by educating each patient on the risk factors predisposing him/her to ETW, with a referral for psychoeducation and dietary advice. Eating disorder patients should take the following precautions [83].

When possible, 'bite-guards' should be worn while vomiting. The inside (tooth surface) of the guard should be coated with a small amount of Sodium Bicarbonate suspension, or Milk of Magnesia, to neutralize any gastric acid pooling in it.

Patients must avoid toothbrushing immediately after each episode of vomiting or bingeing on acidic food or drink; rather, patients should use any of the following to freshen their mouth and wait for at least 60 min before toothbrushing [104]: fluoride mouthwash, fluoride tablets/lozenges or dairy products (e.g., milk) to enhance the rapid remineralization of the softened tooth surface, sugar-free chewing gum or lozenges to increase saliva flow to neutralize the acidity and provide an alkaline environment to facilitate the rapid remineralization of the softened tooth tissue, and sugar-free antacid tablets or a pinch of sodium bicarbonate (or baking soda) dissolved in some water to neutralize the acidic oral fluid.

They must use toothpaste containing either high fluoride concentrations or stabilized stannous fluoride, as well as fluoride mouth rinses, for their routine daily oral hygiene practice.

The intake of acidic dietary products with added calcium reduces the erosive effect of the drink [105].

5.5. Steps to Protect Affected Oral Soft and Hard Tissues

Current recommendations for oral management of patients with EDs are based on the need for early intervention without waiting for the remission/curing of eating symptoms, which sometimes takes years [106]. Oral treatment is part of the comprehensive, multidisciplinary, and personalized care project that is coordinated by the general practitioner and the psychiatrist. After making a differential diagnosis with the usual dental pathologies (carious lesions, fractures, cracks, defective restorative care), the positive diagnosis of dentinal hypersensitivity associated with gingival recessions will be confirmed via the visual inspection of dentinal exposure sites, most often at the cervical level that is sensitive to contact (tactile probe test) and/or air using the Schiff test, which is graded from 0 to 3 [107]. Non-invasive treatment is preferred and may include a prescription of desensitizing agents in the form of toothpaste and/or mouthwash. However, in many situations, in patients with an ED, the advanced stage of the hard and soft tissue lesions will indicate restorative treatment and/or periodontal plastic surgery as part of comprehensive management [108]. However, no study has explored the safety and efficacy of root coverage procedures in ED patients. Tobacco cessation has a beneficial impact on root coverage success and ED patients who smoke have to be encouraged to quit before periodontal plastic surgeries [109].

Eroded tooth surfaces in ED patients can be protected from further erosive damage and deterioration in appearance by using any of the following procedures. Dentin bonding agents can be applied to protect erosively exposed dentinal tissues to reduce the rate of tooth wear [110]. Adhesively retained resin veneers and crowns both improve appearance as well as provide protection against further damage [111,112]. These restorations are easy to place in advanced cases with dentinal exposure when the eating symptoms are not fully controlled [113,114]. Porcelain restorations may be preferred when the patient recovers from the ED.

5.6. Recall Visits for Continued Care to Maintain Compliance and Oral Health

A continued care regime matched to the patient's requirements should be established to check patient compliance, monitor wear, reinforce advice, and provide encouragement to maintain changed behavior. Failure to monitor the patient over time may result in a relapse of the condition (Figure 6).

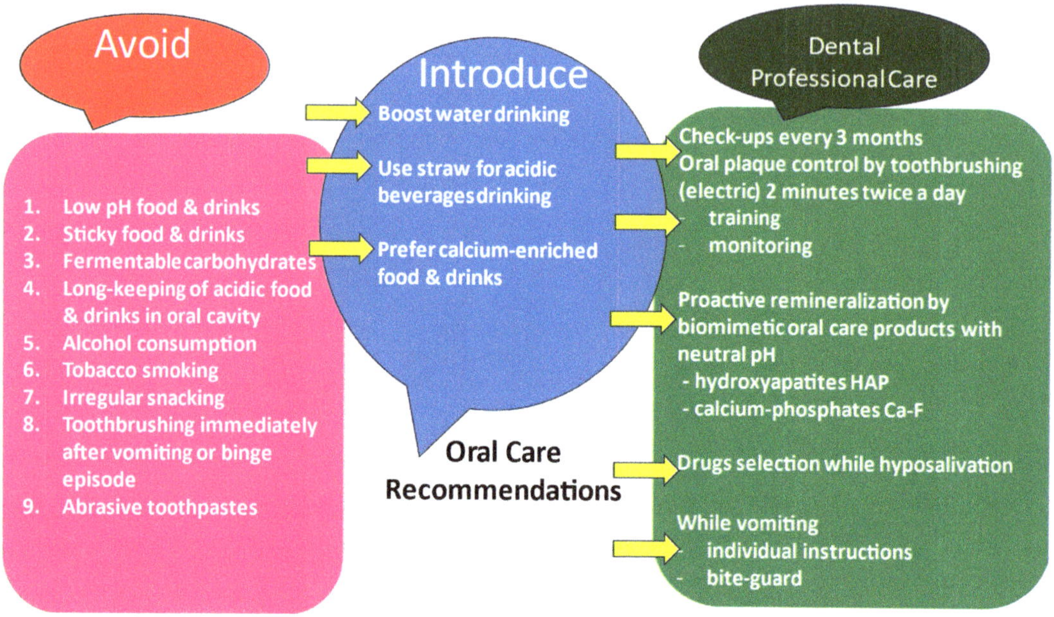

Figure 6. Counseling relating to the prevention of effects on oral hard and soft tissues (what to avoid but introduce for the patient and professional care).

6. Conclusions

Based on the specific psychopathological ED symptoms associated with food intake, it is possible to identify EDs that lead to a number of somatic complications and changes in the functioning of the oral cavity. The present review covers various directions in which dietary habits and oral health may be altered during the course of an ED. However, oral complications in ED patients are of multifactorial origin, and as such, their management requires multidisciplinary approaches. Current concepts in oral disease prevention should focus on tooth remineralization and the control of all the key factors involved in oral complications. The results of this review highlight the role that should be played in this regard by groups of nutritionists, dental professionals, and medical practitioners. Therefore, it should be considered as a serious and complex health problem. The present study uncovers analysis that future investigations may be forwarded. Further studies, preferably multi-center longitudinal studies involving larger populations with EDs, are necessary to establish, precisely, the association between dietary patterns in ED patients and their oral factors, e.g., microbiota. To inhibit oral repercussions, we recommend tailored interventions consisting of oral plaque control, moisturizing oral mucosa, intensive remineralization based not only on fluoride but also hydroxyapatites and calcium-phosphate products, and preventive advice and bite-guard application while vomiting.

Author Contributions: Conceptualization, E.P.; data curation, E.P., M.P., H.R., B.T.A., G.S.G. and A.H.; formal analysis, E.P., M.P., H.R., B.T.A., G.S.G. and A.H.; funding acquisition, E.P. and M.P.; investigation, E.P., M.P., H.R., B.T.A., G.S.G. and A.H.; methodology, E.P.; project administration, E.P.; software, E.P., M.P., H.R., B.T.A., G.S.G. and A.H.; supervision, E.P.; writing original draft, E.P., M.P., H.R., B.T.A., G.S.G. and A.H. All authors have read and agreed to the published version of the manuscript.

Funding: This research was funded by the Poznan University of Medical Sciences and Medical University in Bialystok.

Institutional Review Board Statement: Not applicable.

Informed Consent Statement: This research methodology assumed non-invasive methods of analysis of the scientific literature.

Data Availability Statement: Data associated with the paper are not publicly available but are available from the corresponding author on reasonable request.

Acknowledgments: Particular thanks go to Magdalena Karbowniczek, who supported us with editing and data preparation.

Conflicts of Interest: The authors declare no conflict of interest. The funders (Poznan University of Medical Sciences and Medical University in Bialystok) had no role in the study's design; in the collection, analysis, or interpretation of data; in the writing of the manuscript; or in the decision to publish the results.

References

1. Romanos, G.E.; Javed, F.; Romanos, E.B.; Williams, R.C. Oro-facial manifestations in patients with eating disorders. *Appetite* **2012**, *59*, 499–504. [CrossRef] [PubMed]
2. American Psychiatric Association. *Diagnostic and Statistical Manual of Mental Disorders*, 5th ed.; American Psychiatric Publishing: Arlington, TX, USA, 2013.
3. International Classification of Diseases, Eleventh Revision (ICD-11), World Health Organization (WHO) 2019/2021. Licensed under Creative Commons Attribution-NoDerivatives 3.0 IGO Licence (CC BY-ND 3.0 IGO). Available online: https://icd.who.int/browse11 (accessed on 1 July 2023).
4. Reed, G.M.; First, M.B.; Kogan, C.S.; Hyman, S.E.; Gureje, O.; Gaebel, W.; Maj, M.; Stein, D.J.; Maercker, A.; Tyrer, P.; et al. Innovations and changes in the ICD-11 classification of mental, behavioural and neurodevelopmental disorders. *World Psychiatry* **2019**, *18*, 3–19. [CrossRef] [PubMed]
5. Özbaran, N.B.; Yılancıoğlu, H.Y.; Tokmak, S.H.; YuluğTaş, B.; Çek, D.; Bildik, T. Changes in the psychosocial and clinical profiles of anorexia nervosa patients during the pandemic. *Front. Psychiatry* **2023**, *14*, 1207526. [CrossRef] [PubMed]
6. Jagielska, G.; Kacperska, I. Outcome, comorbidity and prognosis in anorexia nervosa. *Psychiatr. Pol.* **2017**, *51*, 205–218. [CrossRef] [PubMed]
7. Grzelak, T.; Dutkiewicz, A.; Paszynska, E.; Dmitrzak-Weglarz, M.; Slopien, A.; Tyszkiewicz-Nwafor, M. Neurobiochemical and psychological factors influencing the eating behaviors and attitudes in anorexia nervosa. *J. Physiol. Biochem.* **2017**, *73*, 297–305. [CrossRef] [PubMed]
8. Bardone-Cone, A.M.; Fitzsimmons-Craft, E.E.; Harney, M.B.; Maldonado, C.R.; Lawson, M.A.; Smith, R.; Robinson, D.P. The inter-relationships between vegetarianism and eating disorders among females. *J. Acad. Nutr. Diet.* **2012**, *112*, 1247–1252. [CrossRef] [PubMed]
9. Ferguson, C.J.; Muñoz, M.E.; Garza, A.; Galindo, M. Concurrent and prospective analyses of peer, television and social media influences on body dissatisfaction, eating disorder symptoms and life satisfaction in adolescent girls. *J. Youth Adolesc.* **2014**, *43*, 1–14. [CrossRef]
10. Antczak, A.J.; Brininger, T.L. Diagnosed eating disorders in the U.S. Military: A nine year review. *Eat. Disord.* **2008**, *16*, 363–377. [CrossRef]
11. Filaire, E.; Rouveix, M.; Pannafieux, C.; Ferrand, C. Eating Attitudes, Perfectionism and Body-esteem of Elite Male Judoists and Cyclists. *J. Sports Sci. Med.* **2007**, *6*, 50–57.
12. Penniment, K.J.; Egan, S.J. Perfectionism and learning experiences in dance class as risk factors for eating disorders in dancers. *Eur. Eat. Disord. Rev.* **2012**, *20*, 13–22. [CrossRef]
13. Lindberg, L.; Hjern, A. Risk factors for anorexia nervosa: A national cohort study. *Int. J. Eat. Disord.* **2003**, *34*, 397–408. [CrossRef] [PubMed]
14. McClelland, L.; Crisp, A. Anorexia nervosa and social class. *Int. J. Eat. Disord.* **2001**, *29*, 150–156. [CrossRef] [PubMed]
15. Dmitrzak-Weglarz, M.; Moczko, J.; Skibinska, M.; Slopien, A.; Tyszkiewicz, M.; Pawlak, J.; Zaremba, D.; Szczepankiewicz, A.; Rajewski, A.; Hauser, J. The study of candidate genes related to the neurodevelopmental hypothesis of anorexia nervosa: Classical association study versus decision tree. *Psychiatry Res.* **2013**, *30*, 117–121. [CrossRef]
16. Dmitrzak-Weglarz, M.; Szczepankiewicz, A.; Slopien, A.; Tyszkiewicz, M.; Maciukiewicz, M.; Zaremba, D.; Twarowska-Hauser, J. Association of the glucocorticoid receptor gene polymorphisms and their interaction with stressful life events in Polish adolescent girls with anorexia nervosa. *Psychiatr. Danub.* **2016**, *28*, 51–57.
17. Baker, J.H.; Schaumberg, K.; Munn-Chernoff, M.A. Genetics of Anorexia Nervosa. *Curr. Psychiatry Rep.* **2017**, *19*, 84. [CrossRef] [PubMed]
18. Boraska, V.; Franklin, C.S.; Floyd, J.A.; Thornton, L.M.; Huckins, L.M.; Southam, L.; Rayner, N.W.; Tachmazidou, I.; Klump, K.L.; Treasure, J. A genome-wide association study of anorexia nervosa. *Mol. Psychiatry* **2014**, *19*, 1085–1094. [CrossRef]
19. Janas-Kozik, M.; Stachowicz, M.; Krupka-Matuszczyk, I.; Szymszal, J.; Krysta, K.; Janas, A.; Rybakowski, J.K. Plasma levels of leptin and orexin a in the restrictive type of anorexia nervosa. *Regul. Pept.* **2011**, *168*, 5–9. [CrossRef]

20. Pałasz, A.; Tyszkiewicz-Nwafor, M.; Suszka-Świtek, A.; Bacopoulou, F.; Dmitrzak-Węglarz, M.; Dutkiewicz, A.; Słopień, A.; Janas-Kozik, M.; Wilczyński, K.M.; Filipczyk, Ł.; et al. Longitudinal study on novel neuropeptides phoenixin, spexin and kisspeptin in adolescent inpatients with anorexia nervosa—Association with psychiatric symptoms. *Nutr. Neurosci.* **2021**, *24*, 896–906. [CrossRef]
21. Grzelak, T.; Tyszkiewicz-Nwafor, M.; Dutkiewicz, A.; Mikulska, A.A.; Dmitrzak-Weglarz, M.; Slopien, A.; Czyzewska, K.; Paszynska, E. Neuropeptide B and Vaspinas New Biomarkers in Anorexia Nervosa. *Biomed. Res. Int.* **2018**, *10*, 9727509.
22. Tyszkiewicz-Nwafor, M.; Jowik, K.; Paszynska, E.; Dutkiewicz, A.; Słopien, A.; Dmitrzak-Weglarz, M. Expression of immune-related proteins and their association with neuropeptides in adolescent patients with anorexia nervosa. *Neuropeptides* **2022**, *91*, 102214. [CrossRef]
23. Vos, T.; Allen, C.; Arora, M.; Barber, R.M.; Bhutta, Z.A.; Brown, A.; Carter, A.; Casey, D.C.; Charlson, F.J.; Chen, A.Z.; et al. Global, regional, and national incidence, prevalence, and years lived with disability for 310 diseases and injuries, 1990–2015: A systematic analysis for the Global Burden of Disease Study 2015. *Lancet* **2016**, *388*, 1545–1602. [CrossRef] [PubMed]
24. Keski-Rahkonen, A.; Mustelin, L. Epidemiology of eating disorders in Europe: Prevalence, incidence, comorbidity, course, consequences, and risk factors. *Curr. Opin. Psychiatry* **2016**, *29*, 340–345. [CrossRef] [PubMed]
25. van Eeden, A.E.; van Hoeken, D.; Hoek, H.W. Incidence, prevalence and mortality of anorexia nervosa and bulimia nervosa. *Curr. Opin. Psychiatry* **2021**, *34*, 515–524. [CrossRef]
26. van Noort, B.M.; Lohmar, S.K.; Pfeiffer, E.; Lehmkuhl, U.; Winter, S.M.; Kappel, V. Clinical characteristics of early onset anorexia nervosa. *Eur. Eat. Disord. Rev.* **2018**, *26*, 519–525. [CrossRef]
27. Bryant-Waugh, R. Feeding and Eating Disorders in Children. *Psychiatr. Clin. N. Am.* **2019**, *42*, 157–167. [CrossRef]
28. Johansson, A.K.; Norring, C.; Unell, L.; Johansson, A. Eating disorders and oral health: A matched case-control study. *Eur. J. Oral Sci.* **2012**, *120*, 61. [CrossRef]
29. Schebendach, J.E.; Mayer, L.E.; Devlin, M.J.; Attia, E.; Contento, I.R.; Wolf, R.L.; Walsh, B.T. Dietary energy density and diet variety as predictors of outcome in anorexia nervosa. *Am. J. Clin. Nutr.* **2008**, *87*, 810–816, Erratum in *Am. J. Clin. Nutr.* **2012**, *96*, 222. [CrossRef]
30. Raatz, S.K.; Jahns, L.; Johnson, L.K.; Crosby, R.; Mitchell, J.E.; Crow, S.; Peterson, C.; Le Grange, D.; Wonderlich, S.A. Nutritional adequacy of dietary intake in women with anorexia nervosa. *Nutrients* **2015**, *15*, 3652–3665. [CrossRef] [PubMed]
31. Monteiro, C.A.; Cannon, G.; Lawrence, M.; Costa Louzada, M.; Pereira Machado, P. *Ultra-Processed Foods, Diet Quality, and Health Using the NOVA Classification System*; Food and Agriculture Organization of the United Nations: Rome, Italy, 2019.
32. Casarin, M.; da Silveira, T.M.; Bezerra, B.; Pirih, F.Q.; Pola, N.M. Association between different dietary patterns and eating disorders and periodontal diseases. *Front. Oral Health* **2023**, *22*, 1152031. [CrossRef]
33. Dommisch, H.; Kuzmanova, D.; Jönsson, D.; Grant, M.; Chapple, I. Effect of micronutrient malnutrition on periodontal disease and periodontal therapy. *Periodontol. 2000* **2018**, *78*, 129–153. [CrossRef]
34. Kumar, P.S. Interventions to prevent periodontal disease in tobacco-, alcohol-, and drug-dependent individuals. *Periodontol. 2000* **2020**, *84*, 84–101. [CrossRef] [PubMed]
35. Aggarwal, K.; Gupta, J.; Kaur, R.K.; Bansal, D.; Jain, A. Effect of anxiety and psychologic stress on periodontal health: A systematic review and meta-analysis. *Quintessence Int.* **2022**, *53*, 144–154. [PubMed]
36. Liew, V.P.; Frisken, K.W.; Touyz, S.W.; Beumont, P.J.; Williams, H. Clinical and microbiological investigations of anorexia nervosa. *Aust. Dent. J.* **1991**, *36*, 435–441. [CrossRef] [PubMed]
37. Hellström, I. Oral complications in anorexia nervosa. *Scand. J. Dent. Res.* **1977**, *85*, 71–86. [CrossRef] [PubMed]
38. Hanachi, M.; Dicembre, M.; Rives-Lange, C.; Ropers, J.; Bemer, P.; Zazzo, J.F.; Poupon, J.; Dauvergne, A.; Melchior, J.C. Micronutrients Deficiencies in 374 Severely Malnourished Anorexia Nervosa Inpatients. *Nutrients* **2019**, *5*, 792. [CrossRef]
39. Godart, N.T.; Flament, M.F.; Perdereau, F.; Jeammet, P. Comorbidity between eating disorders and anxiety disorders: A review. *Int. J. Eat. Disord.* **2002**, *32*, 253–270. [CrossRef]
40. Godart, N.T.; Perdereau, F.; Rein, Z.; Berthoz, S.; Wallier, J.; Jeammet, P.H.; Flament, M.F. Comorbidity studies of eating disorders and mood disorders. Critical review of the literature. *J. Affect. Disord.* **2007**, *97*, 37–49. [CrossRef]
41. Murakami, S.; Mealey, B.L.; Mariotti, A.; Chapple, I.L.C. Dental plaque-induced gingival conditions. *J. Periodontol.* **2018**, *89*, 17–27. [CrossRef]
42. Papapanou, P.; Sanz, M.; Buduneli, N.; Dietrich, T.; Feres, M.; Fine, D.H.; Flemmig, T.F.; Garcia, R.; Giannobile, W.V.; Graziani, F.; et al. Periodontitis: Consensus report of workgroup 2 of the 2017 World Workshop on the Classification of Periodontal and Peri-Implant Diseases and Conditions. *J. Clin. Periodontol.* **2018**, *45*, 162–170. [CrossRef]
43. Silness, J.; Loe, H. Periodontal Disease in Pregnancy. II. Correlation between Oral Hygiene and Periodontal Condtion. *Acta Odontol. Scand.* **1964**, *22*, 121–135. [CrossRef]
44. Altschuler, B.D.; Dechow, P.C.; Waller, D.A.; Hardy, B.W. An investigation of the oral pathologies occurring in bulimia nervosa. *Int. J. Eat. Disord.* **1990**, *9*, 191–199. [CrossRef]
45. Milosevic, A.; Slade, P.D. The orodental status of anorexics and bulimics. *Br. Dent. J.* **1989**, *22*, 66–70. [CrossRef] [PubMed]
46. Philipp, E.; Willershausen-Zönnchen, B.; Hamm, G.; Pirke, K.M. Oral and dental characteristics in bulimic and anorectic patients. *Int. J. Eat. Disord.* **1991**, *10*, 423–431. [CrossRef]
47. Pallier, A.; Karimovaa, A.; Boillota, A.; Colon, P.; Ringuenete, D.; Bouchard, P. Dental and periodontal health in adults with eating disorders: A case-control study. *J. Dent.* **2019**, *84*, 55–59. [CrossRef]

48. Touyz, S.; Liew, V.; Tseng, P.; Frisken, K.; Williams, H.; Beumont, P. Oral and Dental Complications in Dieting Disorders. *Int. J. Eat. Disord.* **1993**, *14*, 341–347. [CrossRef]
49. Kisely, S.; Baghaie, H.; Lalloo, R.; Johnson, N. Association between poor oral health and eating disorders: Systematic review and meta-analysis. *Br. J. Psychiatry* **2015**, *207*, 299–305. [CrossRef]
50. Chapple, I.L.; Bouchard, P.; Cagetti, M.G.; Campus, G.; Carra, M.C.; Cocco, F.; Nibali, L.; Hujoel, P.; Laine, M.l.; Lingstrom, P.; et al. Interaction of lifestyle, behaviour or systemic diseases with dental caries and periodontal diseases: Consensus report of group 2 of the joint EFP/ORCA workshop on the boundaries between caries and periodontal diseases. *J. Clin. Periodontol.* **2017**, *44*, 39–51. [CrossRef]
51. Chapple, I.L.; Mealey, B.L.; Van Dyke, T.E.; Bartold, P.M.; Dommisch, H.; Eickholz, P.; Geisinger, M.L.; Genco, R.J.; Glogauer, M.; Goldstein, M.; et al. Periodontal health and gingival diseases and conditions on an intact and a reduced periodontium: Consensus report of workgroup 1 of the 2017 World Workshop on the Classification of Periodontal and Peri-Implant Diseases and Conditions. *J. Clin. Periodontol.* **2018**, *45*, 68–77. [CrossRef]
52. Garrido-Martínez, P.; Domínguez-Gordillo, A.; Cerero-Lapiedra, R.; Burgueño-García, M.; Martínez-Ramírez, M.J.; Gómez-Candela, C.; Cebrián-Carretero, J.L.; Esparza-Gómez, G.G. Oral and dental health status in patients with eating disorders in Madrid, Spain. *Med. Oral Patol. Oral Cir. Bucal.* **2019**, *24*, 595–602. [CrossRef]
53. Lourenço, M.; Azevedo, A.; Brandão, I.; Gomes, P. Orofacial manifestations in outpatients with anorexia nervosa and bulimia nervosa focusing on the vomiting behaviour. *Clin. Oral Investig.* **2018**, *22*, 1915–1922. [CrossRef]
54. Loe, H.; Silness, J. Periodontal Disease in Pregnancy. I. Prevalence and Severity. *Acta Odontol. Scand.* **1963**, *21*, 533–551. [CrossRef]
55. Cutress, T.W.; Ainamo, J.; Sardo-Infirri, J. The community periodontal index of treatment needs (CPITN) procedure for population groups and individuals. *Int. Dent. J.* **1987**, *37*, 222–333.
56. Chiba, F.Y.; Sumida, D.H.; Moimaz, S.A.S.; Chaves Neto, A.H.; Nakamune, A.; Garbin, A.J.I.; Garbin, C.A.S. Periodontal condition, changes in salivary biochemical parameters, and oral health-related quality of life in patients with anorexia and bulimia nervosa. *J. Periodontol.* **2019**, *90*, 1423–1430. [CrossRef]
57. Cortellini, P.; Bissada, N.F. Mucogingival conditions in the natural dentition: Narrative review, case definitions, and diagnostic considerations. *J. Periodontol.* **2018**, *45*, 190–198.
58. Range, H.; Colon, P.; Godart, N.; Kapila, Y.; Bouchard, P. Eating disorders through the periodontal lens. *Periodontol. 2000* **2021**, *87*, 17–31. [CrossRef] [PubMed]
59. Stanislav, N.; Tolkachjov, M.D.; Bruce, A.J. Oral manifestations of nutritional disorders. *Clin. Dermatol.* **2017**, *35*, 441–452.
60. Panico, R.; Piemonte, E.; Lazos, J.; Gilligan, G.; Zampini, A.; Lanfranchi, H. Oral mucosal lesions in Anorexia Nervosa, Bulimia Nervosa and EDNOS. *J. Psychiatr. Res.* **2018**, *96*, 178–182. [CrossRef] [PubMed]
61. Lesar, T.; Vidović Juras, D.; Tomić, M.; Čimić, S.; Kraljević Šimunković, S. Oral Changes in Pediatric Patients with Eating Disorders. *Acta Clin. Croat.* **2022**, *61*, 185–192. [CrossRef] [PubMed]
62. Mascitti, M.; Coccia, E.; Vignini, A.; Aquilanti, L.; Santarelli, A.; Salvolini, E.; Sabbatinelli, J.; Mazzanti, L.; Procaccini, M.; Rappelli, G. Anorexia, Oral Health and Antioxidant Salivary System: A Clinical Study on Adult Female Subjects. *Dent. J.* **2019**, *7*, 60. [CrossRef]
63. Paszyńska, E.; Słopień, A.; Slebioda, Z.S.; Dyszkiewicz-Konwińska, M.; Monika Weglarz, M.; Rajewski, A. Macroscopic evaluation of the oral mucosa and analysis of salivary pH in patients with anorexia nervosa. *Psychiatr. Pol.* **2014**, *48*, 453–464.
64. Back-Brito, G.N.; da Mota, A.J.; de Souza Bernardes, L.Â.; Takamune, S.S.; Prado Ede, F.; Cordás, T.A.; Balducci, I.; da Nobrega, F.G.; Koga-Ito, C.Y. Effects of eating disorders on oral fungal diversity. *Oral Surg. Oral Med. Oral Pathol. Oral Radiol.* **2012**, *113*, 512–517. [CrossRef] [PubMed]
65. Hedman, A.; Breithaupt, L.; Hübel, C.; Thornton, L.M.; Tillander, A.; Norring, C.; Birgegård, A.; Larsson, H.; Ludvigsson, J.F.; Sävendahl, L.; et al. Bidirectional relationship between eating disorders and autoimmune diseases. *J. Child Psychol. Psychiatry* **2019**, *60*, 803–812. [CrossRef]
66. Yadav, K.; Prakash, S. Dental Caries: A review. *Asian J. Biomed. Pharm. Sci.* **2016**, *6*, 1–7.
67. Moynihan, P.; Petersen, P.E. Diet, nutrition and the prevention of dental diseases. *Public Health Nutr.* **2004**, *7*, 201–226. [CrossRef]
68. Johansson, A.K.; Norring, C.; Unell, L.; Johansson, A. Diet and behavioral habits related to oral health in eating disorder patients: A matched case-control study. *J. Eat. Disord.* **2020**, *8*, 7. [CrossRef]
69. Jung, E.H.; Jun, M.K. Relationship between risk factors related to eating disorders and subjective health and oral health. *Children* **2022**, *9*, 786. [CrossRef] [PubMed]
70. Sheetal, A.; Hiremath, V.K.; Patil, A.G.; Sajjansetty, S.; Kumar, S.R. Malnutrition and its oral outcome—A review. *J. Clin. Diagn. Res.* **2013**, *7*, 178–180. [CrossRef]
71. Johansson, A.K.; Norring, C.; Unell, L.; Johansson, A. Eating disorders and biochemical composition of saliva: A retrospective matched case–control study. *Eur. J. Oral Sci.* **2015**, *123*, 158–164. [CrossRef]
72. Paszynska, E.; Schlueter, N.; Slopien, A.; Dmitrzak-Weglarz, M.; Dyszkiewicz-Konwinska, M.; Hannig, C. Salivary enzyme activity in anorexic persons-a controlled clinical trial. *Clin. Oral Investig.* **2015**, *19*, 1981–1989.
73. Dynesen, A.W.; Bardow, A.; Petersson, B.; Nielsen, L.R.; Nauntofte, B. Salivary changes and dental erosion in bulimia nervosa. *Oral Surg. Oral Med. Oral Pathol. Oral Radiol. Endodontology* **2008**, *106*, 696–707. [CrossRef]
74. Ximenes, R.; Couto, G.; Sougey, E. Eating disorders in adolescents and their repercussions in oral health. *Int. J. Eat. Disord.* **2010**, *43*, 59–64. [CrossRef] [PubMed]

75. Paszynska, E.; Hernik, A.; Slopien, A.; Roszak, M.; Jowik, K.; Dmitrzak-Weglarz, M.; Tyszkiewicz-Nwafor, M. Risk of dental caries and erosive tooth wear in 117 children and adolescents anorexia nervosa population-a case-control study. *Front. Psychiatry* **2022**, *13*, 874263. [CrossRef] [PubMed]
76. Szupiany-Janeczek, T.; Rutkowski, K.; Pytko-Polończyk, J. Oral cavity clinical evaluation in psychiatric patients with eating disorders: A case-control study. *Int. J. Environ. Res. Public Health* **2023**, *20*, 4792. [CrossRef] [PubMed]
77. Manevski, J.; Stojšin, I.; Vukoje, K.; Janković, O. Dental aspects of purging bulimia. *Vojnosanit. Pregl.* **2020**, *77*, 300–307. [CrossRef]
78. Shaughnessy, B.F.; Feldman, H.A.; Cleveland, R.; Sonis, A.; Brown, J.N.; Gordon, C.M. Oral health and bone density in adolescents and young women with anorexia nervosa. *J. Clin. Pediatr. Dent.* **2008**, *33*, 87–92. [CrossRef]
79. Hermont, A.P.; Pordeus, I.A.; Paiva, S.M.; Abreu, M.H.; Auad, S.M. Eating disorder risk behavior and dental implications among adolescents. *Int. J. Eat. Disord.* **2013**, *46*, 677–683. [CrossRef]
80. Brandt, L.M.T.; Freitas Fernandes, L.H.; Silva Aragão, A.; Costa Aguiar, Y.P.; Auad, S.M.; Dias de Castro, R.; D'Ávila Lins Bezerra, B.; Leite Cavalcanti, A. Relationship between risk behavior for eating disorders and dental caries and dental erosion. *Sci. World J.* **2017**, *2017*, 1656417. [CrossRef]
81. Jugale, P.V.; Pramila, M.; Murthy, A.K.; Rangath, S. Oral manifestations of suspected eating disorders among women of 20–25 years in Bangalore City, India. *J. Health Popul. Nutr.* **2014**, *32*, 46–50.
82. De Moor, R.J.G. Eating disorder-induced dental complications: A case report. *J. Oral Rehabil.* **2004**, *31*, 725–732. [CrossRef]
83. Amaechi, B.T.; Higham, S.M. Dental erosion: Possible approaches to prevention and control. *J. Dent.* **2005**, *33*, 243–252. [CrossRef]
84. The Erosive Tooth Wear Foundation. What Is Erosive Toothwear? Available online: www.erosivetoothwear.com (accessed on 13 August 2023).
85. Schlueter, N.; Luka, B. Risk groups. *Br. Dent. J.* **2018**, *224*, 364–370. [CrossRef] [PubMed]
86. Schlueter, N.; Tveit, A.B. Prevalence of erosive tooth wear in risk groups. *Monogr. Oral Sci.* **2014**, *25*, 74–98. [PubMed]
87. Petersson, C.; Svedlund, A.; Wallengren, O.; Swolin-Eide, D.; Paulson Karlssonet, G.; Ellegard, L. Dietary Intake and Nutritional States in Adolescents & Young Adults with Anorexia Nervosa: A 3-Year Follow-Up Study. *Clin. Nutr.* **2021**, *40*, 5391–5398.
88. McMaster, C.M.; Fong, M.; Franklin, J.; Hart, S. Dietetic Intervention for Adult Outpatients with Eating Disorders: A Systemic Review and Assessment of Evidence Quality. *Nutr. Rev.* **2021**, *79*, 914–930. [CrossRef] [PubMed]
89. Yang, Y.; Conti, J.; McMaster, C.M.; Hay, P. Beyond Refeeding: The Effects of Including a Dietitian in Eating Disorder Treatment. A Systemic Review. *Nutrients* **2021**, *13*, 4490. [CrossRef]
90. Sterling, W.; Crosbie, C.; Shaw, N.; Martin, S. The Use of the Plate-by-Plate Approach for Adolescents Undergoing Family Based Treatment. *J. Acad. Nutr. Diet.* **2019**, *119*, 1075–1084. [CrossRef]
91. Heruc, G.; Hart, S.; Stiles, G.; Fleming, K.; Casey, A.; Sutherland, F.; Jeffrey, S.; Roberton, M.; Hurst, K. ANZAED practice and training standards for dietitians providing eating disorder treatment. *J. Eat. Disord.* **2018**, *8*, 77. [CrossRef]
92. Butera, A.; Maiorani, C.; Morandini, A.; Simonini, M.; Morittu, S.; Trombini, J.; Scribante, A. Evaluation of Children Caries Risk Factors: A Narrative Review of Nutritional Aspects, Oral Hygiene Habits, and Bacterial Alterations. *Children* **2022**, *9*, 262. [CrossRef]
93. Pitts, N.B.; Ismail, A.I.; Martignon, S.; Ekstrand, K.; Douglas, G.V.A.; Longbottom, C. International Caries Classification and Management System (ICCMS™) Guide for Practitioners and Educators. 2014. Available online: www.iccms-web.com (accessed on 16 August 2023).
94. Walsh, T.; Worthington, H.V.; Glenny, A.M.; Marinho, V.C.C.; Jeroncic, A. Fluoride toothpaste of different concentrations for preventing dental caries. *Cochrane Database Syst. Rev.* **2019**, *3*, CD007868. [CrossRef]
95. Paszynska, E.; Pawinska, M.; Gawriolek, M.; Kaminska, I.; Otulakowska-Skrzynska, J.; Marczuk-Kolada, G.; Rzatowski, S.; Sokolowska, K.; Olszewska, A.; Schlagenhauf, U.; et al. Impact of toothpaste with microcrystalline hydroxyapatite on the occurrence of early childhood caries: A 1-year randomized clinical trial. *Sci. Rep.* **2021**, *11*, 2650. [CrossRef]
96. Paszynska, E.; Pawinska, M.; Enax, J.; Meyer, F.; Schulze zur Wiesche, E.; May, T.W.; Amaechi, B.T.; Limeback, H.; Hernik, A.; Otulakowska-Skrzynska, J.; et al. Caries-preventing effect of a hydroxyapatite-toothpaste in adults: An 18-month double-blinded randomized clinical trial Front. *Front. Public Health* **2023**, *11*, 1199728. [CrossRef] [PubMed]
97. Grocholewicz, K.; Matkowska-Cichocka, G.; Makowiecki, P.; Drozdzik, A.; Ey-Chmielewska, H.; Dziewulska, A.; Tomasik, M.; Trybek, G.; Janiszewska-Olszowska, J. Effect of nano-hydroxyapatite and ozone on approximal initial caries: A randomized clinical trial. *Sci. Rep.* **2020**, *10*, 11192. [CrossRef] [PubMed]
98. Robertson, M.A.; Kau, C.H.; English, J.D.; Lee, R.P.; Powers, J.; Nguyen, J.T. MI Paste Plus to prevent demineralization in orthodontic patients: A prospective randomized controlled trial. *Am. J. Orthod. Dentofac. Orthop.* **2011**, *140*, 660–668. [CrossRef] [PubMed]
99. Karabekiroglu, S.; Ünlü, N.; Küçükyilmaz, E.; Sener, S.; Botsali, M.S.; Malkoç, S. Treatment of post-orthodontic white spot lesions with CPP-ACP paste: A three year follow up study. *Dent. Mater. J.* **2017**, *36*, 791–797. [CrossRef] [PubMed]
100. Limeback, H.; Enax, J.; Meyer, F. Biomimetic hydroxyapatite and caries prevention: A systematic review and meta-analysis. *Can. J. Dent. Hyg.* **2021**, *55*, 148–159. [PubMed]
101. Paszyńska, E.; Jurga-Krokowicz, J.; Shaw, H. The use of parotid gland activity analysis in patients with gastroesophageal reflux disease (GERD) and bulimia nervosa. *Adv. Med. Sci.* **2006**, *51*, 208–213.
102. Paszynska, E.; Linden, R.W.; Slopien, A.; Rajewski, A. Flow rates and inorganic composition of whole saliva in purging bulimic patients treated with a fluoxetine. *World J. Biol. Psychiatry* **2011**, *12*, 282–287. [CrossRef]

103. Paszyńska, E.; Słopień, A.; Węglarz, M.; Linden, R.W. Parotid salivary parameters in bulimic patients—A controlled clinical trial. *Psychiatr. Pol.* **2015**, *49*, 709–720. [CrossRef]
104. Nehme, M.; Parkinson, C.R.; Zero, D.T.; Hara, A.T. Randomised study of the effects of fluoride and time on in situ remineralisation of acid-softened enamel. *Clin. Oral Investig.* **2019**, *23*, 4455–4463. [CrossRef]
105. Hughes, J.A.; West, N.X.; Parker, D.M.; Newcombe, R.G.; Addy, M. Development and evaluation of a low erosive blackcurrant juice drink 3. Final drink and concentrate, formulae comparisons in situ and overview of the concept. *J. Dent.* **1999**, *27*, 345–350. [CrossRef]
106. Haute Autorité de Santé. Detection and Management of Dental Problems by Dentists. Available online: www.has-sante.fr (accessed on 16 August 2023).
107. Schiff, T.; Wachs, G.N.; Petrone, D.M.; Chaknis, P.; Kemp, J.H.; DeVizio, W. The efficacy of a newly designed toothbrush to decrease tooth sensitivity. *Compend. Contin. Educ. Dent.* **2009**, *30*, 238–240.
108. Antezack, A.; Ohanessian, R.; Sadowski, C.; Faure-Brac, M.; Brincat, A.; Etchecopar-Etchart, D.; Monnet-Corti, V. Effectiveness of surgical root coverage on dentin hypersensitivity: A systematic review and meta-analysis. *J. Clin. Periodontol.* **2022**, *49*, 840–851. [CrossRef] [PubMed]
109. Erley, K.J.; Swiec, G.D.; Herold, R.; Bisch, F.C.; Peacock, M.E. Gingival recession treatment with connective tissue grafts in smokers and non-smokers. *J. Periodontol.* **2006**, *77*, 1148–1155. [CrossRef] [PubMed]
110. Wiegand, A.; Lechte, C.; Kanzow, P. Adhesion to eroded enamel and dentin: Systematic review and meta-analysis. *Dent. Mater.* **2021**, *37*, 1845–1853. [CrossRef] [PubMed]
111. King, P.A. Tooth surface loss: Adhesive techniques. *Br. Dent. J.* **1999**, *186*, 371–376. [CrossRef]
112. Milosevic, A. Use of porcelain veneers to restore palatal tooth loss. *Restor. Dent.* **1990**, *6*, 15–18.
113. Carvalho, T.S.; Colon, P.; Ganss, C.; Huysmans, M.C.; Lussi, A.; Schlueter, N.; Schmalz, G.; Shellis, R.P.; Tveit, A.B.; Wiegand, A. Consensus report of the European Federation of Conservative Dentistry: Erosive tooth wear—Diagnosis and management. *Clin. Oral. Investig.* **2015**, *19*, 1557–1561. [CrossRef]
114. Colon, P.; Lussi, A. Minimal intervention dentistry: Part 5. Ultra-conservative approach to the treatment of erosive and abrasive lesions. *Br. Dent. J.* **2014**, *216*, 463–468. [CrossRef]

MDPI
St. Alban-Anlage 66
4052 Basel
Switzerland
www.mdpi.com

Nutrients Editorial Office
E-mail: nutrients@mdpi.com
www.mdpi.com/journal/nutrients

www.ingramcontent.com/pod-product-compliance
Lightning Source LLC
Chambersburg PA
CBHW061456210926
51850CB00008B/1732
9783725813032